Cyanobacteria Biotechnology

Muhammad Aamer Mehmood • Sana Malik •
Syed Ghulam Musharraf • Ramaraj Boopathy
Editors

Cyanobacteria Biotechnology

Sustainability of Water-Energy-Environment Nexus

Springer

Editors
Muhammad Aamer Mehmood
Department of Bioinformatics and Biotechnology
Government College University Faisalab
Faisalabad, Pakistan

Syed Ghulam Musharraf
HEJ Research Institute of Chemistry
International Center of Chemical and Biological Sciences, University of Karachi
Karachi, Pakistan

Sana Malik
Climate Change Cluster
University of Technology Sydney
Ultimo, NSW, Australia

Ramaraj Boopathy
Department of Biological Sciences
Nicholls State University
Thibodaux, LA, USA

ISBN 978-3-031-70697-4 ISBN 978-3-031-70698-1 (eBook)
https://doi.org/10.1007/978-3-031-70698-1

This Springer imprint is published by the registered company Springer Nature Switzerland AG
The registered company address is: Gewerbestrasse 11, 6330 Cham, Switzerland

Dedicated to the all Mothers who ever lived on Earth, for their unwavering love, boundless support, and endless contributions to society's upliftment, whose amazing strength and matchless compassion inspire us all.

With most profound respect and gratitude!

Editors

Preface

Cyanobacteria have emerged as versatile organisms with immense potential in biotechnology, offering solutions to global challenges ranging from environmental sustainability to human health. The book *Cyanobacteria Biotechnology—Sustainability of Water-Energy-Environment Nexus* delves into the multifaceted applications of these fascinating microorganisms. It is a pinnacle of contributions from leading experts in the field, offering a comprehensive overview of the current state and prospects of cyanobacteria in biotechnological applications. The book explores a diverse array of topics, including carbon capture, resource recovery and recycling, bioprocessing of wastewater-grown biomass into bioproducts, metabolic pathway engineering of cyanobacteria for industrial and environmental applications, and leveraging the potential of cyanobacteria for CO_2-to-O_2 conversion in global space expansion programs. Each chapter focuses on specific aspects of cyanobacterial biotechnology, discussing innovative methodologies, cutting-edge research findings, and practical applications. This compilation showcases scientific advancements and underscores cyanobacteria's potential to address pressing global challenges sustainably.

As editors, we aim to offer researchers, students, entrepreneurs, and industry professionals a valuable resource that informs and inspires continued exploration and innovation in cyanobacterial biotechnology. We envision this book fostering interdisciplinary collaborations and sparking fresh ideas to advance the field.

We extend our sincere gratitude to all the contributors for their insightful chapters and to the readers for their interest in this rapidly evolving field. Together, let us dive into the exciting world of cyanobacteria and pave the way for a more sustainable and greener future.

Faisalabad, Pakistan — Muhammad Aamer Mehmood
Ultimo, Australia — Sana Malik
Karachi, Pakistan — Syed Ghulam Musharraf
Thibodaux, LA — Ramaraj Boopathy

Acknowledgments

I would like to thank Springer Nature and its representatives, particularly *Ms. Susanne Dathe* and *Mr. Bibhuti Bhusan Sharma*, for their patience and technical support throughout this process. I sincerely appreciate the time and sincere efforts of all the contributing authors whose valuable work made completing this book possible.

I am grateful for the scientific and technical guidance of my *Co-Editors*, Dr. Sana Malik, Prof. Dr. Syed Ghulam Musharraf, and Prof. Dr. Ramaraj Boopathy. A special note of appreciation goes to my dedicated and passionate team members at the BioEcoTech Research Cluster for their essential input during the preparation of this document.

Faisalabad, Pakistan — Muhammad Aamer Mehmood

Contents

About the Editors

Muhammad Aamer Mehmood is Professor and Group Leader of the BioEcoTech Research Cluster at the Department of Bioinformatics and Biotechnology, Government College University Faisalabad, Pakistan. He is also Founding Executive Board Member and the Secretary-General of the Asian Federation of Biotechnology, Pakistan Chapter. His research team is dedicated to optimizing processes for the sustainable utilization of bio-resources, including wastewater recycling, resource recovery, and the biotransformation of biomass into bioproducts such as biofertilizers, industrial enzymes, biopolymers, and functional foods. They achieve this by using microalgae, cyanobacteria, and fungi as microbial cell factories within a multiproduct biorefinery framework.

Sana Malik is a distinguished researcher at the Climate Change Cluster (C3) at the University of Technology Sydney (UTS) in Australia. With previous tenures at Government College University Faisalabad in Pakistan and Johns Hopkins University in the USA, Dr. Malik brings a wealth of experience to her current role. Her research is dedicated to advancing carbon-neutral biorefinery pathways for the algae industry within a circular bioeconomy framework. At present, she leverages phenomics to identify elite algal strains for industrial and environmental applications. Dr. Malik leads several innovative projects focusing on strain enhancement through adaptive laboratory evolution, metabolic pathway engineering, and phenomics, integrating cutting-edge technologies such as AI, machine learning, and robotics.

Syed Ghulam Musharraf is Professor at the International Center for Chemical and Biological Sciences at the University of Karachi. He is a distinguished authority in metabolomics, having pioneered numerous innovative applications for industrial, environmental, human health, and food-related purposes. Dr. Musharraf is a member of the Global Young Academy and the Pakistan Academy of Sciences, among various other national and international organizations. His exceptional contributions have earned him several prestigious awards and honors, including one of Pakistan's highest civil awards, the Tamgha-e-Imtiaz, bestowed on him by the Government of Pakistan.

Ramaraj Boopathy is Fulbright Scholar, Fulbright Senior Specialist, World Class Professor—Government of Indonesia, Honorary Visiting Professor, ITB, Indonesia, Alcee Fortier Distinguished Service Professor, and John Brady Jr. and John Brady Sr. Endowed Professor in Biological Sciences, working at Nicholls State University, USA. His research focuses on bioremediation of hazardous chemicals, anaerobic microbiology, antibiotic-resistant bacteria/genes in the environment, and the design and development of biological reactor systems for wastewater treatment. In addition, he has also been working on clean energy through bioprocessing. During his academic career, he has received more than 50 million USD in funding to support his academic and research activities.

Chapter 1
Emerging Trends in Cyanobacterial Biotechnology for Sustainable Development

Ayesha Shahid, Muhammad Mubashar, Rabail Zulekha, Muhammad Muzamil Sultan, Jing Li, and Xuezhi Zhang

Abstract The ecological crisis has been detrimental to our ecosystem, where poor resource management and accumulated aquatic and environmental pollutants are the key interlinking factors for several global issues. It has necessitated adopting a pragmatic approach to natural resource management to mitigate this crisis. Sustainable management and conservation are pertinent to alleviating the atrocious effects of pollution and climate change. In this regard, circular bioeconomy is the core of sustainability to minimize the response to critical environmental challenges by cutting emissions and minimizing material loss. Cyanobacteria are considered fundamental renewable sustainability sources due to their unique physiochemical profile. This is the opening chapter of the book, which captures the multifaceted role of newly emerging cyanobacterial biotechnology in reducing CO_2 emissions and aquatic pollution, along with the provision of biomass for sustainable production and processing value chains. Integrating environmental contaminants (wastewater and CO_2) reclamation for value addition provides economic and ecological benefits. The novel product profile of cyanobacteria, including biopolymers, 3D printing, and carbon-negative pigments production, has indicated their immense potential for responsible production and consumption, which is discussed here. Besides it also provided an overview of the environmental and economic impact of cyanobacterial biotechnology, challenges encountered in the industrial inclusion and public acceptance of cyanobacterial products, and possible solutions to set out the path to greener, renewable, and sustainable future communities.

Keywords Environmental mitigation · Net-zero emissions · Wastewater treatment · *Spirulina* · 3D printing

---

A. Shahid
School of Chemical Engineering, Zhengzhou University, Zhengzhou, People's Republic of China

M. Mubashar · R. Zulekha · M. M. Sultan · J. Li · X. Zhang (✉)
Key Laboratory of Algal Biology, Institute of Hydrobiology, Chinese Academy of Sciences, Wuhan, People's Republic of China
e-mail: zhangxuezhi@ihb.ac.cn

M. A. Mehmood et al. (eds.), *Cyanobacteria Biotechnology*,
https://doi.org/10.1007/978-3-031-70698-1_1

1.1 Introduction

Ecological crises, including aquatic and environmental pollution, fossil fuel-driven climate change, and unsustainable production and consumption of everyday commodities, are detrimental to the ecosystem. These are among the most pressing challenges of the twenty-first century. By agreeing to the United Nations' "Sustainability Development Goals (SDG)," the world intends to have a prosperous, sustainable, and socially inclusive future for all. SDG, together with the Paris Climate Agreement, aims to promote sustainable developmental patterns (Van Vuuren et al. 2022) by focusing on circular and net-zero carbon emission technologies. Circular processing is a vital enabler of the sustainable utilization of natural resources to meet the sustainability criteria outlined by the UN. In this regard, "Green Growth" provides the framework and impetus to achieve sustainability by 2030. It fosters renewable means of processing that promote sustainable production and consumption technologies to achieve a healthy environment through a greener economy (Abid et al. 2022). Endeavoring to sustainable development, the European Union set the goal to reduce 40% greenhouse emissions with a 32% increase in renewable energy-efficient technologies to become the first climate-neutral continent by 2050 (Filipović et al. 2022).

Cyanobacteria are envisioned as critical players in the sustainable bio-based circular economy owing to their highest CO_2 sequestration rates, efficient nutrient recovery and conversion, higher productivities, robust photosynthesis capacity, dynamic biochemical profile, elevated adaptability to varying biotic/abiotic conditions, and vast application range (Shahid et al. 2023; Wang et al. 2024). The nutrient-recycling capabilities of cyanobacteria make them pivotal in fighting climate change. Cyanobacteria mitigate the effects of climate change by carbon, nitrogen, phosphorus, and sulfur cycling, where CO_2 is captured and converted to O_2 and biomass while all other nutrients are transformed into bioavailable forms, which are rendered to biota, ensuring the healthy balance of the ecosystem (Bhardwaj et al. 2024). The acclaimed benefits of cyanobacteria drive their use in developing modern-day sustainable and greener societies. Currently, biofuel production in America, Asia, and Europe has been powered by annual tonnage of cyanobacterial biomass. Due to their color and nutraceutical properties, cyanobacterial pigments (especially phycobilin) are important food, cosmetic, and supplementation ingredients. Their toxins are ideal pharmacological material to design novel anticancerous, antibacterial, antimicrobial, anti-inflammatory, and immune boisterous medical compounds. High protein, vitamin, and mineral content, along with antioxidant and immune-protectant properties of cyanobacterial biomass and secondary metabolites, promotes their inclusion in the food chain as an alternative protein source. Cyanobacterial biomass also provides active materials for bioplastics, nanoparticles, and carbon-neutral dyes/inks (Xu et al. 2023).

Despite the environmental and societal benefits of cyanobacterial systems, their scale-up to industrial processes has met with certain restrictions. This chapter summarizes the latest intervention of cyanobacterial systems for environmental and water pollution management and the provision of biomass for sustainable production and processing. These emerging trends directly contribute to SDGs 6, 11, 12, and 13 while indirectly targeting SDGs 2, 3, 7, 8, 9, 14, and 15.

1.2 Atmospheric CO_2 Sequestration Using Cyanobacteria

Climate change driven by environmental pollution has threatened biodiversity in general. The adverse effects of environmental pollution could be rectified by focusing on decarbonization technologies. Various mitigation strategies, including carbon offset by bioenergy production and direct carbon capture and storage (CCS), are some of the viable options. Still, legal and institutional barriers to bioenergy and the logistical challenges of CCS limit their applicability (Steg et al. 2022).

Alternatively, cyanobacteria offer the possibility to not only store CO_2 but also transform it into sustainable bioproducts through carbon capture and utilization (CCU) technology. It is estimated that cyanobacteria could fix 25 gigatons of CO_2 globally. However, the large-scale implementation of CCU plants is challenged by low sequestration efficiency, high/unaffordable energy consumption, photosynthetic efficiency, and light-energy dissipation rates. Theoretically, 100 tons of cyanobacteria could sequester 183 kg CO_2 with 440–550 g m^{-2} day^{-1} or 1800 tons ha^{-1} $year^{-1}$ of carbon fixation rate while practically it could only fix 14.67–18.33 gCO_2 $m^{-2}day^{-1}$ (Wang et al. 2023b). This trend is attributed to the narrowing down of the photosynthetically active radiative band of cyanobacterial and high energy losses that lead to 6% photosynthetic efficiency in controlled conditions and <2% in outdoor (uncontrolled) conditions as compared to the estimated value of 13% in photoautotrophic conditions. Secondly, energy-intensive cultivation and processing escalated the cost of CCU (Wu et al. 2023). Recent technologies, including genetic engineering, chemical adsorption, material hybrids, and nanoparticles, are being investigated to enhance the carbon availability, solubility, and capturing capacity of cyanobacterial cell factories (Table 1.1). Regardless of the adopted method, strain selection is the foremost prerequisite to attain the desired results. A locally resourced strain that exhibited the highest fitness to 20% inlet CO_2 was isolated; in another attempt, a strain with a 15% CO_2 (flue gas) fixation rate was obtained by genetic mutation-adaptive technology. Similarly, overexpression of RUBISCO and CA (carbonic anhydrase) dramatically enhanced biomass production, carbon capture rate, and lipid accumulation (Wu et al. 2023).

The downstream processing of produced biomass allows valorizing carbon fixation biorefineries with the additional benefit of reducing waste by-product generation. As an emerging intervention, scientists explored nature's biomineralization process to develop cyanobacterial-material hybrids based on microscale interactions to impart novel/new properties of the photosynthetic system. Through this process, immobilization, or co-association of cyanobacteria with SiO_2, TiO_2, CeO_2, glucose oxide, polyethylene glycol, tannic acid, graphite, carbon nanotubes, sodium alginate, etc., has been explored to produce semi-natural photosynthetic systems. These systems increase the CO_2 fixation and photosynthetic rate for environmental

Table 1.1 Emerging technologies to attain cyanobacteria-mediated carbon neutrality

Technology	Process	Species	Base material (gene, chemical, compound)	Findings	References
Chemical adsorption	Adsorption of CO_2 onto chemical solvents like potassium carbonate, phase change solvents, blending solvents	–	300 ppm monoethanolamine	6-times increase in carbon solubility, 63% improvement in CO_2 capture, and 31.4% higher biomass production	You et al. (2022)
Genetic engineering	CRISPR-mediated modification	*Synechocystis* sp. PCC 6803	*gltA*, *ackA*, *phaEC*, *ndhD1–2*	Butanol production by improved carbon partitioning	Shabestary et al. (2018)
Genetic engineering	Glass bead transformation	*Chlamydomonas reinhardtii*	Heterologous expression of carbonic anhydrase *Mesorhizobium loti* and *Sulfurihydrogenibium yellowstonense*	Carbon capture rate of 2.75 g CO_2 g^{-1} DCW with twofold higher biomass and eightfold higher lipids	Lin et al. (2022)
Nanoparticle-mediated capture	Nanosheets	*Nostoc sphaeroides*	Molybdenum disulfide nanoparticles	Enhance CO_2 fixation with the accumulation of primary and secondary metabolites	Chen et al. (2021)
Nanoparticle-mediated capture	Reactive oxygen quenching by intracellular nanoparticles	*Synechocystis* sp. PCC 6803	Gold nanoparticles	1.6-fold increased biomass, 21% increase in apparent oxygen evolution, 40% increase in glyceraldehyde-3-phosphate activity	Li et al. (2023)
Microalgae material hybrid	Cell immobilization	Mutant *Synechocystis* sp.	Carbon felt and FTO	CO_2 to ethylene conversion	Li et al. (2021)

protection and offer the possibility of water protection, biomedical therapy, and chemical/bioenergy production (Xiong et al. 2023). Although not explored extensively, a circular biorefinery outlook to achieve carbon neutrality by cyanobacteria was proposed where industrial algal CCU plants can consume flue gases from industry (mainly cement and petrochemical) as carbon sources to transform environmental pollutants into valuable biomass photosynthetically. Co-palletization of produced biomass with conventional biomass is viable for producing solid biofuels that provide the necessary energy and electricity to drive cultivation ponds or industrial processes (Yang et al. 2022).

However, cyanobacterial CCU requires intensive maintenance and colossal capital to create feasible and realistic CO_2-utilizing systems. In this regard, researchers recommended (1) improvement of the CO_2 dissolution process through genetic engineering or process optimization, (2) enhancement of light distribution uniformity, (3) investigation of abiotic factors and cultivation parameters on CO_2 conversion rate, and (4) designing of novel photobioreactor designs with better gas flow.

1.3 Water Pollution Management by Cyanobacteria

Water shortage and anthropogenic activities-driven aquatic pollution are the driving factors for developing sustainable technologies for wastewater treatment as endorsed by the UN SDG 6 (Clean Water and Sanitation). According to recent estimates, 80% of wastewater remains untreated and poses severe threats to aquatic and terrestrial ecosystems by promoting eutrophication and public health diseases like diarrhea, kidney failure, and cancer (Amin et al. 2022). Cyanobacteria-remediated wastewater is viable for reducing water footprint and associated problems, but toxins, pathogens, and unpleasant odor limit its industrial applicability.

1.3.1 Selective Cyanobacteria for Effluents Nutrient Removal

Cyanobacteria play an essential role in the decontamination of water and can be employed in various treatment systems. However, in freshwater ecosystems like lakes, nutrient enrichment from multiple sources contributes to eutrophication, leading to harmful algal blooms dominated by cyanobacteria (Bibi et al. 2024). Certain cyanobacterial species create toxins that are detrimental to aquatic life and human health. Eutrophication and toxins significantly enhance the costs of protecting aquatic bodies or drinking water treatment. This global issue can be turned into a win-win situation by using cyanobacterial species that engage in beneficial interactions with the environment (Nowruzi et al. 2021). Cyanobacteria absorb and assimilate contaminants, which helps to remediate nutrient-rich waters (Czyżewska et al. 2020). Researchers proposed using cyanobacteria for nutrient removal as a green detoxification technology. Using cyanobacteria in soil and water bioremediation is a popular study area (Zanganeh et al. 2022). Several species, including *Anabaena*, *Nostoc*, *Synechocystis*, and *Cyanothece*, demonstrated the potential to remove hazardous heavy metals from wastewater (Bloch and Ghosh 2022). Bolatkhan et al. (2020) investigated that cultivating *Cyanobacterium* sp. IPPASB-1200 on reservoir wastewater decreases 68–100% organic pollutants, improving water's physicochemical qualities. This indicates a significant likelihood of using wastewater to produce cyanobacteria strains for biomass while also recycling water.

Implementing selective cultivation of beneficial cyanobacterial species in ponds presents a promising approach to mitigate water pollution by sequestering excess nutrients (Pathak et al. 2018). Cyanobacteria biosorption is well-acknowledged as a practical, rapid, low-cost, and environmentally friendly treatment approach (Singh 2020). Figure 1.1 shows cyanobacterial cultivation of selective species from effluents of wastewater treatment plants for removing remaining nutrients (N, P) and sequestering CO_2 from the atmosphere. These beneficial cyanobacteria effectively utilize nutrients for growth, thereby suppressing algal blooms and reducing the production of toxins (Kollmen and Strieth 2022). Furthermore, the cyanobacteria could sequester CO_2 from the atmosphere while removing nutrients, and the biomass of beneficial cyanobacteria could serve as a valuable resource, providing opportunities for sustainable practices, such as biofertilizer production to improve soil fertility (Chanda et al. 2024). Biomass can be utilized as feedstock for biodiesel production, contributing to renewable energy initiatives (Mishra et al. 2023). This integrated approach not only addresses water pollution and CO_2 sequestration but also promotes the sustainable utilization of cyanobacterial biomass for environmental and economic benefits.

1.3.2 *Waste Stabilization Pond (WSP)*

WSP is a passive wastewater treatment alternative that utilizes the natural processes of algal-bacteria symbiosis for tertiary treatment of sludge, septate, and wastewater. Effluents from this simple-to-construct and easy-to-operate economical solution

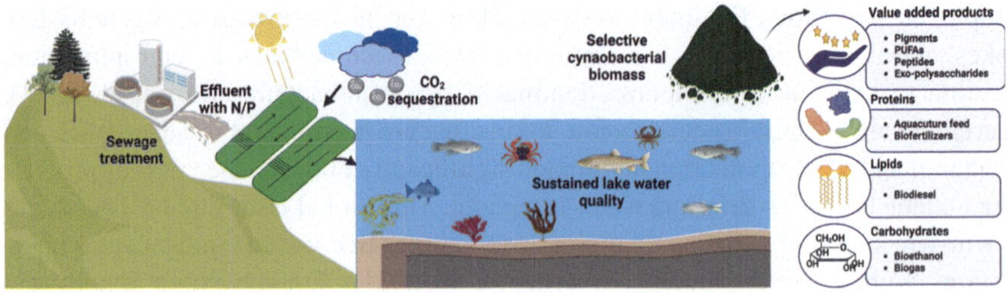

Fig. 1.1 A conceptual drawing showing the cultivation of selective cyanobacteria to remove nutrients and sequester CO_2 from the atmosphere for maintenance in aquatic environmental protection and resource recycling

met the juristic and WHO discharge guidelines for reuse of treated wastewater and thus have potential use for irrigation, watering, and recreational activities due to the high removal rate of pathogenic microbes and helminthic pathogens. In WSP, algae/cyanobacteria (*Microcystis*, *Euglena*, *Chlorella*, etc.) promote the photosynthetic oxygen production that enables nutrient and waste degradation by organic matter oxidizing bacteria (*Pseudomonas, Proteus, Rhodothecae*, *Achromobacter*, and *Thiospirillum*) into CO_2, ammonia, nitrate, and H_2O that in turn are used by algae as growth resources.

Although WSP can remove 90–95% of pathogens, 75% of organic pollutants, and 85% of total suspended solids, and is not the method of choice to reduce nutrient (nitrogen and phosphorous) load in treated wastewater (Achag et al. 2023). The treatment efficiency of WSP has greatly been influenced by the number of fecal microbes, hydraulic and solid retention time, pond depth, temperature, sunlight, pH, DO, and nutrient availability. Therefore, WSP is usually preferred as a post-treatment system to reduce the organic load and pathogens. An upgradation of WSP by combining it with micro-sieve and upstream anaerobic sludge blanket reactor resulted in 25–50% better load management for TN with reduction from 70 to 39 mgL^{-1}. TP removal remains unaffected as its concentration in the effluent (10.6 mgL^{-1}) exceeds the influent concentration of 9.5 mgL^{-1}. Based on load-based removal, up to 72% removal of TN and COD, along with a 36% reduction in daily TP load, was observed (Sinn et al. 2023). Similarly, the integration of construction wetlands with WSP resulted in 48% TN, 80–83% COD, and 67% total suspended solids reduction from municipal wastewater (Arslan et al. 2023).

In the future, investigation of suitable wild-type or engineered indigenous cyanobacterial strains and optimization of operating parameters along with WSP modeling by artificial intelligence and machine learning can help ensure standard effluent quality. Investigation of cost-effective pond management schemes and integration of WSP with fuel production will ensure the economic viability of this system (Liu et al. 2018; Mahapatra et al. 2022).

1.3.3 Cyanobacterial Biomass to Biochar

Biochar, a partially combusted carbon-rich biomass, is an emerging carbon-negative material of interest for aquatic pollution remediation (Table 1.2). It plays a pivotal role in the water-energy-environment nexus by providing innovative means to treat domestic and industrial wastewater through adsorption, flocculation, catalytic degradation, and surface-redox reactions. Engineered biochar, modulated by customized surface physiochemical properties (driven by biomass selection and pyrolysis conditions), boosts nutrient recovery, sludge settlement/management, biological treatment, odor mitigation, and circular resource loops (He et al. 2022). Zhu and Zou produced 5 gL^{-1} of biochar from the phycocyanin-extracted *Spirulina* residues which was able to remove 66.7% of methyl orange from the industrial wastewater, where methyl orange forms uniform biofilm monolayer on the surface of biochar

Table 1.2 Cyanobacterial biochar as an effective option to treat industrial, pharmaceutical, and textile wastewater

Cyanobacterial species	Biomass/biochar condition	Optimum pyrolysis	Targeted pollutant	Removal efficiency	References
Spirulina	Lipid-free biomass	600 °C for 2 h	Lead (Pb^{2+})	74.4%	Yang et al. (2021)
Spirulina	–	200 °C	Heavy metals (Cd, Pb, Cu) and bisphenol A (BPA)	95.2% Cd 97.58% Pb 96.02% Cu 60.6% BPA	Moon et al. (2023a)
Spirulina	$ZnCl_2$-modified biochar	400 °C for 2 h	Acetaminophen (ACT) and BPA	67.7% ACT 96.0% BPA	Moon et al. (2023b)
Synechocystis sp. 6803	Acid-washed triphenyl phosphate-doped biochar	950 °C	Norfloxacin	79%	Wang et al. (2023a)
Spirulina	Phycocyanin-extracted high-salinity residues	350 °C	Hg (II)	75–88%	Ge et al. (2020)
Chlamydomonas sp. Tai-03	–		p-nitrophenols	106 mg g^{-1}	Zheng et al. (2017)

(Zhu and Zou 2021). Similarly, phycocyanin-extracted *Microcystis aeruginosa* biochar successfully absorbs 85 mgg^{-1} malachite green dye (Zhang et al. 2017). However, the potential of cyanobacterial biochar for wastewater treatment has been scrutinized in pristine laboratory conditions, for its industrial inclusion testing should be performed in real-time settings with diverse contaminant profiles. Improvement in adsorption capacity and reusability are the future areas of research.

1.3.4 Nanomaterials and Fabricated Nanocomposites

Cyanobacterial nanomaterials displayed diverse capabilities for disinfection and wastewater treatment. They have proven beneficial for antifouling, nutrient recovery, pollutant biosensing, heavy metal mitigation, and microbial removal in aquatic systems. ZnO nanoparticles of *C. reinhardtii* effectively degraded methyl orange, while the Ag-ZnO nanoparticles of the same species additionally degraded methylene blue and 4-nitrophenols (Khan et al. 2022). Synthesis of CdS nanoparticles from wastewater-cultivated *S. platensis* not only improved the bioremediation by degrading toxic cadmium (Cd) to less toxic CdS nanoparticles but also facilitated the malachite green removal from industrial wastewater and antifouling activity (Mandal et al. 2016), while silver (Ag) nanoparticles of *Anabaena* and *S. platensis* successfully removed 81–82% and 88–93% of malachite green dye (Ismail et al. 2021). Recently, a unique combination of graphene and silver was employed for nanoparticle synthesis that combines the antimicrobial properties of Ag nanoparticles and the strong support of graphene with antifouling properties. 99% decrease in biofilm formation and GAg nanoparticles with >80% antifouling activity against *Isochrysis* sp. was observed (Yee et al. 2016).

Alternatively, nanocomposite fabrication has been tested to develop carbon-neutral and low-cost absorbents. An innovative dual-functional approach has been applied based on the Spirogyra biochar-based Ag nanocomposite (AgBC). 0.5 g L^{-1} of AgBC efficiently adsorbed 95.92% Congo red dye from industrial wastewater in 60 min with a reusability potential of up to 5 cycles (Shaikh et al. 2022). The fabrication of *Spirulina* residual biochar into "Emmental cheese"-inspired highly porous structure improved surface area by 81-fold and enhanced adsorption capacity for sulfathiazole removal from 25.78 to 218.4 mg g^{-1} (Wang et al. 2022). It is to be noted that although cyanobacterial nanoparticle synthesis is relatively more straightforward and less time-consuming than metal nanoparticle synthesis, the composition and properties of green nanoparticles are greatly affected by selected species, biochemical profile, dosage, fabrication material, and production conditions.

1.3.5 Immobilized Polymeric Substances

Integrating wastewater treatment with biorefinery (Fig. 1.2) is integral to developing sustainable commercialized closed-looped systems that are economically competitive, reduce carbon/water footprint, alleviate climate change, and encourage resource recovery (Shahid et al. 2020). However, separating biomass from wastewater is daunting, increasing operational and maintenance costs. Immobilization (in comparison to free cyanobacteria) offers easy harvesting, high cell density for reduced lag phase, less sensitivity to environmental factors, high chloroplast formation (essential for photosynthesis), and improved oxygen production. Natural materials like carrageenan, sodium/calcium alginate, chitosan, and polyvinyl foam are commonly used for cyanobacterial immobilization as they are biodegradable, are nontoxic, and provide high protection against biotic and abiotic factors (Lee et al. 2020). *C. reinhardtii* (immobilized on sodium alginate) beads with 3.5 mm pore size with 20% photobioreactor volume fraction and 3 mm depth successfully removed 95% of TN and 100% TP (Lee et al. 2020). Due to high density, alginate algal beads tend to settle at the bottom of the photobioreactor, which prevents light penetration and reduces treatment efficiency. In a proof-of-concept study, alginate algal beads were used for actual food wastewater treatment where suspended cultivation in the annular photobioreactor doubled cell growth and improved COD removal from 78% to 83% and TN removal from 68% to 84.4% (Hu et al. 2021). Similarly, a novel wastewater treatment bioreactor based on the immobilized activity of nitrogen-doped zinc oxide nanoparticles and algal cells successfully adsorbs and degrades azo dyes and toxic by-products from methylene blue industrial wastewater (Leung Edwin et al. 2024). Polymeric substances such as cellulose acetate, polyethylene, polypropylene, polysulfone, polyethersulfone, polyacrylonitrile,

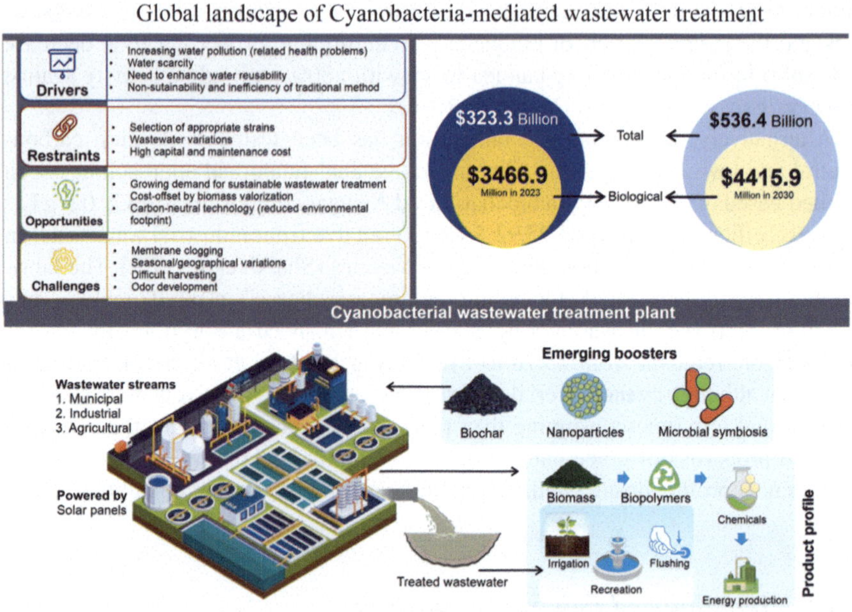

Fig. 1.2 Overview of cyanobacteria mediated wastewater treatment in terms of (i) key driving factors, (ii) market share, and (iii) biorefinery landscape

polycarbonates, and polyvinylidene fluoride are grafted on the membrane surface as antifouling agents.

1.4 Sustainable Production and Processing Profile

1.4.1 Biopolymers

Plastics are an indispensable commodity, but their non-degradable nature atrociously affects the climate by contributing to plastic pollution (Fig. 1.3). The world generates 535 million tons of plastic waste annually, out of which only 9% is recycled, 19% incinerated, 22% discarded, and 50% goes to landfills. GHG emissions associated with the production and disposal of traditional plastics are forecasted to reach 19% of the global carbon budget by 2040 (OECD 2022; Progaramme 2024). With 115 million metric tons of production, single-use packaging plastics are the major contributors (46%) of plastic pollution. The contribution of microplastics from degradation contaminates and physically damages organs/systems of aquatic organisms, disrupts aquatic ecosystem, alters aquatic biota, contaminates the food chain, contributes to metal/chemical-induced cellular toxicity, disrupts metabolic pathways, promotes carcinogenicity and genetic defects, and impairs immune function. Moreover, exceeding levels of dangerous gases like CO, NOx, hydrogen cyanide, dioxins, etc., are released into the environment during plastic production.

These problems threaten the natural ecosystem, the environment, and aquatic and human health (Ali et al. 2023; Thacharodi et al. 2024).

Algal plastics can potentially replace 1540–2200 USD per metric ton of commodity thermoplastics, 2640–5500 USD per metric ton of biodegradable resins, and 1540–8800 USD per metric ton of engineered resins market. However, the share of bioplastics is only 1% of total plastic production. Prominent multinational companies like Carrefour, Hofer, Walmart, Spar, and Sainsbury, as well as automotive giants like Mercedes, Toyota, and Ford, seek green bioplastic alternatives to incorporate into their production and packaging pipeline. Including these alternatives has been reported to reduce CO_2 and energy emissions by 11% and 18%, respectively, along with 25% lesser energy consumption. Food and beverage giant Nestle has committed to a 33% reduction in synthetic plastic consumption by 2050, whereas Ecover is reducing its annual CO_2 emissions to 2500 tons by utilizing 95% bioplastic as packaging material. Shoe soles of PUMA are now being produced by 60% of bioplastics (Nanda and Bharadvaja 2022). Bioplastic startup "Algenesis Materials" in collaboration with a group of researchers from UC-San Diego developed a relatively sturdy but rapidly biodegradable flip-flop by formulating polyurethane foams (standard foot-ware plastic) from algal oils (Gunawan et al. 2020).

Cyanobacterial polylactic acid (PLA), polyhydroxyalkanoates (PHA), and polyhydroxybutyrate (PHB) are gaining attention as promising biodegradable biopolymers to replace traditional non-degradable petroleum-based plastics (Chia et al. 2020) with the capability to hold 55.5% of total bioplastic market share. *Spirulina*, *Nostoc*, *Synechococcus*, and *Synechocystis* can produce up to 78% of PHA or PHB under varying cultivation conditions. *Nostoc* produced 19–22% PHB under photoautotrophic conditions, which improved to 35% under glucose/acetate-supplemented mixotrophic conditions, with further improvement to 40% under gas limitation. *Nostoc* cultivation with propionate or valerate supplementation under mixotrophic and heterotrophic conditions resulted in 60–78% production of PHA co-polymers. *Synechococcus* MA19 has been reported to produce 55% PHA under photoautotrophic conditions, while *Spirulina* could only achieve 6% PHA production. Additionally, species of *Oscillatoria*, *Calothrix*, *Gloeothece*, *Gloeocapsa*, *Aulosira*, and *Chlorogloea* are investigated for bioplastics production (Afreen et al. 2021).

PLA, due to its high mechanical strength, biocompatibility, and biodegradability, is the most promising "green plastic" that is intensively used in disposable products, 3D printing, and medical applications (Fig. 1.3). Recently, *S. elongatus* PCC7942 was metabolically engineered with D-lactic dehydrogenase, polyhydroxyalkanoate synthase, and propionate CoA-transferase for the direct biosynthesis of PLA from CO_2, resulting in 19-fold higher PLA as compared to control with the final titer of 108 mg L^{-1} by employing high-density cultivation strategy (Tan et al. 2022). Cyanophycin, a specialized biopolymer with L-aspartate backbone and L-arginine side chains (first discovered in cyanobacterium *Anabaena*), is now used to replace polyacrylate in surface coatings and paints (Schniete et al. 2024).

In an attempt to scale up production, cultivation of PHB-producing *Synechococcus leopoliensis* in a thin-layer open photobioreactor indicated that optimization of initial cultivation conditions is pertinent for the PHB accumulation (Mariotto et al.

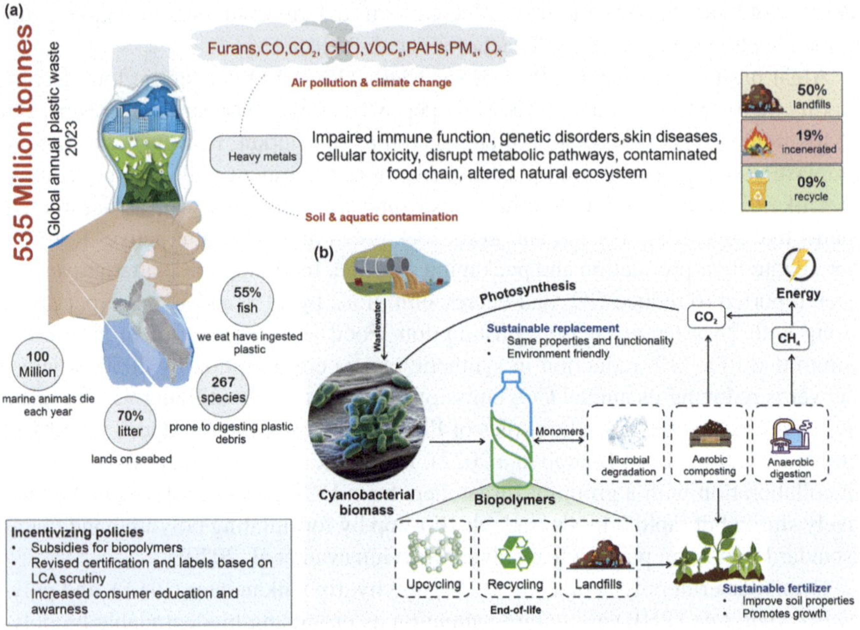

Fig. 1.3 Impact assessment of (**a**) plastic waste on the aquatic, terrestrial, and environmental ecosystems and (**b**) bioplastics/biopolymers in circular bioeconomy

2023). Similarly, the pilot-scale trial of mutant *Synechocystis* sp. PCC 6714 was able to produce 0.35 g L^{-1} PHB, indicating the reassessment of cultivation conditions for pilot-scale production (Kamravamanesh et al. 2019), while a two-section reactor configuration (respectively dedicated to biomass production under optimal conditions and PHB accumulation) has been proposed for the continuous PHB production from engineered *Synechocystis* PCC6803 (Carpine et al. 2020). The production of 4.8 g L^{-1} of PHB in a pilot-scale photobioreactor from secondary effluent-fed mix cyanobacterial biomass demonstrated the feasibility of wastewater as a low-cost alternative for such systems (Senatore et al. 2023). LCA of PHB production from wastewater-cultivated *Synechocystis* sp. indicated that wastewater cultivation contributed to 20% of environmental impact reduction while 67–75% of ecological impacts reduction is possible by improving PHB accumulation to 50% dry weight (Rueda et al. 2023). For economical PHB production, a roll-to-roll winding machine/heat-sealer hybrid system was developed to improve PHB accumulation by 29% and biomass yield by 21% (Lee et al. 2022). Improvement in biomass productivity up to 810 mg L^{-1} day^{-1} has been recommended to reduce PHB selling price from 135 € kg^{-1} PHB to the market price of 4 € kg^{-1} PHB. In contrast, wastewater cultivation could further reduce it to 2.9 € kg^{-1} PHB. Implementing better reactor design with a lower volume: surface ratio and selecting robust/genetically modified high-yielding cyanobacteria strains could further contribute to economic feasibility (Rueda et al. 2023). Recovery and purification of produced PHB are significant bottlenecks in the commercialization of bioplastic as cost-effective (€1.02–1.11 kg^{-1}) chemical treatments are not environment friendly. Although alternatives like ionic liquids, non-chlorinated solvents, and solvent-free extraction are tested (Carpine et al. 2020), their inclusion in industrial processes is hindered by large-scale production, process optimization, by-product release after product degradation, and techno-economic and environmental impact assessments.

1.4.2 Cyanobacterial 3D Bioprinting

Green bioprinting is a rapid bio-fabrication prototyping technology where cyanobacterial cells are embedded/immobilized into polymeric hydrogels through 3D printing. It is the recent intervention to ensure a stable carrier formulation aiming toward a high survival rate and experimental performance for multifactorial applications. In a proof-of-concept study, *C. reinhardtii* was embedded in methylcellulose biomaterial ink, ensuring the steady growth and high stability of immobilized material during 3D printing. In contrast, in a follow-up study, high viability of immobilized cells in response to 26–37 °C temperature and continuous irradiance was observed.

Co-immobilization of *C. reinhardtii* and human cells in a single scaffold has been tested as a novel therapeutic intervention to improve the tissue oxygenation and delivery of secondary metabolites to mammalian cells as immobilization of *C. reinhardtii* into 3D-alginate scaffolds improved cell number (as indicated by chlorophyll content) by 16-fold and oxygen production from 0.05 $mgL^{-1}\ h^{-1}$ to 0.25 $mgL^{-1}\ h^{-1}$ within 24–48 h (Lode et al. 2015). *Synechococcus elongatus* PCC 7942 gel patch has been developed for chronic skin wound regeneration in diabetic patients where alginate-embedded cells improve the oxygen supply to the wound area 100-fold by enhancing fibroblast proliferation and angiogenesis (Chen et al. 2020). In a commercial bio-scaffold "HULK," *C. reinhardtii* immobilization on polysulfone nanofibrous web improved the local oxygen supply and was found effective for tissue engineering (Obaíd et al. 2021).

3D-bioprinted alginate *Chlorella* hydrogel has also been produced to enhance wastewater treatment efficiency and shorten settling time. The resultant hydrogel was found stable for >25 days with 88% and 52% removal efficiency for phosphorous and nitrogen, respectively (Yoon et al. 2024). Similarly, calcium alginate-immobilized *C. reinhardtii* biodegrade 1300 $\mu molL^{-1}$ of phenol within ten days of cultivation (Nazos and Ghanotakis 2021), while co-immobilized consortium of *Synechococcus* sp. and *Chroococcus* sp. efficiently remove 71.8% TN, 91–97% TP, 70–77% COD along with 55–70% sulfamethoxazole antibiotic (Fang et al. 2024).

Cyanobacterial 3D-bioprinted living building materials (LBM) are gaining interest in reducing construction-relevant adverse environmental effects. Recently, a 3D-printed construct of *Synechococcus* sp. PCC 7002 was developed using alginate, sea sand, and methylcellulose as bio-ink. The produced LBM could produce biocement through calcium carbonate mineralization (Reinhardt et al. 2023). Similarly, encapsulation of *Synechococcus* on a loofah scaffold as a biolatex binder has been adopted to produce an innovative biocomposite that could fix 5–10 times more CO_2 than in suspension mode. In another attempt, 3D-printed photovoltaic electrode pillars were constructed using indium tin oxide particles as printing ink to harness the photocurrent activity of *Nostoc* and *Synechocystis* (Goodchild-Michelman et al. 2023). Being a relatively newer addition to the cyanobacterial industry, 3D printing has been challenged by the feasibility and effectiveness of immobilization materials and 3D scaffolds in real-time multi-dynamic settings. Therefore, the search for better fitting immobilization materials, exploitation of entrapment methods, optimization of experimental conditions, clarification of nutrient conversion and metabolite production mechanisms, and evaluation of integration technologies along with life-cycle and environmental impact assessment analysis are recommended to enhance the inclusion of 3D cyanobacterial biomaterials in the ecosystem (Han et al. 2023). 4D bioprinting is a recent theoretical intervention where cyanobacteria could be investigated to develop responsive hydrogels that have the ability to bend, swell, twist, etc., in response to certain stimuli of temperature, moisture, light, chemical, electricity, and magnet to enhance the use of cyanobacterial LBM in real settings (Grira et al. 2023).

1.4.3 Cyanobacterial Biomass Pigments

Petroleum-derived traditional inks and colorants are notorious for health and the environment. Potential perils of these colorants impel the industries (particularly textile) to parade toward sustainable alternatives to eliminate the pollution and associated risks. Biogenic dyes/pigments have been an important research area since the 1960s, where, in earlier stages, they were mainly produced for the food and pharmaceutical industry as a colorant, but later grew their application to the textile, cosmetics, and paint industries. The pigment profile of cyanobacteria, consisting of phycocyanin (blue pigment), allophycocyanin (blue-green pigment), and carotenoids (yellow, orange, red, and purple pigment) and their health-promoting properties (antioxidant, antimicrobial, cytotoxic), makes them suitable for the biogenic dye production (Fried et al. 2022; Saini et al. 2018). Sustainable production and amenability to higher yield by genetic engineering or environmental modulation (Rana et al. 2021) are additional attractive drivers.

Food colorants are an industry with a market share of USD4.1 billion in 2023, of which natural food colorants account for USD1.7 billion. As sustainable, nontoxic, and biodegradable, cyanobacterial pigments are gaining prominence with changed consumer perception (7% increase in the demand for natural pigments). A San Diego-based food colorant startup, "Triton Algae Innovations," has developed highly nutritious strains of algae (mainly *C. reinhardtii*) by precision fermentation. The startup intends to soon launch the yellow, red, and white versions of algae along with the traditional green variant to meet the colorant and nutritional needs of the food industry. Triton, in collaboration with Chinese company Touyun Tech and an anonymous US food company, started commercial-scale production in 2021 (Pacific).

However, it must be noted that producing food colorants is challenging due to molecular instability and higher cost. Moreover, they must be compatible with nutritional value, food flavors, and safety requirements. Microbial engineering, CRISPR-based genetic modulation, and environmental stress modulations are commonly adopted methods for strain improvement, while ionic liquids, supercritical CO_2, and green solvents achieve improvement in pigment extraction. Pigment preservation is critical, and recently, milder and non-thermal approaches like ohmic heating, high-pressure CO_2, cold plasma, high hydrostatic pressure, and high-pulse field electricity along with solid-state fermentation are implemented to reduce thermal and enzymatic degradation (Rana et al. 2021; Sandybayeva et al. 2022; Sen et al. 2019).

Natural pigment-embedded intelligent food packaging is an innovative approach to reducing food wastage through sustainable monitoring of food's freshness from manufacturing to consumption. This packaging (also known as clever or smart packaging) acts as a freshness indicator to monitor highly perishable foods' pH, temperature, and oxygen variation in real time. The combination of cyanobacterial food colorants like phycocyanin, lutein, curcumin, anthocyanin, β-carotene, and chlorophylls with polyvinyl alcohols, starch, and glycerol composite films or with polylactic acid (PLA)/polyethylene oxide (PEO) nanofibers has been used for intelligent packaging. Anthocyanin (water-soluble flavonoid) and phycocyanin (water-soluble phycobilin) are commonly applied as colorimetric indicators of pH variation in packaged food (Bhargava et al. 2020; Yu et al. 2023). Anthocyanin shifts from red to purple in response to pH changes, while phycocyanin appears greenish blue at acidic pH (<pH 6) and discolored at alkaline pH. Whereas 3% phycocyanin containing PLA/PEO fibers produces red, blue, and green hues in response to acidic to alkaline pH variations, while the use of 2% phycocyanin as pH indicator in food packaging produces brown color at acid pH (1–3 range), green at acidic to neutral pH (4–7 range), and discolored green at alkaline pH (Echegaray et al. 2023).

Carbon black, a commonly employed petroleum-derived pigment, has vast industrial applications, but its large carbon footprint and safety issues motivate researchers to search for alternatives. In recent years, the company "Living Ink" has used algal biomass to generate Algal Black™ pigment as the sustainable and easy-to-use ink option for the screen-printing process. Produced flexographic, spray, and screen inks have high color density and do not lose color upon UV exposure (Living Ink Technologies). For large-scale extrusion 3D cyanobacterial bioprinting, investigation of eight alginate-based algae-laden bio-ink formulations indicated the suitability of carrageenan and methylcellulose as bio-ink due to their excellent shear thinning properties (Malik et al. 2020).

Although an exciting arena to explore, biogenic dye production is challenged by the selection of suitable production conditions that could be implemented for industrial production. Moreover, biogenic dyes offer a limited and patchy color profile instead of a vast and continuous spectrum of synthetic colorants. The unavailability of colorimetric data on biogenic dyes also limits their application as a textile commodity. As a comparable measurement depending on fabric texture and dye concentration, it is impossible to give dyed clothes the final appearance. The provision of efficient downstream processing, undesired protective properties, low water solubility, and limited stability are additional bottlenecks that impede the application of biogenic dyes as textile colorants. Genetic engineering technologies could help to address these issues (Fried et al. 2022).

1.5 Environmental and Economic Impact

Life-cycle and techno-economic assessment of various processes of the cyanobacterial biotechnology pipeline provides information regarding SDG compliance and cost minimization guidance (Table 1.3). Integrative processing (combining

Table 1.3 Economic and environmental impact of cyanobacterial biotechnology for CO_2 fixation, wastewater treatment, and bioproduct production

Perspective	Cultivation conditions	Environmental impact (carbon footprint)	Economic impact	References
Phycocyanin production	*Spirulina platensis*, raceway pond, coal-fired electricity	1259 tCO_2 eq emission reduction	Cost 4239 USD carbon credits to offset emissions	Dias et al. (2022)
Biological CCU for phycocyanin production	*S. platensis*, raceway pond, biological energy	2.86 tCO_2 eq emission reduction	Cost 51 USD carbon credits to offset emissions, increase product selling price by 1.6%	Dias et al. (2022)
Use of flue gas as a carbon source	Open raceway pond, batch cultivation	612 CO_2-eq GHG emissions	34.46 $gm^{-2}\,d^{-1}$ of areal productivity	Wu et al. (2023)
Use of photovoltaic system for energy requirements	Industrial production plant	791 metric tons of CO_2 per year	Per annum saving of 123,000 USD	CyanoTech (2014)
Biomass cultivation and wastewater treatment	Primary treated municipal wastewater with urine solution	1.4089×10^5 CO_2-eq. GHG emissions per kg year^{-1}	3.22 USD bio-oil selling cost at zero net present value	Wu et al. (2023)
Biomass cultivation and cyclic wastewater treatment	Recycled and compensated wastewater from the harvesting stage	99.9% reduction in aquatic ecotoxicity and 99.6% reduction in eutrophication	6.88% increase in net present value and 9.14% increase in internal rate of return	Wu et al. (2023)

environmental and aquatic pollution mitigation with industrial processes) is believed to address the obstacles of cyanobacterial biorefinery.

Global warming potential (GWP) of pilot-scale and industrial-scale cyanobacterial cultivation systems has been estimated to be <1 kg CO_2 kg^{-1} of biomass, whereas energy consumption varied from 4 to 800 MJ kg^{-1}. A significant offset of >50% in greenhouse gas emissions is possible by replacing 75% of traditional energy (mainly electricity) with renewable bioenergy. Nutrients (carbon, nitrogen, phosphorous, minerals, etc.) are the major contributors of GWP. By leveraging the CO_2 fixation and wastewater treatment efficiency of cyanobacterial systems (wastewater and flue gas), GWP could be reduced to 0.33 kg CO_2 kg^{-1} biomass. At the same time, a negative value of −9.0 GWP kg^{-1} is possible for symbiotic algal-bacterial systems by consuming wastewater as a nutrient source and gasoline as combustion gas (Xu et al. 2023). Similarly, GWP as low as −56.6 kg CO_2 eq. MJ^{-1} biocrude is possible by replacing traditional electricity with bioelectricity (produced by the combustion of cyanobacterial biomass) during the cultivation process. Conclusively, cyanobacterial biorefineries are effective for wastewater remediation, GHG reductions, and energy economy, but their environmental impacts are only partially positive. Careful curation of the processing value chain and more sensitive assessments of the cyanobacteria life cycle regarding strain characteristics, cultivation parameters, steady-state processing, and climate suitability/sustainability are required to evaluate LCA thoroughly.

1.6 Conclusion and Prospects

Green growth is pervasive in attaining sustainable development in a country. Aiming toward this, a massive shift toward bio-based sustainable circular economies is anticipated in the next two decades. Cyanobacteria possess a wide range of metabolic abilities and distinctive biochemical pathways that hold great potential for tackling different environmental, agricultural, and industrial issues. Integration of cyanobacterial biotechnology into various sectors promotes resource efficiency and ecosystem resilience and fosters economic growth and social well-being. As an environmentally friendly industrial resource, cyanobacteria is the imperative choice for circular economies. Cyanobacterial systems need to attain the promised environmental benefits due to high energy consumption and exhaustive downstream processing. Substantial efforts and investment are required to identify and address the bottlenecks of the cyanobacterial value chain. Complete resource utilization, switching to energy alternatives, simplifying the production-to-processing pipeline, and process regulation and automation are crucial for higher efficiencies and economic viability. This chapter reviews the recent vistas of cyanobacterial biotechnology for circular bioeconomy by utilizing ecological services. It will strengthen the realm of circular bioeconomy by provoking an out-of-box mindset for developing novel ideas and processes to enhance climate protection and economic development for SDG attainment by 2030.

References

Abid N, Ceci F, Ikram M (2022) Green growth and sustainable development: dynamic linkage between technological innovation, ISO 14001, and environmental challenges. Environ Sci Pollut Res 29:25428–25447

Achag B, Mouhanni H, Bendou A (2023) Performance of municipal waste stabilization ponds in the Saharan desert (Morocco). J Water Climate Change 14:1741–1761

Afreen R, Tyagi S, Singh GP et al (2021) Challenges and perspectives of Polyhydroxyalkanoate production from microalgae/cyanobacteria and bacteria as microbial factories: an assessment of hybrid biological system. Front Bioeng Biotechnol 9

Ali SS, Abdelkarim EA, Elsamahy T et al (2023) Bioplastic production in terms of life cycle assessment: a state-of-the-art review. Environ Sci Ecotechnol 15:100254

Amin M, Tahir F, Ashfaq H et al (2022) Decontamination of industrial wastewater using microalgae integrated with biotransformation of the biomass to green products. Energy Nexus 6:100089

Arslan M, Wilkinson S, Naeth MA et al (2023) Performance of constructed floating wetlands in a cold climate waste stabilization pond. Sci Total Environ 880:163115

Bhardwaj A, Singh P, Gupta N et al (2024) Cyanobacteria: a key player in nutrient cycling. In: Mishra AK, Singh SS (eds) Cyanobacteria. Academic Press

Bhargava N, Sharanagat VS, Mor RS et al (2020) Active and intelligent biodegradable packaging films using food and food waste-derived bioactive compounds: a review. Trends Food Sci Technol 105:385–401

Bibi S, Saadaoui I, Bibi A et al (2024) Applications, advancements, and challenges of cyanobacteria-based biofertilizers for sustainable agro and ecosystems in arid climates. Bioresour Technol Reports 25:101789

Bloch K, Ghosh S. (2022). Cyanobacteria mediated toxic metal removal as complementary and alternative wastewater treatment strategy. In Elsevier eBooks, pp. 533–548

Bolatkhan K, Sadvakasova AK, Zayadan BK et al (2020) Prospects for the creation of a waste-free technology for wastewater treatment and utilization of carbon dioxide based on cyanobacteria for biodiesel production. J Biotechnol 324:62–170

Carpine R, Olivieri G, Hellingwerf KJ, et al (2020) Industrial production of poly-β-hydroxybutyrate from CO_2: can cyanobacteria meet this challenge? Processes [Online] 8

Chanda S, Dattamudi S, Jayachandran K et al (2024) The application of cyanobacteria as a biofertilizer for okra (abelmoschus esculentus) production with a focus on environmental and ecological sustainability. Environment 11:45

Chen H, Cheng Y, Tian J et al (2020) Dissolved oxygen from microalgae-gel patch promotes chronic wound healing in diabetes. Sci Adv 6:eaba4311

Chen S, Shi N, Huang M et al (2021) MoS_2 nanosheets–cyanobacteria interaction: reprogrammed carbon and nitrogen metabolism. ACS Nano 15:16344–16356

Chia WY, Ying Tang DY, Khoo KS et al (2020) Nature's fight against plastic pollution: algae for plastic biodegradation and bioplastics production. Environ Sci Ecotechnol 4:100065

Cyanotech (2014) CyanoTech-our history [Online]

Czyżewska W, Piontek M, Łuszczyńska K et al (2020) The occurrence of potential harmful cyanobacteria and cyanotoxins in the Obrzyca River (Poland), a source of drinking water. Toxins 12:284

Dias RR, Deprá MC, Zepka LQ et al (2022) Roadmap to net-zero carbon emissions in commercial microalgae-based products: environmental sustainability and carbon offset costs. J Appl Phycol 34:1255–1268

Echegaray N, Guzel N, Kumar M et al (2023) Recent advancements in natural colorants and their application as coloring in food and in intelligent food packaging. Food Chem 404:134453

Fang Y, Lin G, Liu Y et al (2024) Advanced treatment of antibiotic-polluted wastewater by a consortium composed of bacteria and mixed cyanobacteria. Environ Poll 344:123293

Filipović S, Lior N, Radovanović M (2022) The green deal – just transition and sustainable development goals Nexus. Renew Sust Energy Rev 168:112759

Fried R, Oprea I, Fleck K et al (2022) Biogenic colourants in the textile industry – a promising and sustainable alternative to synthetic dyes. Green Chem 24:13–35

Ge Y, Zhu S, Chang J-S et al (2020) Immobilization of Hg(II) on high-salinity *Spirulina* residue-induced biochar from aqueous solutions: sorption and transformation mechanisms by the dual-mode isotherms. Environ Poll 265:115087

Goodchild-Michelman IM, Church GM, Schubert MG et al (2023) Light and carbon: synthetic biology toward new cyanobacteria-based living biomaterials. Mat Today Bio 19:100583

Grira S, Khalifeh HA, Alkhedher M et al (2023) 3D printing algae-based materials: pathway towards 4D bioprinting. Bioprint 33:e00291

Gunawan NR, Tessman M, Schreiman AC et al (2020) Rapid biodegradation of renewable polyurethane foams with identification of associated microorganisms and decomposition products. Bioresour Technol Rep 11:100513

Han M, Zhang C, Ho S-H (2023) Immobilized microalgal system: an achievable idea for upgrading current microalgal wastewater treatment. Environ Sci Ecotechnol 14:100227

He M, Xu Z, Hou D et al (2022) Waste-derived biochar for water pollution control and sustainable development. Nat Rev Earth Environ 3:444–460

Hu X, Meneses YE, Hassan AA et al (2021) Application of alginate immobilized microalgae in treating real food industrial wastewater and design of annular photobioreactor: a proof-of-concept study. Algal Res 60:102524

Ismail GA, Allam NG, El-Gemizy WM et al (2021) The role of silver nanoparticles biosynthesized by *Anabaena variabilis* and *Spirulina platensis* cyanobacteria for malachite green removal from wastewater. Environ Technol 42:4475–4489

Kamravamanesh D, Kiesenhofer D, Fluch S et al (2019) Scale-up challenges and requirement of technology-transfer for cyanobacterial poly (3-hydroxybutyrate) production in industrial scale. Int J Bio Plast 1:60–71

Khan F, Shahid A, Zhu H et al (2022) Prospects of algae-based green synthesis of nanoparticles for environmental applications. Chemosphere 293:133571

Kollmen J, Strieth D (2022) The beneficial effects of cyanobacterial Co-Culture on plant growth. Life 12:223

Lee H, Jeong D, Im S et al (2020) Optimization of alginate bead size immobilized with *Chlorella vulgaris* and *Chlamydomonas reinhardtii* for nutrient removal. Bioresour Technol 302:122891

Lee JS, Sung YJ, Kim DH et al (2022) Development of a limitless scale-up photobioreactor for highly efficient photosynthesis-based polyhydroxybutyrate (PHB)-producing cyanobacteria. Bioresour Technol 364:128121

Leung Edwin HW, Lam W, Lo CM et al (2024) Novel nitrogen-doped zinc oxide (ND-ZnO) algal-bioreactor system for methylene blue wastewater treatment. J Environ Eng 150:04023102

Li Z, Wu C, Gao X et al (2021) Exogenous electricity flowing through cyanobacterial photosystem I drives CO_2 valorization with high energy efficiency. Energy Environ Sci 14:5480–5490

Li D, Yao S, Cao X et al (2023) Enhancing cyanobacterial photosynthetic carbon fixation via quenching reactive oxygen species by intracellular gold nanoparticles. ACS Sust Chem Eng 11:11140–11148

Lin J-Y, Sri Wahyu Effendi S, Ng IS (2022) Enhanced carbon capture and utilization (CCU) using heterologous carbonic anhydrase in *Chlamydomonas reinhardtii* for lutein and lipid production. Bioresour Technol 351:127009

Liu L, Hall G, Champagne P (2018) Disinfection processes and mechanisms in wastewater stabilization ponds: a review. Environ Rev 26:417–429

Lode A, Krujatz F, Brüggemeier S et al (2015) Green bioprinting: fabrication of photosynthetic algae-laden hydrogel scaffolds for biotechnological and medical applications. Eng Life Sci 15:177–183

Mahapatra S, Samal K, Dash RR (2022) Waste stabilization pond (WSP) for wastewater treatment: a review on factors, modelling and cost analysis. J Environ Manag 308:114668

Malik S, Hagopian J, Mohite S et al (2020) Robotic extrusion of algae-laden hydrogels for large-scale applications. Global Chall 4:1900064

Mandal RP, Sekh S, Sarkar NS et al (2016) Algae mediated synthesis of cadmium sulphide nanoparticles and their application in bioremediation. Mater Res Exp 3:055007

Mariotto M, Egloff S, Fritz I et al (2023) Cultivation of the PHB-producing cyanobacterium *Synechococcus leopoliensis* in a pilot-scale open system using nitrogen from waste streams. Algal Res 70:103013

Mishra S, Kumari N, Singh V et al (2023) Cyanobacterial biofuel: a platform for green energy. Adv Env Eng Res 04:1–42

Moon S, Lee Y-J, Choi M-Y et al (2023a) Adsorption of heavy metals and bisphenol A from wastewater using *Spirulina* sp.-based biochar as an effective adsorbent: a preliminary study. J Appl Phycol 35:2257–2269

Moon S, Lee Y-J, Park S-J et al (2023b) Enhanced removal of micropollutants from water using $ZnCl_2$-modified *spirulina* sp.-based biochar. J Appl Phycol 36:167

Nanda N, Bharadvaja N (2022) Algal bioplastics: current market trends and technical aspects. Clean Technol Environ Pol 24:2659–2679

Nazos TT, Ghanotakis DF (2021) Biodegradation of phenol by alginate immobilized *Chlamydomonas reinhardtii* cells. Arch Microbiol 203:5805–5816

Nowruzi B, BouaïCha N, Metcalf JS et al (2021) Plant-cyanobacteria interactions: beneficial and harmful effects of cyanobacterial bioactive compounds on soil-plant systems and subsequent risk to animal and human health. Phytochemistry 192:112959

Obaíd ML, Camacho JP, Brenet M et al (2021) A first in human trial implanting microalgae shows safety of photosynthetic therapy for the effective treatment of full thickness skin wounds. Front Med 8

OECD (2022) Global plastics outlook

Pacific TGFIA Triton Algae Innovations: Ushering in the Algae-based Alt-Protein Movement

Pathak J, Maurya PK, Singh SP et al (2018) Cyanobacterial farming for environment friendly sustainable agriculture practices: innovations and perspectives. Front Environ Sci 6:311475

Progaramme UE (2024) Beat plastic pollution-our planet is choking on plastic [Online]. United Nations

Rana B, Bhattacharyya M, Patni B et al (2021) The realm of microbial pigments in the food color market. Front Sust Food Syst 5

Reinhardt O, Ihmann S, Ahlhelm M et al (2023) 3D bioprinting of mineralizing cyanobacteria as novel approach for the fabrication of living building materials. Front Bioeng. Biotechnol 11

Rueda E, Senatore V, Zarra T et al (2023) Life cycle assessment and economic analysis of bioplastics production from cyanobacteria. Sust Mater Technol 35:e00579

Saini DK, Pabbi S, Shukla P (2018) Cyanobacterial pigments: perspectives and biotechnological approaches. Food Chem Toxicol 120:616–624

Sandybayeva SK, Kossalbayev BD, Zayadan BK et al (2022) Prospects of cyanobacterial pigment production: biotechnological potential and optimization strategies. Biochem Eng J 187:108640

Schniete JK, Brüser T, Horn MA et al (2024) Specialized biopolymers: versatile tools for microbial resilience. Curr Opin Microbiol 77:102405

Sen T, Barrow CJ, Deshmukh SK (2019) Microbial pigments in the food industry-challenges and the way forward. Front Nutri 6

Senatore V, Rueda E, Bellver M et al (2023) Production of phycobiliproteins, bioplastics and lipids by the cyanobacteria *Synechocystis* sp. treating secondary effluent in a biorefinery approach. Sci Total Environ 857:159343

Shabestary K, Anfelt J, Ljungqvist E et al (2018) Targeted repression of essential genes to arrest growth and increase carbon partitioning and biofuel titers in cyanobacteria. ACS Synth Biol 7:1669–1675

Shahid A, Malik S, Zhu H et al (2020) Cultivating microalgae in wastewater for biomass production, pollutant removal, and atmospheric carbon mitigation; a review. Sci Total Environ 704:135303

Shahid A, Khan AZ, Jabeen F et al (2023) Cyanobacteria-based biorefineries for a sustainable future of bioindustry. In: Oncel SS (ed) A sustainable green future: perspectives on energy, economy, industry, cities and environment. Springer International Publishing, Cham

Shaikh WA, Chakraborty S, Islam RU et al (2022) Fabrication of biochar-based hybrid Ag nanocomposite from algal biomass waste for toxic dye-laden wastewater treatment. Chemosphere 289:133243

Singh SB (2020) Biosorption of heavy metals by cyanobacteria: potential of live and dead cells in bioremediation. In Springer eBooks:409–423

Sinn J, Agrawal S, Orschler L et al (2023) Upgrade of waste stabilization ponds to improve effluents for reuse purposes. H2Open J 6:315–330

Steg L, Veldstra J, De Kleijne K et al (2022) A method to identify barriers to and enablers of implementing climate change mitigation options. One Earth 5:1216–1227

Tan C, Tao F, Xu P (2022) Direct carbon capture for the production of high-performance biodegradable plastics by cyanobacterial cell factories. Green Chem 24:4470–4483

Thacharodi A, Meenatchi R, Hassan S et al (2024) Microplastics in the environment: a critical overview on its fate, toxicity, implications, management, and bioremediation strategies. J Environ Manag 349:119433

Van Vuuren DP, Zimm C, Busch S et al (2022) Defining a sustainable development target space for 2030 and 2050. One Earth 5:142–156

Wang K, Wang Y, Zhang S et al (2022) Tailoring a novel hierarchical cheese-like porous biochar from algae residue to boost sulfathiazole removal. Environ Sci Ecotechnol 10:100168

Wang C, Holm PE, Andersen ML et al (2023a) Phosphorus doped cyanobacterial biochar catalyzes efficient persulfate oxidation of the antibiotic norfloxacin. Bioresour Technol 388:129785

Wang Y, Yang S, Liu J et al (2023b) Realization process of microalgal biorefinery: the optional approach toward carbon net-zero emission. Sci Total Environ 901:165546

Wang N, Xiang Y, Ma Y, Zhang P, Zhou X, Zhu H, Zhang Z, Li Z, Xiao X, He M, Mehmood MA, Zhu H (2024) Fermented *Spirulina platensis* alleviated DSS-induced ulcerative colitis by regulating gut microbiota and MyD88/TLR4 signaling pathway. Asian J Agric Biol 1(1):2023232. https://doi.org/10.35495/ajab.2023.232

Wu W, Tan L, Chang H et al (2023) Advancements on process regulation for microalgae-based carbon neutrality and biodiesel production. Renew Sust Energy Rev 171:112969

Xiong W, Peng Y, Ma W et al (2023) Microalgae–material hybrid for enhanced photosynthetic energy conversion: a promising path towards carbon neutrality. Nat Sci Rev 10:nwad200

Xu P, Li J, Qian J et al (2023) Recent advances in CO_2 fixation by microalgae and its potential contribution to carbon neutrality. Chemosphere 319:137987

Yang Z, Hou J, Wu J et al (2021) The effect of carbonization temperature on the capacity and mechanisms of Pb(II) adsorption by microalgae residue-derived biochar. Ecotoxicol Environ Saf 225:112750

Yang J, Zhao T, Cui X et al (2022) New insights into the carbon neutrality of microalgae from culture to utilization: a critical review on the algae-based solid biofuels. Biomass Bioenergy 166:106599

Yee MS-L, Khiew P-S, Chiu WS et al (2016) Green synthesis of graphene-silver nanocomposites and its application as a potent marine antifouling agent. Coll Surf B: Biointerf 148:392–401

Yoon S-W, Kim S-Y, Jeon J-S et al (2024) 3D-printed *Chlorella vulgaris* biocarriers: a novel approach to wastewater treatment. J Water Process Eng 57:104711

You X, Yang L, Zhou X et al (2022) Sustainability and carbon neutrality trends for microalgae-based wastewater treatment: a review. Environ Res 209:112860

Yu Z, Boyarkina V, Liao Z et al (2023) Boosting food system sustainability through intelligent packaging: application of biodegradable freshness indicators. ACS Food Sci Technol 3:199–212

Zanganeh F, Heidari A, Sepehr A, Rohani A et al (2022) Bioaugmentation and bioaugmentation–assisted phytoremediation of heavy metal contaminated soil by a synergistic effect of cyanobacteria inoculation, biochar, and purslane (*Portulaca oleracea L.*). Environ Sci Pollut Res 29:6040–6059

Zhang H, Zhang F, Huang Q (2017) Highly effective removal of malachite green from aqueous solution by hydrochar derived from phycocyanin-extracted algal bloom residues through hydrothermal carbonization. RSC Adv 7:5790–5799
Zheng H, Guo W, Li S et al (2017) Adsorption of p-nitrophenols (PNP) on microalgal biochar: analysis of high adsorption capacity and mechanism. Bioresour Technol 244:1456–1464
Zhu H, Zou H (2021) Characterization of algae residue biochar and its application in methyl orange wastewater treatment. Water Sci Technol 84:3716–3725

Chapter 2
Algae Species of Industrial, Environmental, and Food Importance

Muhammad Rehan Khalid, Irum Yousaf, Mahwish Amin, Musaab Sohail, and Muhammad Nawaz

Abstract Exploring nonconventional alternatives for food, feed, and environmental sustainability has become essential. Among several nontraditional resources, microalgae and cyanobacteria have come forward for their diverse and sustainable applications. However, selecting elite strains followed by optimizing their pilot-scale cultivation for industrial application faces a number of challenges. This chapter explores previously explored elite microalgal strains for diverse applications including environmental remediation, feed/food, and biofuel production. It has been found that algae species including *Spirulina platensis*, *Chlorella vulgaris*, *Scenedesmus* sp., *Dunaliella salina*, *Haematococcus pluvialis*, *Tolypothrix distorta*, *Scytonema hyalinum*, *Nostoc commune*, *Nostoc*, *Anabaena* spp., *Synechococcus* spp., *Synechocystis* spp., *Oscillatoria* spp., and *Phormidium* spp. have come forward as the most promising species for industrial, environmental, and food/feed applications. However, most of the algal species are still unexplored, which could be one of the priority research areas in the future. Different optimization approaches to increase the bioactive compounds from different microalgal strains are also described. Some challenges and prospects of microalgal products are also discussed.

Keywords Algae · Alternative food · Environmental mitigation · Sustainability · Carbon neutral

---

M. R. Khalid · I. Yousaf · M. Sohail · M. Nawaz (✉)
Department of Botany, Government College University Faisalabad, Faisalabad, Pakistan
e-mail: muhammadnawaz@gcuf.edu.pk

M. Amin
Bioenergy Research Center, Department of Bioinformatics and Biotechnology, Government College University Faisalabad, Faisalabad, Pakistan

M. A. Mehmood et al. (eds.), *Cyanobacteria Biotechnology*,
https://doi.org/10.1007/978-3-031-70698-1_2

2.1 Introduction

To ensure the sustainability of food security, alternative and sustainable food sources must be explored. Food deficiency is expected to go on increasing due to increasing population (Wolkers et al. 2011). Several algal species have been recognized as GRAS (Generally Regarded as Safe) and are being considered as alternate food to feed the resources. There are more than 7000 microalgae species contributing toward the primary food chain in the ecosystem (Prabhakaran et al. 2014). Algae have shown numerous applications including food, fuel (Fig. 2.1), wastewater treatment/recycling, soil amender, and nutraceuticals. GRAS microalgal species include *Spirulina* sp., *Anabaena* sp., *Nostoc* sp., *Synechococcus* sp., *Haematococcus* sp., *Prototheca* sp., *Chlamydomonas* sp., *Crypthecodinium* sp., *Chlorella* sp., *Scenedesmus* sp., and *Haematococcus* sp. These strains of microalgae produce various types of beneficial biomolecules like proteins, amino acids, and vitamins that are necessary for growth. Furthermore, these edible microalgae include a high concentration of essential elements such as calcium, sodium, magnesium, phosphorus, sulfur, and nitrogen along with trace elements such as cobalt, iodine, selenium, copper, manganese, zinc, and molybdenum. Algae are also rich in essential amino acids and omega-3 and omega-6 fatty acids. Due to a wide range of therapeutic and nutritional contents, the market for algal products including nutraceuticals is rising across the globe. A wide range of algal metabolites are not well characterized and yet to be explored. Among various industrial applications of algae, biofertilizer is an

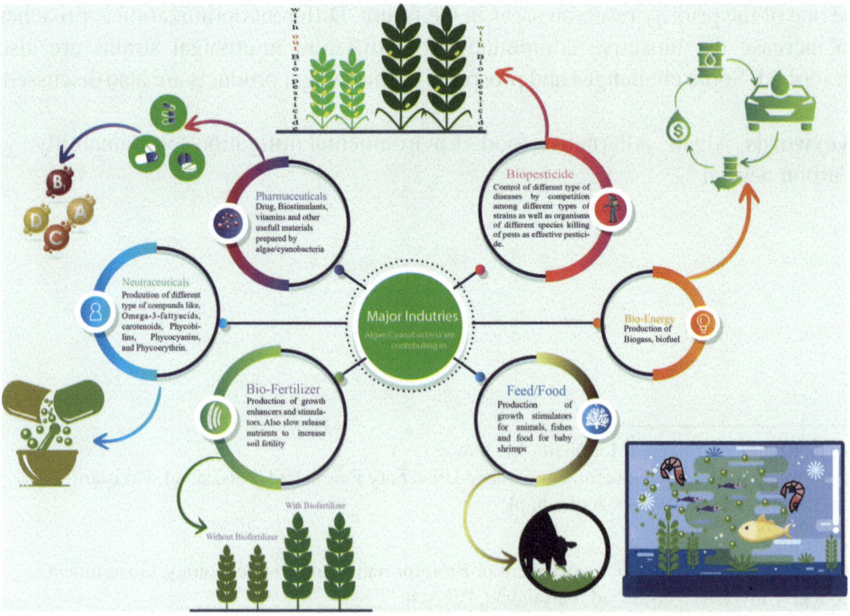

Fig. 2.1 Algal products and their association with different industries

important use which restricts the use of pesticide and chemical fertilizers where the latter pose a threat to environment like nutrient deficiency, soil deterioration, and eutrophication (Calvo et al. 2014; Sharma et al. 2014; Bulgari et al. 2015).

Rice (*Oryza sativa* L.), wheat (*Triticum aestivum* L.), cotton (*Gossypium* spp.), maize (*Zea mays* L.), and other leguminous plants have been shown to benefit from biofortification with useful molecules derived from algae in terms of restoring soil fertility in both flooded and harsh terrestrial environments with varying ecological conditions such as temperature, soil pH, and moisture content (Karthikeyan et al. 2007; Prasanna et al. 2013, 2014, 2015). Because they boost plant defense mechanisms and increase output, algae and cyanobacteria are utilized as biofertilizer in agriculture (Kholssi et al. 2019; Ronga et al. 2019; Castro et al. 2020). This new wave of biostimulants, biofertilizers, and biopesticides is upending the agriculture sector by encouraging plant growth, flower/fruit set, fruit production, and seed germination (Bulgari et al. 2015; Singh et al. 2016; Colla and Rouphael 2020). Their application is also expanding the defense mechanism of species for abiotic stresses like salinity, temperature, drought, light, and frost as well as biotic stresses like insects and other microorganism as pathogens (Lesueur et al. 2016; Costa et al. 2019a; Kaur and Purewal 2019). Other fertilizers only give a boost of nutrients but don't increase the soil fertility; in contrast, biofertilizers have a slow-release mechanism and increase the fertility of the soil by absorbing excess nutrients in flooded lands (Coppens et al. 2016).

When compared to traditional crop plants, microalgae biomass production has lower water footprint, can tolerate harsh environments, and can be produced on barren lands which might help in preserving fertility for agriculture purpose. Besides, their high photosynthetic activity and production potential in wastewater add more value to their selection for biomass production (Brennan et al. 2010; Kuddus et al. 2013). One of many such microalgae strains, *Arthrospira platensis* commonly called "*Spirulina*," has been studied and characterized using open pond cultures and photobioreactors (Jiménez et al. 2003; Brennan et al. 2010). Globally, *Spirulina* has been declared as Generally Regarded as Safe (GRAS) and has been approved for its use as dried food products (DFPs) (Thomassen et al. 2016; Torres-Tiji et al. 2020). *Spirulina* profile includes functional biomolecules in excess including carbohydrates, lipids, proteins, and fatty acids, which offers diverse opportunities for various industrial products (Van Iersel and Flammini 2010; Wang et al. 2024).

Microalgae provide a wide range of biochemicals/metabolites, some of which have been listed in Table 2.1, but their nature and their production conditions are yet to be identified and characterized (Khan et al. 2009; Górka et al. 2015; Colla and Rouphael 2020). A thorough investigation from cultivation to production is a key requirement for the use of strains for any industrial application (Costa et al. 2019b; Vaz et al. 2019a, b).

Table 2.1 Industrially important microalgae/cyanobacteria species for diverse applications

Species	Products/Industry	Findings	References
Synechocystis sp.	Phycobiliprotein/ Bioplastics/Biofuel	Resource recovery and water treatment of effluent	Senatore et al. (2023)
Synechocystis sp. PCC6803	Nutrient/Resource Recovery	Resource recovery from human urine	Canizales et al. (2023)
Chlamydomonas reinhardtii CC124, *Chlorella* sp. MACC-360	Cosmetics	Anti-aging biomolecules and reduction in hyperpigmentation	Favas et al. (2022)
Picochlorum maculatum, Nannochloris atomus	Bio-stimulant/ Biofertilizer	The algae strains were cultivated in tris-acetate-phosphate (TAP) media, at pH 7	Gitau et al. (2022)
Chlorella sorokiniana Chlorella–Micractinium clade	Bioresource Conservation/Feed Aquaculture	Feed production	Rasheed et al. (2022)
Chlorella sp.	Nutraceuticals	High production of biomolecules and secondary metabolites and pigments	Chiellini et al. (2022)
Spirulina platensis LB 2340	Biofertilizer	Granular Nitrogen biofertilizer production	Slinksienė et al. (2022)
Nostoc commune	Probiotics/ Biostimulants/Feed/ Food	Used as probiotics in aquaculture for white shrimp	Kim et al. (2021)
Tetradesmus obliquus, C. protothecoides, Chlorella vulgaris	Pharmaceuticals/ Cosmetics	Polysaccharide extract production for cosmetics	Tseng et al. (2021)
Anabaena sp.	Biostimulant and biopesticide	Diluted wastewater (1:20) to treat piggery wastewater (PWW)	Ferreira et al. (2021)
Rhodomonas sp.	Feed/Supplement	Feed supplement for *Nile tilapia*	Fadl et al. (2020)
Phormidium autumnale	Aquaculture	Aquaculture Feed for fishes and shrimps	Oostlander et al. (2020)
Scenedesmus spp., *Neochloris* spp., *Nitzschia* spp., *Chlorella* spp., *Desmodesmus* spp., *Cosmarium* spp.	Bioenergy/Biofuel	By Maintaining biofuel quality and sustainability	Jacob-Lopes et al. (2020)
C. vulgaris, C. sorokiniana, C. regularis, S. platensis	Bioremediation	MA was used in heavy metal polluted water to reduce its toxicity	Chakdar et al. (2022)
Phormidium autumnale UTEX1580 *Synechococcus* sp. PCC7942, *Anabaena flosaquae* UTCC64, *Sorokiniana* XJK *S. platensis*	MA as a future food source	Production of beneficial food grade biomolecules	Torres-Tiji et al. (2020)

(continued)

Table 2.1 (continued)

Species	Products/Industry	Findings	References
Tolypothrix distorta, *Scytonema hyalinum*, *Nostoc commune*	Dye removal from textile industry effluent	Nutrient sequestration from textile dye contaminated wastewater	Brar et al. (2019)
Nostoc, *Anabaena* spp., *Synechococcus* spp., *Synechocystis* spp., *Oscillatoria* spp., *Phormidium tenue*	Biofilm formation	Desalinization of saline soil	Román et al. (2018)
Microcoleus vaginatus, *Scytonema javanicum*, *Phormidium tenue*	Bioremediation	Wastewater treatment and recycling	Safi et al. (2014)
S. platensis, *S. maxima*, *Aphanizomenon flosaquae*, *Chlorella* spp.	Ecosystem restoration	Biocrust formation	Lan et al. (2014)
S. platensis	Dietary supplement	Source of vitamin B12, riboflavin, beta-carotenes	Khan et al. (2005)

2.2 Industrially Important Algal Species

2.2.1 Skincare Products and Cosmetics Industry

Cosmetics are getting popular for their capacity to safeguard the skin from external chemical aggressors such as xenobiotics, UV radiation, and dryness, which are the fundamental causes of skin aging and loss of integrity. The aging process of the skin is driven by natural physiological factors, but it may also be the result of biotic and abiotic stresses, and these factors add more to this process (Mukherjee et al. 2011). Different types of MMPs (matrix metalloproteinases) are produced which play a main role in the breakdown of different types of bodily proteins like elastin and collagen critically involved in skin formation and some hyaluronic acids. All these form the extracellular matrix of cells (Freitas-Rodríguez et al. 1864; Zhuang et al. 2021). People who desire to live a healthier and attractive lifestyle have therefore made the use of skincare products a routine (Morone et al. 2019). Production of cosmetic components that are derived from algae and cyanobacteria are more friendly toward the skin and environment, consequently committing to sustainability (Morone et al. 2019). Due to their exceptional secondary metabolite production and plasticity, cyanobacteria have developed a large diversity of biomolecules which demonstrates therapeutic, nutraceutical, and food-conservation potential (Wiegand et al. 2005; Maschek et al. 2008; Nunnery et al. 2010; Nowruzi et al. 2020). They can block the UV radiations and disintegrate the enzymes responsible for the degradation of extracellular products to prevent the aging of skin (Favas et al. 2021).

Production of such beneficial chemicals might open a path to use them in anti-inflammatory creams, sunscreens, and moisturizers (Morone et al. 2019). Different algal species including *Synechocystis, Leptolyngbya, Nostoc*, and *Arthrospira*

(Nanni et al. 2001; Basheva et al. 2018) have been explored for various biomolecules including phycocyanin, phycobilin, allophycocyanin, phycoerythrin, and phenolic compounds which are being used as antioxidant in cosmetics (Freile-Pelegrín et al. 2013; Jerez-Martel et al. 2017). Besides, algal species, namely *Coleofasciculus* spp., *Nostochopsis lobatus*, and *Schizothrix* sp., produce biomolecules which act as inhibitor of enzymes like elastases, collagenases, and hyaluronidases (Volkmann et al. 2006; Oren et al. 2007; Yamaguchi et al. 2016). Furthermore, it has been established recently that various aqueous extracts of cyanobacteria could postpone the loss of dermal density (Favas et al.). Cyanobacterial pigments, such as carotenoids, phenolic compounds, phycobiliproteins (PBPs), mycosporine-like amino acids (MAAs), and scytonemin (SCY), have been listed among the leading skincare compounds (Gao et al. 2021).

2.2.2 *Biopolymer Production*

The circular economy model (CEM) has attracted a lot of interest as a potential remedy for the problems raised by increased production and consumption demands. It focuses on developing strategies that ensure resource conservation, recycling, and minimum or no waste generation (Ness. 2008). A biorefinery is essential to the shift to a society with net-zero emissions (Madadi et al. 2021a); therefore, using biomass and renewable resources has gained more attention in recent years to achieve the goals set by CEM, while microalgae are considered as the most sustainable source of biomass (Fig. 2.3). Approximately one-third of the world's annual plastic production, more than 400 million tons, ends up as plastic debris in landfills, freshwater bodies, and seas. The manufacturing of synthetic polymers increased significantly, resulting in increased energy usage and greenhouse gas emissions. Furthermore, several toxic substances are released during the manufacture of these polymers (Singh et al. 2017). Therefore, finding sustainable and ecofriendly materials has become essential to replace the synthetic plastics (Madadi et al. 2021b).

Polyhydroxyalkanoates (PHAs) are bioplastics derived from cyanobacteria and microalgae using sugar and lipids as their internal carbon source. Given their similar physical properties, PHAs are expected to be appropriate alternatives for petrochemical polymers (Lambert and Wagner 2017). PHAs are made up of repeated units of ester with two oxygen atoms and a carbon chain linked to alpha carbon's R group. Their extreme hydrophobicity explains the limited solubility of these polymers in water, and they are also inert and stable in the environment. In addition to being biodegradable and biocompatible, PHAs may have low solvent resistance, exceptionally high purity within the cell, excellent resistance to UV radiation degradation, and thermoplastic and elastomeric properties (da Costa et al. 2019). These properties make them suitable materials for a wide range of applications, including wound dressing, vein valves, bone marrow scaffolds, replacements, tendon repair molecules, and surgical devices (bone plates, orthopedic pins, etc.). PHA concentrations have been observed to range from 1 to 10% in a variety of cyanobacteria, including *Calothrix* sp., *Synechocystis* sp., *Nostoc* sp., *Oscillatoria* sp., and *Spirulina* sp. (Zhang et al. 2019).

P(3HB) is one of the most extensively studied and commercially available short-chain-length PHAs (scl-PHA) produced by cyanobacteria (Balaji et al. 2013). *Calothrix scytonemicola, Spirulina* spp., *Gloeothece* spp., and *Synechococcus* spp., and *Chlorella* spp., are some of the species which are reported to produce P(3HB) using CO_2 as a carbon source (Balaji et al. 2013). The P(3HB) generated from algae has been demonstrated to be more recyclable, biocompatible, biodegradable, plasticizing capable, and usable than petroleum-based polymers (Abdo and Ali 2019). Das et al. (2018) have reported the production of bioplastics from *Chlorella pyrenoidosa*.

2.2.3 *Agriculture and Animal Feed*

Since the introduction of fertilizers and insecticides, the yield and production have increased significantly. But their extensive usage has poisoned the soil, reduced soil fertility, and decreased productivity; it was a plague masquerading as a gift (Cooper et al. 2007; Fenner et al. 2013; Chiaiese et al. 2018; Gitau et al. 2022). The synthesis of biofertilizer was put on hold and limited to the symbiotic connections between a few of rhizobacteria and the plant (Madani et al. 2018; Fayzi et al. 2020). Different microorganisms have been used to synthesize biofertilizers, but use of microalgae is considered a more environmentally friendly and sustainable approach (Fig. 2.2). The emergence of bio-filmed formulations is predicated on the fact that diazotrophic filamentous cyanobacteria possess extracellular mucilage, providing a nutrient-rich habitat for the bacteria. This approach has been used to create advantageous mixtures of cyanobacteria with beneficial bacteria for agriculture while maintaining optimal growing conditions (Prasanna et al. 2011, 2013). In several crops, including

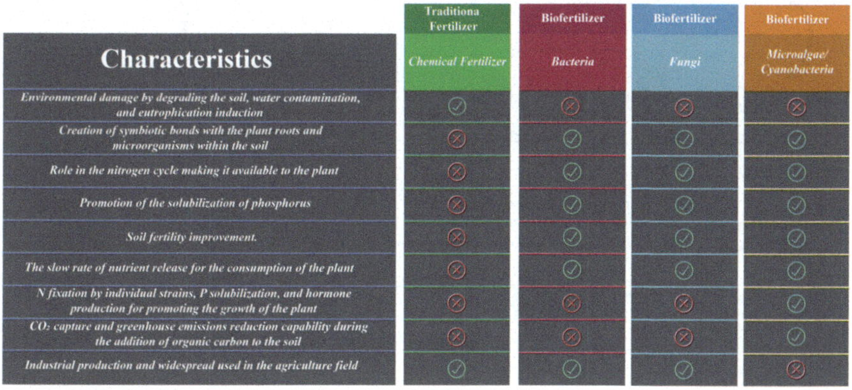

Fig. 2.2 Comparison of microalgae/cyanobacteria and other microorganisms used as biofertilizers (Osorio-Reyes et al. 2023)

rice (*Oryza sativa*), wheat (*Triticum aestivum*), cotton (*Gossypium hirsutum*), mungbean (*Vigna radiata*), soybean (*Glycine max*), tomato (*Solanum lycopersicum*), leguminous crops, and spices, these biofilms displayed plant growth-promoting activity and increased nutrient sequestration (Karthikeyan et al. 2009; Kumar et al. 2013; Prasanna et al. 2014, 2015). Biofortification of chickpea (*Cicer arietinum* L.) using biofilms is reported by Bidyarani et al. (2016). The biostimulant effect of *Chlorophyta* sp. is also reported on tomato (*Solanum lycopersicum*) by Gitau et al. (2022).

A peculiar non-photosynthetic, non-dividing cell known as a heterocyst protects the oxygen-labile nitrogenase complex, which is responsible for fixing nitrogen into a bioavailable form in *Anabaena*, a multicellular, filamentous freshwater cyanobacterium (Videau et al. 2019), and this is extensively used in fish diet. For instance, when *Labeo gonius* was provided with different algal diets, its growth performance was boosted and its lipid, protein, and carbohydrate contents were improved when compared to the control. Thus, the use of microalgae as fish feed in the current aquaculture industry is expanding fast as a realistic replacement source (Sirakov et al. 2015; Fadl et al. 2020). Optimization of *Rhodomonas* sp. in aquaculture as live feed as industrial application for crustaceans is going on at many farms (Oostlander et al. 2020). Figure 2.3 shows a schematic workflow for the strain selection of industrial importance. Although promising, the use of microalgae in agriculture as biostimulant and biofertilizers is still in its infancy owing to the absence of recognized and defined procedures for processing and applying the algal material.

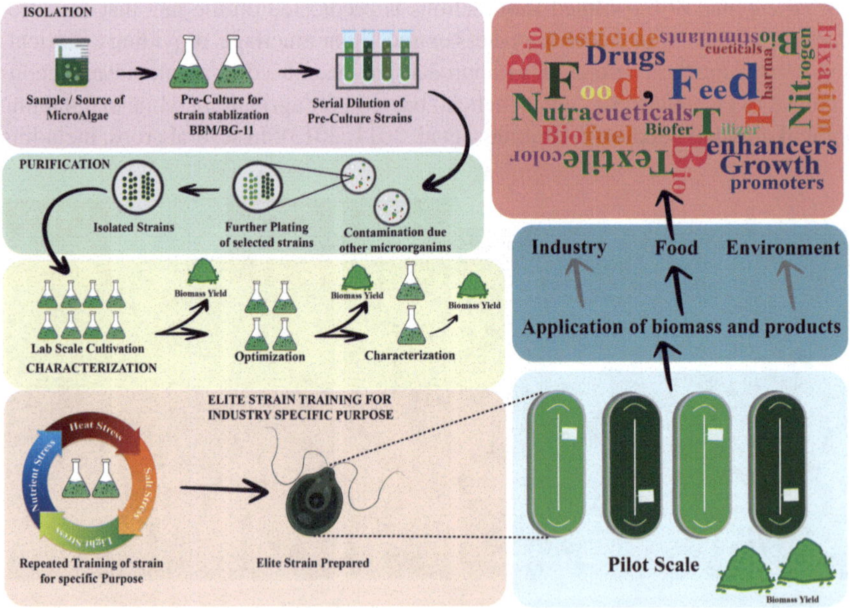

Fig. 2.3 A schematic workflow for the strain selection of industrial applications

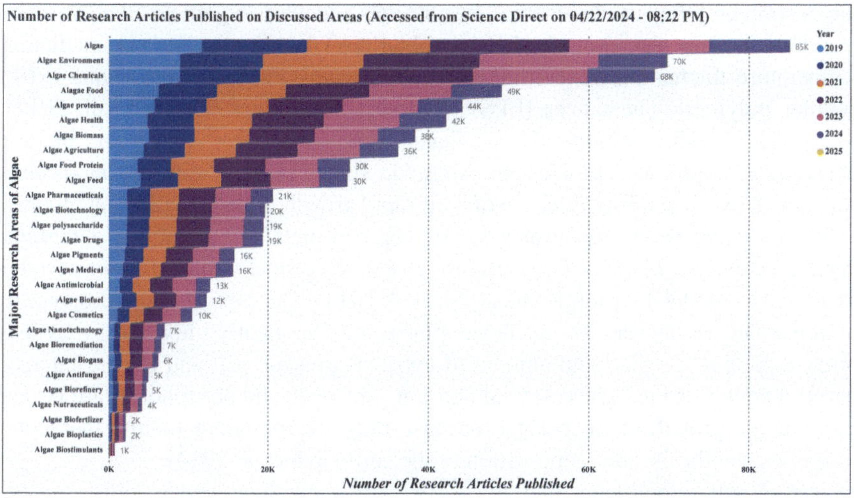

Fig. 2.4 Number of articles published from 2019 to 2025 about algal research from ScienceDirect accessed on 22/04/2024 at 08:22 PM

2.3 Algae Species in Environmental Applications

For their diverse application, there has been a rising trend in algae research for environmental applications in recent years (Fig. 2.4).

2.3.1 Bioremediation

Currently available chemical and physical approaches for removing contaminants from soil and water include ion exchange, filtration, oxidation-reduction, membrane-based procedures, and immobilization techniques (Ramzan et al. 2024). These conventional techniques are costly, inefficient, and labor intensive. Bioremediation is a substitute for chemical and artificial remediation. The use of microorganisms to absorb or extract contaminants from the environmental bodies (fields, lakes, rivers, and oceans) is known as bioremediation. Some of the features that make microorganisms suitable for efficient bioremediation of heavy metals are their tiny size, existence of robust heavy metals sensing apparatus, and high surface to volume ratio (Potnis et al. 2021). The microalgae/cyanobacteria have the ability to degrade contaminants along with significant environmental benefits (Agüera et al. 2020).

Microalgal systems are receiving attention as the most exciting approaches for bioremediation of wastewater. These species are selected for wastewater treatment based on rapid and consistent growth rate, as well as their potential to extract pollutants effectively. Algal species are capable of efficiently removing waste or harmful substances from household and industrial effluent. Biosorption is one of the major strategies they use to resorb extra nutrients and waste organic matter (Mujtaba et al. 2018). They also control other microbial growth by increasing pH, mediating temperature fluctuations, excreting extracellular substances, and producing extracellular polymeric substances (EPS) to protect themselves (García et al. 2017). Microalgal species commonly evaluated for wastewater treatment include *Scenedesmus* spp., *Neochloris* spp., *Nitzschia* spp., *Chlorella* spp., *Desmodesmus* spp., and *Cosmarium* spp. Because of their rapid growth, linkage with algal blooms, resilience under harsh wastewater conditions, and metabolic flexibility, *Chlorella* spp., especially *Chlorella vulgaris*, are extensively investigated for wastewater treatment as model microorganisms (Safi et al. 2014). The capacity of microalgae to bioremediate wastewater and its development may be greatly impacted by its features, including the kind and amount of nutrients present, pH, and the presence of harmful contaminants. The essential nutrients, nitrogen and phosphorus, are necessary for the growth of microalgae because they are important for the metabolic processes, synthesis, and composition of the cells (Safi et al. 2014).

Intriguingly, employing microalgae for wastewater treatment would also produce algal biomass, which might be a source of value-added products for the feed, energy, nutraceutical, and agricultural industries in addition to the phytoremediation impact (Chandrasekhar et al. 2022). However, all algal species are not well adapted to wastewater and are not capable of removing nitrogen and phosphorus from contaminated water; thus, selection of microalgae strain is very important (Kim et al. 2019).

Cyanobacteria participate in the process of biotransformation of more toxic forms of heavy metals into less toxic forms known as biosorption (Chakdar et al. 2022). *Synechocystis, Phormidium, Nostoc, Oscillatoria*, and *Anabaena* are few of the members of phylum cyanobacteria that have been commonly isolated from heavy metal (Hg, Cd, and As) polluted sites, reflecting their resilience for heavy metal stress (Anjana et al. 2007). Despite photoautotrophic life, cyanobacteria have also developed distinct mechanisms to diminish stress caused by reactive oxygen species (ROS). These mechanisms help them to tolerate high heavy metals concentration in their native environment. Some species of microalgae have good potential for bioremediation of different contaminants like xenobiotics, textile dyes, heavy metals, pesticides, phenols, and naphthalene (Kesaano and Sims 2014). Cyanobacteria have the capacity to remove pollutants from soil and water, helping to build sustainable agriculture practices. Species like *Lyngbya* sp., *Microcystis* sp., *Anabaena* sp., and *Nostoc* sp. can be extremely helpful to eliminate the herbicide effect of glyphosate from agricultural sites.

2.3.2 Removal of Dyes from Textile Effluent

On global scale textile industries produce almost 17–20% of dye containing wastewater (Brar et al. 2019). Hence, the release of dye containing wastewater into environment would result in several challenges, such as eutrophication, negative effect on aesthetics, reduction in photosynthetic activity, dissolved chemical oxygen of water bodies, and bioaccumulation of toxins in aquatic environment (Dellamatrice et al. 2017). Industries use numerous conventional physical and chemical techniques to treat wastewater. But they end up with a number of drawbacks, including the need to handle sludge, high cost, ineffective dye removal, and formation of hazardous degradation products (El-Kassas and Mohamed 2014). Microalgae can use textile dye wastewater as cultivation media, in which dye molecules are used as carbon source, whereas phosphorus, nitrogen, and mineral components are used as macronutrients and micronutrients needed for cell proliferation (Cai et al. 2013). As a result, textile dye water would remediate and become decolorized. The mechanism by which microalgae removes dyes from wastewater can be classified into two steps. The first step involves bioabsorption, which binds dye molecules to surface. In the second step bioconversion, microalgae enzymes like azoreductase and peroxidase break down dyes into simple compounds (Varjani et al. 2021).

Textile dyes like black sulfur, indigo, and Remazol Brilliant Blue can be degraded by using cyanobacterial species. Species like *Phormidium autumnale* UTEX1580, *Synechococcus* sp. PCC7942, and *Anabaena flosaquae* UTCC64 can degrade dyes from wastewater of textile industries. *Spirulina* can extract nitrogen from wastewater. Indigo dye can be degraded completely by *Phormidium autumnale* UTEX1580 (Dellamatrice et al. 2017). These applications and studies will solve a major problem of industrial and textile effluent.

2.3.3 Microalgal Potential for Ecosystem Recuperation

The ability of cyanobacteria to thrive in extreme environmental conditions makes them an excellent candidate for ecosystem restoration (Li et al. 2019). Cyanobacteria are pioneers of low fertile and poorly textured soils where they form symbiotic assemblage with fungi, plants, lichens, mosses, and other microorganisms and cover the upper layer of soil. This assemblage is called biocrust (Giraldo-Silva et al. 2019). Cyanobacteria boost biocrust formation, which is crucial for soil stability and better functioning of ecosystem. Biocrust regulates ecosystem restoration and modulate the physio-biochemical properties of soil, hydrological cycle, and nutrient cycle (Lan et al. 2014).

These photosynthetic bacteria improve soil microbial community activity and diversity by this symbiotic association (Rodriguez-Caballero et al. 2018). In degraded soils, their inoculation has shown amazing results in terms of soil erosion resistance, soil biodiversity, moisture content, and organic matter. They do so as they exhibit unique characteristics such as increasing carbon content, fix nitrogen, and produce extracellular polysaccharide (EPS) that assists in the restoration of degraded soils (Potnis et al. 2021). Cyanobacteria produce a matrix of lipids, protein, water, EPS, and other substances that improve soil aggregation and lessen moisture content (Abinandan et al. 2019). In erosion-prone soils cyanobacteria accelerate nitrogen and carbon fixation, suppress runoff, and enhance aggregate formation by production of EPS (Sepehr et al. 2019) so they induce soil colonization. *Microcoleus vaginatus, Scytonema javanicum*, and *Phormidium tenue* have been used as inoculants for improvement and stabilizing sandy soils (Lan et al. 2014). Cyanobacteria have the extraordinary ability to survive in drought conditions and quickly resume their photosynthetic activity upon hydration. This capacity is essential for their survival in desert areas, as they experience drought periodically (Singh et al. 2016). Upon rehydration they can quickly restore their metabolic processes and rebuild cellular components. Cyanobacteria are desirable choices to restore the desert ecosystem because they have short wet cycle (Perera et al. 2018). Cyanobacteria species vary in their ability to endure abiotic stress conditions. So, it is necessary to select appropriate species for the restoration of desert areas (Smaoui et al. 2023).

A thorough strategy is needed for the restoration of post-mining sites because topsoil disturbance involves soil microbial populations that have the potential to change ecosystem function. Cyanobacteria have the capacity to enhance soil function in post-mining sites by carbon sequestration in soil. Chua et al. (2020) highlighted that bio-priming of plant seeds with cyanobacteria enhances seedling growth in soils that are frequently used in mine restoration sites. Because polysaccharides and protein receptors are present on the surface of cyanobacteria, they can accumulate toxins, making them important remediation agents for hydrocarbon and heavy metal polluted soils (Rocha et al. 2020).

2.3.4 Reclamation of Saline Soils

Cyanobacteria species such as *Anabaena* and *Nostoc* have special characteristics that make them useful for reclamation of saline soil (Li et al. 2019). They have capability to extract Na^+ ion from saline soil through trichome-based biosorption (Kirsch et al. 2019). Cyanobacteria secrete EPS (exopolysaccharides) in soil that binds with Na^+ ion and make biofilms. Biofilms protect plants from harmful effects of salinity (Deng et al. 2020). The secretion of EPS in soil assists in stability and aggregate formation, which promotes plant establishment in saline soils. Besides, EPS secretion cyanobacteria also release various kinds of vitamins, acids, amino acids, and hormones that in turn enhance plant growth and development (Rai et al. 2019). As compared to other bacteria, cyanobacteria raise the soil biomass and water holding capacity.

Recent studies showed that direct de-salinization by use of cyanobacteria have significantly enhanced the overall growth and yield of some major crops like wheat, rice, and maize (Li et al. 2019). Better soil structure facilitates salt to move in deeper layers and reduces crop damage (Benidire et al. 2020). The usefulness of cyanobacteria for restoration of salt-affected soils highly depends on climatic conditions and soil properties. As soil texture is a key factor in colonization of cyanobacteria in an ecosystem. According to recent studies *Tolypothrix distorta* and *Scytonema hyalinum* are dominant components of microbial community of low fertile soil of arid zones (Román et al. 2018).

2.3.5 CO_2 Sequestration and O_2 Enrichment

The changing global environment and elevated levels of CO_2 have caused global warming to reach a destructive level by the end of this century. As our industries rely on different carbon sources like natural gas, oil, and coal to meet their energy demands, they are major contributors of CO_2 in the atmosphere (Hansen et al. 2000). It is needed to reduce CO_2 emission from industries. Methods used for sequestration of CO_2 include physical and biological methods. Physical strategies (physical trapping, solubility trapping, ionic trapping, and mineral/chemical trapping) are costly; it is important to develop cost-effective and sustainable strategies for CO_2 sequestration (Intergovernmental Panel on Climate Change, Working Group III 2000). Biological methods are suitable alternative to physical methods. Biological methods include the use of cyanobacteria and microalgae (Shreyash et al. 2021). Cyanobacteria and microalgae absorb atmospheric CO_2 and use it for their photosynthetic activity. Terrestrial plants can capture a massive amount of CO_2 from the air. But carbon fixation capacity of microalgae and cyanobacteria is 10–50 times greater as compared to terrestrial plants (Li et al. 2008). The use of microalgae and cyanobacteria for CO_2 has several advantages: in addition to CO_2 sequestration, they also produce beneficial by-products, co-products, and secondary metabolites. It is estimated that 1 kg of dry cells algae utilize 1.83 kg of carbon dioxide (Kumar et al. 2011).

2.4 Use of Cyanobacteria as Functional Foods

Globally, food safety is a major issue that affects both developing and developed countries (Savary et al. 2012). By 2050, the world's population will demand a 60% increase in agriculture productivity. But current and future crop production is at risk due to climate change (Rosenzweig et al. 2014). The two main leading causes of environmental degradation are rising human population and expansion of agriculture. Approximately 40% of the world's land is made up of pastures and croplands, which constitutes one of main terrestrial biomes. Agronomic practices represent the only significant, widespread type of land use on the planet (Foley et al. 2005). Microalgae and cyanobacteria are important producers that can be found in many terrestrial and aquatic ecosystems. These do not require agricultural land to grow, making them sustainable resources for the revolution in agriculture and food products (Hopes and Mock 2015). An extensive range of nutritionally rich molecules such as proteins, lipids, carbohydrates, antioxidants, vitamins, and different pigments is produced by microalgae and cyanobacteria. Since the Chinese first used microalgae 2000 years ago to survive starvation, they have been an essential source of sustenance (Mobin et al. 2019).

Cyanobacteria are grown for diverse products which are utilized in the food and feed industries. Many nations have also exploited cyanobacteria such as *Spirulina* as food source. The cyanobacterium was first discovered in Lake Texcoco in the Mexico basin by the Aztecs, who were among the first people to use Spirulina as functional food (Barrios et al. 2017). *Spirulina* is high in proteins, valuable unsaturated fatty acids like omega-3, and iron and is one of the best natural green sources of vitamin B_{12}. *Spirulina* also has many beneficial effects on human health, such as the ability to lower blood pressure, prevent renal failure, and promote growth of beneficial *Lactobacillus* bacteria in the intestine (Beheshtipour et al. 2013).

2.4.1 Natural Food Colors

There has also been a noticeable shift in consumer preferences toward naturally occurring pigments made through biological synthesis for use in food and medicinal applications. Therefore, food and nutraceutical producers are exploring the extraction and industrial manufacturing of naturally occurring high-value biochemicals in response to this propensity (Aschemann-Witzel et al. 2021). Since most of the commercially available food colors are now derived from chemical synthesis, their molecular structures or organic reactions that occur during processing sometime lead to concerns about their safety. Instead, natural pigments function as nutritional enhancers with a variety of biological activities and are nontoxic when compared to their synthetic equivalents (Patel et al. 2022).

Microalgae pigments are expected to reach a market value of USD452.4 million by 2025, growing at a compound annual growth rate (CAGR) of 4%. By then, microalgae products will be sold on a bigger scale (Sun et al. 2023). Vital components found in microalgae include proteins, lipids, polysaccharides, minerals vitamins, colors, and polyunsaturated fatty acids (PUFAs), which are highly valuable for industry and health (Silva et al. 2020). Algae's photosynthetic processes and cell development are driven by three primary groups of pigments: phycobiliproteins (PBPs), which make up 8% of DW; chlorophylls, which make up 0.5–0.1% of DW; and carotenoids, which typically make up 0.1–0.2% of DW (or up to 14% in some species) (Lopez and Hall 2020). It has been demonstrated that pigments produced from microalgae have exceptional physiological properties (such as antibacterial and antioxidant properties) and a broad range of medical applications. Microalgae pigments may display a wide range of attractive tints, hues, and natural tones in food, which gives them a substantial competitive advantage as edible colorants that can replicate the color of natural foods (Ruiz-Domínguez et al. 2021). An inexpensive and attractive source of natural colorants is microalgae. For instance, astaxanthin from *Haematococcus pluvialis* and β-carotene from *Dunaliella salina* are marked as functional foods having the potential to be employed for color-giving qualities (Lafarga 2019). Rich in B-phycoerythrin, extracted from red alga

Porphyridium cruentum have been employed as coloring agents in milk products to imitate the color of strawberry-flavored drinks. These extracts exhibit excellent stability throughout a broad pH range and in storage (García et al. 2021).

2.4.2 Dietary Supplements and Vitamins

The demand of dietary supplements was estimated at USD140.1 billion in 2018 which is projected to reach USD216.3 billion by 2026. Microalgae have emerged as a potentially valuable source of health promoting compounds, including carotenoids, phycobiliproteins, and omega-3 fatty acids (El-Baz et al. 2020).

The market for products drives from *Spirulina* is anticipated to expand at a compound annual growth rate (CAGR) of 18.1% through 2028 (Sotiroudis et al. 2013). Most commercial microalgae products are made in Asia or Australia, with European business making up around 5% of the worldwide microalgae market for food and feed (Czerwonka et al. 2018). Nonetheless, microalgae have long been utilized in food; since the 1960s, Japan has been growing *Chlorella* sp. as protein supplement. More recently, the growth of *Spirulina* and *Chlorella* spp. has spread to numerous additional nations for a variety of goods, including as carotenoids and omega-3 fatty acids (Camacho et al. 2019).

Microalgae can be studied as a dietary supplement under two general perceptions. Firstly, whole cells produce high levels of primary metabolites, including proteins, carbohydrates, and lipids that are like common foods such as cereals, vegetables crops, and nuts. Secondly, by extracting precise secondary metabolites from the cell, such as omega-3 fatty acids and carotenoids (Shao et al. 2018). Both human and animal ingestion of these secondary metabolites, sometimes referred to as bioactive substances, have been linked to favorable health effects. Among these are qualities that are anti-inflammatory, antibacterial, anticancer, and anti-obesity (Khoo et al. 2021).

Microalgal products with high potential to be used as dietary supplements include fucoxanthin, lutein, astaxanthin, β-carotene and fatty acids (such as docosahexaenoic acid (DHA), and eicosapentaenoic acid (EPA) (Úbeda et al. 2017). The antioxidant, anti-inflammatory, and anticancer properties of carotenoids derived from microalgae have been demonstrated (Panis and Carreon 2016). The omega-3 fatty acids (EPA and DHA) are the most widely available nutritional supplements in the market and mostly found in fish oil-based supplements (Katiyar and Arora 2020). Fish cannot, however, produce these substances on their own or obtain omega-3 fatty acids by eating algae. Thus, by directly consuming microalgae, several problems associated with fish oil capsules may be avoided, including fishy aftertaste, indigestion, bloating, poor oxidative stability, toxins accumulation, taste, and vegetarian concerns (Kejžar et al. 2021).

Dried *Spirulina* has 5% water, 8% fats, 51–71% proteins, and 24% carbohydrates contents (Ruiz-Domínguez et al. 2021) in addition to iron, calcium, phosphorus, pigments (carotenoids, c-phycocyanin, chlorophyll-a), and vitamins (B_{12}, E).

Additionally, it is thought to be an excellent source of ω-3 fatty acids, such as docosahexaenoic acid (DHA), eicosapentaenoic acid (EPA), and α-linolic acid, as well as ω-6 fatty acids, such as linoleic and γ-linolenic fatty acids. It has been proposed that this high nutritional content may have a beneficial effect on the management of several pathological disorders, including hyperlipidemia, hepatoxicity, cardiovascular disease, and some types of cancer (Rutar et al. 2022).

2.5 Conclusions and Prospects

Resources depletion is a major issue on Earth nowadays. The underdeveloped countries are facing severe food depletion as well as resource deficit conditions. They require productive as well as sustainable resources for their survival in the future. Algae and cyanobacteria are proposing a promising future and being used in almost every field and industry in the world. Their use as food, feed, fertilizer, biofuel, in tanning industry, and in textile industry requires proper characterization of strains. Although they are well adapted to their natural environment, bulk production and sustainable supply is the major issue. There are many unseen factors affecting the production of specific strains like their competition with other species and contamination, etc. to be exploited on an industrial scale. There are many other issues in scaling up the process including biomass production, collection, quality maintenance, and transportation. For these issues, proper research development is required at present to be more effective in future use. Algae-specific cross-discipline research as well as funding calls must focus on the isolation and characterization of novel algal and cyanobacterial strains for a diverse range of applications.

References

Abdo SM, Ali GH (2019) Analysis of polyhydroxybutrate and bioplastic production from microalgae. Bull Natl Res Cent 43. https://doi.org/10.1186/S42269-019-0135-5

Abinandan S, Subashchandrabose SR, Venkateswarlu K, Megharaj M (2019) Soil microalgae and cyanobacteria: the biotechnological potential in the maintenance of soil fertility and health. Taylor & Francis 39:981–998. https://doi.org/10.1080/07388551.2019.1654972

Agüera A, Plaza-Bolaños P, Fernández FGA (2020) Removal of contaminants of emerging concern by microalgae-based wastewater treatments and related analytical techniques. In Current developments in biotechnology and bioengineering: emerging organic micro-pollutants 503–525. https://doi.org/10.1016/B978-0-12-819594-9.00020-6

Anjana K, Kaushik A, Kiran B, Nisha R (2007) Biosorption of Cr (VI) by immobilized biomass of two indigenous strains of cyanobacteria isolated from metal contaminated soil. J Hazard Mater 148:383–386. https://doi.org/10.1016/J.JHAZMAT.2007.02.051

Aschemann-Witzel J, Gantriis RF, Fraga P, Perez-Cueto FJA (2021) Plant-based food and protein trend from a business perspective: markets, consumers, and the challenges and opportunities in the future. Crit Rev Food Sci Nutr 1–10. https://doi.org/10.1080/10408398.2020.1793730

Balaji S, Gopi K, Research BM-A (2013) A review on production of poly β hydroxybutyrates from cyanobacteria for the production of bio plastics. Elsevier 2:278–285. https://doi.org/10.1016/j.algal.2013.03.002

Barrios CAZ, Nandini S, Sarma SSS (2017) Effect of crude extracts from cyanobacterial blooms in Lake Texcoco (Mexico) on the population growth of *Brachionus calyciflorus* (Rotifera). Toxicon 139:45–53. https://doi.org/10.1016/J.TOXICON.2017.09.013

Basheva D, Basheva D, Moten D et al (2018) Content of phycoerythrin, phycocyanin, alophycocyanin and phycoerythrocyanin in some cyanobacterial strains: applications. Eng Life Sci 12(2):144–154. https://doi.org/10.1002/elsc.201800035

Beheshtipour H, Mortazavian AM, Mohammadi R et al (2013) Supplementation of *Spirulina platensis* and *Chlorella vulgaris* algae into probiotic fermented milks. Compr Rev Food Sci Food Saf 12:144–154. https://doi.org/10.1111/1541-4337.12004

Benidire L, El Khalloufi F, Oufdou K et al (2020) Phytobeneficial bacteria improve saline stress tolerance in *Vicia faba* and modulate microbial interaction network. Sci Total Environ 729:139020. https://doi.org/10.1016/J.SCITOTENV.2020.139020

Bidyarani N, Prasanna R, Babu S et al (2016) Enhancement of plant growth and yields in chickpea (*Cicer arietinum* L.) through novel cyanobacterial and bio filmed inoculants. Microbiol Res 188–189:97–105. https://doi.org/10.1016/J.MICRES.2016.04.005

Brar A, Kumar M, Vivekanand V, Pareek N (2019) Phycoremediation of textile effluent-contaminated water bodies employing microalgae: nutrient sequestration and biomass production studies. Int J Environ Sci Te 16:7757–7768. https://doi.org/10.1007/S13762-018-2133-9

Brennan L, Brennan L, Owende P, Owende P (2010) Biofuels from microalgae-a review of technologies for production, processing, and extractions of biofuels and co-products. Renew Sust Energ Rev 14(2):557–577. https://doi.org/10.1016/j.rser.2009.10.009

Bulgari R, Cocetta G, Trivellini A et al (2015) Biostimulants and crop responses: a review. Biol Agric Hortic 31:1–17. https://doi.org/10.1080/01448765.2014.964649

Cai T, Park SY, Li Y (2013) Nutrient recovery from wastewater streams by microalgae: status and prospects. Renew Sust Energ Rev 19:360–369. https://doi.org/10.1016/J.RSER.2012.11.030

Calvo P, Nelson LM, Calvo P et al (2014) Agricultural uses of plant biostimulants. Plant Soil 383:3–41. https://doi.org/10.1007/s11104-014-2131-8

Camacho F, Macedo A, Malcata F (2019) Potential industrial applications and commercialization of microalgae in the functional food and feed industries: a short review. Mar Drugs 17:312. https://doi.org/10.3390/MD17060312

Canizales S, Chen PH, Temmink H et al (2023) Cyanobacteria cultivation on human urine for nutrients recovery. Algal Res 71:103064. https://doi.org/10.1016/j.algal.2023.103064

Castro J d S, Castro J d S, Calijuri ML et al (2020) Microalgae based biofertilizer: a life cycle approach. Sci Total Environ 724:138138. https://doi.org/10.1016/j.scitotenv.2020.138138

Chakdar H, Thapa S, Srivastava A, Shukla P (2022) Genomic and proteomic insights into the heavy metal bioremediation by cyanobacteria. J Hazard Mater 424:127609. https://doi.org/10.1016/J.JHAZMAT.2021.127609

Chandrasekhar K, Raj T, Ramanaiah SV et al (2022) Algae biorefinery: a promising approach to promote microalgae industry and waste utilization. J Biotechnol 345:1–16. https://doi.org/10.1016/j.jbiotec.2021.12.008

Chiaiese P, Chiaiese P, Colla G et al (2018) Renewable sources of plant biostimulation: microalgae as a sustainable means to improve crop performance. Front Plant Sci 9:430391. https://doi.org/10.3389/fpls.2018.01782

Chiellini C, Serra V, Gammuto L et al (2022) Evaluation of nutraceutical properties of eleven microalgal strains isolated from different freshwater aquatic environments: perspectives for their application as nutraceuticals. Food Secur 11(5):654. https://doi.org/10.3390/foods11050654

Chua M, Erickson TE, Merritt DJ et al (2020) Bio-priming seeds with cyanobacteria: effects on native plant growth and soil properties. Restor Ecol 28:S168–S176. https://doi.org/10.1111/REC.13040

Colla G, Rouphael Y (2020) Microalgae: new source of plant biostimulants. Agronomy 10:1240. https://doi.org/10.3390/AGRONOMY10091240
Cooper J, Cooper J, Dobson H, Dobson H (2007) The benefits of pesticides to mankind and the environment. Crop Prot 26(9):1337. https://doi.org/10.1016/j.cropro.2007.03.022
Coppens J, Grunert O, Van Den Hende S et al (2016) The use of microalgae as a high-value organic slow-release fertilizer results in tomatoes with increased carotenoid and sugar levels. J Appl Phycol 28:2367–2377. https://doi.org/10.1007/S10811-015-0775-2/FIGURES/6
Costa JAV, Freitas BCB, Cruz CG et al (2019a) Potential of microalgae as biopesticides to contribute to sustainable agriculture and environmental development. J Environ Sci Health B 54(5):366–375. https://doi.org/10.1080/03601234.2019.1571366
Costa SS, Costa SS, Miranda AL et al (2019b) Microalgae as source of polyhydroxyalkanoates (PHAs) - a review. Int J Biol Macromol 131:536–547. https://doi.org/10.1016/j.ijbiomac.2019.03.099
Czerwonka A, Kaławaj K, Sławińska-Brych A et al (2018) Anticancer effect of the water extract of a commercial Spirulina (*Arthrospira platensis*) product on the human lung cancer A549 cell line. Biomed Pharmacother 106:292–302. https://doi.org/10.1016/J.BIOPHA.2018.06.116
da Costa TP, Quinteiro P, Tarelho LAC et al (2019) Environmental assessment of valorisation alternatives for woody biomass ash in construction materials. Resour Conserv Recycl 148:67–79. https://doi.org/10.1016/J.RESCONREC.2019.04.022
Das SK, Sathish A, Stanley J (2018) Production of biofuel and bioplastic from Chlorella pyrenoidosa. Mater Today Proc 5:16774–16781. https://doi.org/10.1016/J.MATPR.2018.06.020
Dellamatrice PM, Silva-Stenico ME, de Moraes LAB et al (2017) Degradation of textile dyes by cyanobacteria. Braz J Microbiol 48:25–31. https://doi.org/10.1016/J.BJM.2016.09.012
Deng S, Zhang D, Wang G et al (2020) Biological soil crust succession in deserts through a 59-year-long case study in China: how induced biological soil crust strategy accelerates desertification reversal from decades to years. Soil Biol Biochem 141:107665. https://doi.org/10.1016/J.SOILBIO.2019.107665
El-Baz FK, Aly HF, Abd-Alla HI (2020) The ameliorating effect of carotenoid rich fraction extracted from *Dunaliella salina* microalga against inflammation- associated cardiac dysfunction in obese rats. Toxicol Rep 7:118–124. https://doi.org/10.1016/J.TOXREP.2019.12.008
El-Kassas HY, Mohamed LA (2014) Bioremediation of the textile waste effluent by *Chlorella vulgaris*. Egypt J Aquat Res 40:301–308. https://doi.org/10.1016/J.EJAR.2014.08.003
Fadl SE, Elsadany AY, El-Shenawy AM et al (2020) Efficacy of cyanobacterium *Anabaena* sp. as a feed supplement on productive performance and immune status in cultured Nile tilapia. Aquac Rep 17:100406. https://doi.org/10.1016/j.aqrep.2020.100406
Favas R, Morone J, Martins R et al (2021) Cyanobacteria and microalgae bioactive compounds in skin-ageing: potential to restore extracellular matrix filling and overcome hyperpigmentation. J Enzyme Inhib Med Chem 36(1):1829–1838. https://doi.org/10.1080/14756366.2021.1960830
Favas R, Favas R, Morone J et al (2022) Cyanobacteria secondary metabolites as biotechnological ingredients in natural anti-aging cosmetics: potential to overcome hyperpigmentation, loss of skin density and UV radiation-deleterious effects. Mar Drugs 20(3):183. https://doi.org/10.3390/md20030183
Fayzi L, Dayan MP, Fayzi L et al (2020) Biostimulant effect of four Moroccan seaweed extracts applied as seed treatment and foliar spray on maize. Asian J Plant Sci 419–428. https://doi.org/10.3923/ajps.2020.419.428
Fenner K, Canonica S, Wackett LP et al (2013) Evaluating pesticide degradation in the environment: blind spots and emerging opportunities. Science 341(6147):752–758. https://doi.org/10.1126/science.1236281
Ferreira A, Melkonyan L, Carapinha S et al (2021) Biostimulant and biopesticide potential of microalgae growing in piggery wastewater. Environ Adv 4:100062. https://doi.org/10.1016/j.envadv.2021.100062
Foley J, DeFries R, Asner G et al (2005) Global consequences of land use. Science 309(5734):570–574. https://doi.org/10.1126/science.1111772

Freile-Pelegrín Y, Freile-Pelegrín Y, Robledo D (2013) Bioactive phenolic compounds from algae. Bioactive compounds from marine foods: plant and animal sources 113–29. https://doi.org/10.1002/9781118412893.ch6

Freitas-Rodríguez S, Folgueras AR, Freitas-Rodríguez S et al (1864) The role of matrix metalloproteinases in aging: tissue remodeling and beyond. Biochim Biophys Acta 11:2015–2025. https://doi.org/10.1016/j.bbamcr.2017.05.007

Gao X, Jing X, Gao X et al (2021) Biotechnological production of the sunscreen pigment scytonemin in cyanobacteria: progress and strategy. Mar Drugs 19(3):129. https://doi.org/10.3390/md19030129

García D, Alcántara C, Blanco S et al (2017) Enhanced carbon, nitrogen and phosphorus removal from domestic wastewater in a novel anoxic-aerobic photobioreactor coupled with biogas upgrading. Chem Eng J 313:424–434. https://doi.org/10.1016/J.CEJ.2016.12.054

García A, Longo E, Murillo M et al (2021) Using a B-phycoerythrin extract as a natural colorant: application in milk-based products. Molecules 26(2):297. https://doi.org/10.3390/molecules26020297

Giraldo-Silva A, Nelson C, Barger NN, Garcia-Pichel F (2019) Nursing biocrusts: isolation, cultivation, and fitness test of indigenous cyanobacteria. Restor Ecol 27:793–803. https://doi.org/10.1111/REC.12920

Gitau MM, Farkas A, Ördög V, Maróti G (2022) Evaluation of the biostimulant effects of two Chlorophyta microalgae on tomato (*Solanum lycopersicum*). J Clean Prod 364:132689. https://doi.org/10.1016/J.JCLEPRO.2022.132689

Górka B, Górka B, Lipok J et al (2015) Biologically active organic compounds, especially plant promoters, in algae extracts and their potential application in plant cultivation. Marine Algae Extracts 16:659–680. https://doi.org/10.1002/9783527679577.ch37

Hansen J, Sato M, Ruedy R et al (2000) Global warming in the twenty-first century: an alternative scenario. Proc Natl Acad Sci USA 97(18):9875–9880. https://doi.org/10.1073/PNAS.170278997

Hopes A, Mock T (2015) Evolution of microalgae and their adaptations in different marine ecosystems. Encyclopedia Life Sci 1–9. https://doi.org/10.1002/9780470015902.A0023744

Intergovernmental Panel on Climate Change. Working Group III. (2000) Emissions scenarios. Summary for policymakers: a special report of IPCC Working Group III

Jacob-Lopes E, Santos AB, Severo IA et al (2020) Dual production of bioenergy in heterotrophic cultures of cyanobacteria: process performance, carbon balance, biofuel quality and sustainability metrics. Biomass Bioenergy 142:105756. https://doi.org/10.1016/j.biombioe.2020.105756

Jerez-Martel I, García-Poza S, Rodríguez-Martel G et al (2017) Phenolic profile and antioxidant activity of crude extracts from microalgae and cyanobacteria strains. J Food Qual 1:2924508. https://doi.org/10.1155/2017/2924508

Jiménez C, Jiménez C, Cossío BR et al (2003) The feasibility of industrial production of Spirulina (*Arthrospira Platensis*) in southern Spain. Aquaculture 217(1–4):179–190. https://doi.org/10.1016/s0044-8486(02)00118-7

Karthikeyan N, Prasanna R, Karthikeyan N et al (2007) Evaluating the potential of plant growth promoting cyanobacteria as inoculants for wheat. Eur J Soil Biol 43(1):23–30. https://doi.org/10.1016/j.ejsobi.2006.11.001

Karthikeyan N, Prasanna R, Karthikeyan N et al (2009) Physiological characterization and electron microscopic investigation of cyanobacteria associated with wheat rhizosphere. Folia Microbiol 54:43–51. https://doi.org/10.1007/s12223-009-0007-8

Katiyar R, Arora A (2020) Health promoting functional lipids from microalgae pool: a review. Algal Res 46:101800. https://doi.org/10.1016/J.ALGAL.2020.101800

Kaur P, Purewal SS (2019) Biofertilizers and their role in sustainable agriculture. Biofertilizers for sustainable agriculture and environment, 285–300

Kejžar J, Jagodic Hudobivnik M, Nečemer M et al (2021) Characterization of algae dietary supplements using antioxidative potential, elemental composition, and stable isotopes approach. Front Nutr 7:618503. https://doi.org/10.3389/FNUT.2020.618503/FULL

Kesaano M, Sims RC (2014) Algal biofilm-based technology for wastewater treatment. Algal Res 5:231–240. https://doi.org/10.1016/j.algal.2014.02.003

Khan Z, Bhadouria P, Bisen P (2005) Nutritional and therapeutic potential of Spirulina. Curr Pharm Biotechnol 6:373–379. https://doi.org/10.2174/138920105774370607

Khan WZ, Khan W, Rayirath UP et al (2009) Seaweed extracts as biostimulants of plant growth and development. J Plant Growth Regul 28:386–399. https://doi.org/10.1007/s00344-009-9103-x

Kholssi R, Kholssi R, Marks EAN et al (2019) Biofertilizing effect of Chlorella sorokiniana suspensions on wheat growth. J Plant Growth Regul 15(38):644–649. https://doi.org/10.1007/s00344-018-9879-7

Khoo KS, Chong YM, Chang WS et al (2021) Permeabilization of *Chlorella sorokiniana* and extraction of lutein by distillable CO_2-based alkyl carbamate ionic liquids. Sep Purif Technol 256:117471. https://doi.org/10.1016/J.SEPPUR.2020.117471

Kim I, Yang H-M, Park CW et al (2019) Removal of radioactive cesium from an aqueous solution via bioaccumulation by microalgae and magnetic separation. Sci Rep 9:10149. https://doi.org/10.1038/s41598-019-46586-x

Kim S, Jeon H, Han HS, Hur JW (2021) Evaluation of *Bacillus albus* SMG-1 and *B. safensis* SMG-2 isolated from Saemangeum Lake as probiotics for aquaculture of white shrimp (*Litopenaeus vannamei*). Aquac Rep 20:100743. https://doi.org/10.1016/j.aqrep.2021.100743

Kirsch F, Klähn S, Hagemann M (2019) Salt-regulated accumulation of the compatible solutes sucrose and glucosyl glycerol in cyanobacteria and its biotechnological potential. Front Microbiol 10:482569. https://doi.org/10.3389/FMICB.2019.02139

Kuddus M, Kuddus M, Singh P et al (2013) Recent developments in production and biotechnological applications of c-phycocyanin. Biomed Res Int 1:742859. https://doi.org/10.1155/2013/742859

Kumar K, Dasgupta CN, Nayak B et al (2011) Development of suitable photobioreactors for CO_2 sequestration addressing global warming using green algae and cyanobacteria. Bioresour Technol 102:4945–4953. https://doi.org/10.1016/J.BIORTECH.2011.01.054

Kumar M, Kumar M, Prasanna R et al (2013) Evaluating the plant growth promoting ability of thermotolerant bacteria and cyanobacteria and their interactions with seed spice crops. Sci Hortic 164:94–101. https://doi.org/10.1016/j.scienta.2013.09.014

Lafarga T (2019) Effect of microalgal biomass incorporation into foods: nutritional and sensorial attributes of the end products. Algal Res 41:101566. https://doi.org/10.1016/J.ALGAL.2019.101566

Lambert S, Wagner M (2017) Environmental performance of bio-based and biodegradable plastics: the road ahead. Chem Soc Rev 46:6855–6871. https://doi.org/10.1039/C7CS00149E

Lan S, Zhang Q, Wu L et al (2014) Artificially accelerating the reversal of desertification: cyanobacterial inoculation facilitates the succession of vegetation communities. Environ Sci Technol 48:307–315. https://doi.org/10.1021/ES403785J

Lesueur D, Deaker R, Herrmann L et al (2016) The production and potential of biofertilizers to improve crop yields. Bioformulations Sustain Agric 2016:71–92. https://doi.org/10.1007/978-81-322-2779-3_4

Li Y, Horsman M, Wu N et al (2008) Biofuels from microalgae. Biotechnol Prog 24:815–820. https://doi.org/10.1021/BP070371K

Li H, Zhao Q, Huang H (2019) Current states and challenges of salt-affected soil remediation by cyanobacteria. Sci Total Environ 669:258–272. https://doi.org/10.1016/J.SCITOTENV.2019.03.104

Lopez M, Hall C (2020) Physiology, osmosis. StatPearls Publishing

Madadi R, Madadi R, Zahed MA et al (2021a) Simultaneous phycoremediation of petrochemical wastewater and lipid production by *Chlorella vulgaris*. SN Appl Sci 3:1–0. https://doi.org/10.1007/s42452-021-04511-w

Madadi R, Maljaee H, Serafim LS, Ventura SPM (2021b) Microalgae as contributors to produce biopolymers. Mar Drugs 19:466. https://doi.org/10.3390/MD19080466

Madani NSH, Adorian TJ, Madani NSH et al (2018) The effects of dietary probiotic Bacilli (*Bacillus subtilis* and *Bacillus licheniformis*) on growth performance, feed efficiency, body composition and immune parameters of whiteleg shrimp (*Litopenaeus vannamei*) postlarvae. Aquac Res 49(5):1926–1933. https://doi.org/10.1111/are.13648

Maschek JA, Baker BJ, Maschek JA, et al (2008) The chemistry of algal secondary metabolism. 1–24 https://doi.org/10.1007/978-3-540-74181-7_1

Mobin SMA, Chowdhury H, Alam F (2019) Commercially important bioproducts from microalgae and their current applications – a review. Energy Procedia 160:752–760. https://doi.org/10.1016/J.EGYPRO.2019.02.183

Morone J, Alfeus A, Morone J et al (2019) Revealing the potential of cyanobacteria in cosmetics and cosmeceuticals: a new bioactive approach. Algal Res-Biomass Biofuels Bioproducts 41:101541. https://doi.org/10.1016/j.algal.2019.101541

Mujtaba G, Rizwan M, Kim G, Lee K (2018) Removal of nutrients and COD through co-culturing activated sludge and immobilized *Chlorella vulgaris*. J Chem Eng 343:155–162. https://doi.org/10.1016/J.CEJ.2018.03.007

Mukherjee PK, Mukherjee PK, Maity N et al (2011) Bioactive compounds from natural resources against skin aging. Phytomedicine 19(1):64–73. https://doi.org/10.1016/j.phymed.2011.10.003

Nanni B, Balestreri E, Nanni B et al (2001) Characterisation of a specific phycocyanin-hydrolysing protease purified from *Spirulina platensis*. Microbiol Res 156(3):259–266. https://doi.org/10.1078/0944-5013-00110

Ness D (2008) Sustainable urban infrastructure in China: towards a factor 10 improvement in resource productivity through integrated infrastructure systems. Int J Sustain Dev World Ecol 15(4):288–301. https://doi.org/10.3843/SusDev.15.4:2a

Nowruzi B, Sarvari G, Blanco S (2020) The cosmetic application of cyanobacterial secondary metabolites. Algal Res 49:101959. https://doi.org/10.1016/j.algal.2020.101959

Nunnery JK, Mevers E, Nunnery JK et al (2010) Biologically active secondary metabolites from marine cyanobacteria. Curr Opin Biotechnol 21(6):787–793. https://doi.org/10.1016/j.copbio.2010.09.019

Oostlander PC, van Houcke J, Wijffels RH, Barbosa MJ (2020) Optimization of *Rhodomonas* sp. under continuous cultivation for industrial applications in aquaculture. Algal Res 47:101889. https://doi.org/10.1016/j.algal.2020.101889

Oren A, Gunde-Cimerman N, Oren A, Gunde-Cimerman N (2007) Mycosporines and mycosporine-like amino acids: UV protectants or multipurpose secondary metabolites? FEMS Microbiol Lett 269(1):1–0. https://doi.org/10.1111/j.1574-6968.2007.00650.x

Osorio-Reyes JG, Valenzuela-Amaro HM, Pizaña-Aranda JJP et al (2023) Microalgae-based biotechnology as alternative biofertilizers for soil enhancement and carbon footprint reduction: advantages and implications. Mar Drugs 21:93. https://doi.org/10.3390/MD21020093

Panis G, Carreon JR (2016) Commercial astaxanthin production derived by green alga Haematococcus pluvialis: a microalgae process model and a techno-economic assessment all through production line. Algal Res 18:175–190. https://doi.org/10.1016/J.ALGAL.2016.06.007

Patel AK, Albarico FPJB, Perumal PK et al (2022) Algae as an emerging source of bioactive pigments. Bioresour Technol 351:126910. https://doi.org/10.1016/J.BIORTECH.2022.126910

Perera I, Subashchandrabose SR, Venkateswarlu K et al (2018) Consortia of cyanobacteria/microalgae and bacteria in desert soils: an underexplored microbiota. Appl Microbiol Biotechnol 102:7351–7363. https://doi.org/10.1007/S00253-018-9192-1

Potnis AA, Raghavan PS, Rajaram H (2021) Overview on cyanobacterial exopolysaccharides and biofilms: role in bioremediation. Rev Environ Sci Biotechnol 20:781–794. https://doi.org/10.1007/S11157-021-09586-W

Prabhakaran M, Elumalai S, Santhose BI, Kanna GR (2014) Collection, isolation and identification of *Haematococcus pluvialis* Flotow from high altitude region of Pithoragarh district, Uttarakhand, India. Gold Res Thoughts 3:1–7

Prasanna R, Prasanna R, Pattnaik S et al (2011) Development of cyanobacterium-based biofilms and their in vitro evaluation for agriculturally useful traits. Folia Microbiol 56:49–58. https://doi.org/10.1007/s12223-011-0013-5

Prasanna R, Prasanna R, Sharma E et al (2013) Soil fertility and establishment potential of inoculated cyanobacteria in rice crop grown under non-flooded conditions. Paddy Water Environ 11:175–183. https://doi.org/10.1007/s10333-011-0302-2

Prasanna R, Prasanna R, Triveni S et al (2014) Evaluating the efficacy of cyanobacterial formulations and biofilmed inoculants for leguminous crops. Arch Agron Soil Sci 60(3):349–366. https://doi.org/10.1080/03650340.2013.792407

Prasanna R, Adak A, Prasanna R et al (2015) Cyanobacterial inoculation in rice grown under flooded and SRI modes of cultivation elicits differential effects on plant growth and nutrient dynamics. Ecol Eng 84:532–541. https://doi.org/10.1016/j.ecoleng.2015.09.033

Rai AN, Singh AK, Syiem MB (2019) Plant growth-promoting abilities in cyanobacteria. Cyanobacteria:459–476. https://doi.org/10.1016/B978-0-12-814667-5.00023-4

Ramzan M, Sarwar S, Ahmad MZ et al (2024) Phytoremediation of heavy metal-contaminated soil of Lyari River using bioenergy crops. S Afr J Bot 167:663–670. https://doi.org/10.1016/J.SAJB.2024.02.034

Rasheed R, Thaher M, Younes N et al (2022) Solar cultivation of microalgae in a desert environment for the development of techno-functional feed ingredients for aquaculture in Qatar. Sci Total Environ 835:155538. https://doi.org/10.1016/j.scitotenv.2022.155538

Rocha F, Lucas-Borja ME, Pereira P, Muñoz-Rojas M (2020) Cyanobacteria as a nature-based biotechnological tool for restoring salt-affected soils. Agronomy 10:1321. https://doi.org/10.3390/AGRONOMY10091321

Rodriguez-Caballero E, Belnap J, Büdel B et al (2018) Dryland photoautotrophic soil surface communities endangered by global change. Nat Geosci 11:185–189. https://doi.org/10.1038/S41561-018-0072-1

Román JR, Roncero-Ramos B, Chamizo S et al (2018) Restoring soil functions by means of cyanobacteria inoculation: importance of soil conditions and species selection. Land Degrad Dev 29:3184–3193. https://doi.org/10.1002/LDR.3064

Ronga D, Ronga D, Biazzi E et al (2019) Microalgal biostimulants and biofertilisers in crop productions. Agronomy 9(4):192. https://doi.org/10.3390/agronomy9040192

Rosenzweig C, Elliott J, Deryng D et al (2014) Assessing agricultural risks of climate change in the 21st century in a global gridded crop model intercomparison. Proc Natl Acad Sci USA 111:3268–3273. https://doi.org/10.1073/PNAS.1222463110

Ruiz-Domínguez MC, Fuentes JL, Mendiola JA et al (2021) Bioprospecting of cyanobacterium in Chilean coastal desert, Geitlerinema sp. molecular identification and pressurized liquid extraction of bioactive compounds. Food Bioprod Process 128:227–239. https://doi.org/10.1016/J.FBP.2021.06.001

Rutar JM, Hudobivnik MJ, Nečemer M et al (2022) Nutritional quality and safety of the spirulina dietary supplements sold on the Slovenian market. Food Secur 11:849. https://doi.org/10.3390/FOODS11060849/S1

Safi C, Zebib B, Merah O et al (2014) Morphology, composition, production, processing and applications of Chlorella vulgaris: a review. Renew Sust Energ Rev 35:265–278. https://doi.org/10.1016/J.RSER.2014.04.007

Savary S, Ficke A, Aubertot JN, Hollier C (2012) Crop losses due to diseases and their implications for global food production losses and food security. Food Secur 4:519–537. https://doi.org/10.1007/S12571-012-0200-5/TABLES/3

Senatore V, Rueda E, Bellver M et al (2023) Production of phycobiliproteins, bioplastics and lipids by the cyanobacteria *Synechocystis* sp. treating secondary effluent in a biorefinery approach. Sci Total Environ 857:159343. https://doi.org/10.1016/j.scitotenv.2022.159343

Sepehr A, Hassanzadeh M, Rodriguez-Caballero E (2019) The protective role of cyanobacteria on soil stability in two Aridisols in northeastern Iran. Geoderma Reg 16:e00201. https://doi.org/10.1016/J.GEODRS.2018.E00201

Shao W, Zhang J, Wang K et al (2018) Cocamidopropyl betaine-assisted foam separation of freshwater microalgae *Desmodesmus brasiliensis*. Biochem Eng J 140:38–46. https://doi.org/10.1016/J.BEJ.2018.09.006

Sharma HSS, Sharma HSS, Fleming CC et al (2014) Plant biostimulants: a review on the processing of macroalgae and use of extracts for crop management to reduce abiotic and biotic stresses. J Appl Phycol 26:465–490. https://doi.org/10.1007/s10811-013-0101-9

Shreyash N, Sonker M, Bajpai S et al (2021) The review of carbon capture-storage technologies and developing fuel cells for enhancing utilization. Energies 14:4978. https://doi.org/10.3390/EN14164978

Silva S, Ferreira I, Dias M et al (2020) Microalgae-derived pigments: a 10-year bibliometric review and industry and market trend analysis. Molecules 25(15):3406. https://doi.org/10.3390/molecules25153406

Singh JS, Kumar A, Rai AN, Singh DP (2016) Cyanobacteria: a precious bio-resource in agriculture, ecosystem, and environmental sustainability. Front Microbiol 7:186282. https://doi.org/10.3389/FMICB.2016.00529/BIBTEX

Singh AK, Sharma L, Singh AK et al (2017) Progress and challenges in producing polyhydroxyalkanoate biopolymers from cyanobacteria. J Appl Phycol 29:1213–1232. https://doi.org/10.1007/s10811-016-1006-1

Sirakov I, Velichkova K, Sirakov I et al (2015) The importance of microalgae for aquaculture industry. Rev Int J Fish Aquat Stud 2(4):81–84

Slinksienė R, Sendzikiene E, Mikolaitiene A et al (2022) Use of microalgae biomass for production of granular nitrogen biofertilizers. Green Chem Lett Rev 15(2):416–426. https://doi.org/10.1080/17518253.2022.2071593

Smaoui S, D'Amore T, Tarapoulouzi M et al (2023) Aflatoxins contamination in feed commodities: from occurrence and toxicity to recent advances in analytical methods and detoxification. Microorganisms 11(10):2614. https://doi.org/10.3390/MICROORGANISMS11102614

Sotiroudis T, Chemical, GS-J of the S (2013) Undefined (2013) health aspects of spirulina (*Arthrospira platensis*) microalga food supplement. J Serb Chem Soc 78:395–405. https://doi.org/10.2298/JSC121020152S

Sun H, Wang Y, He Y et al (2023) Microalgae-derived pigments for the food industry. Mar Drugs 21(2):82. https://doi.org/10.3390/MD21020082

Thomassen G, Egiguren Vila U, Van Dael M et al (2016) A techno-economic assessment of an algal-based biorefinery. Clean Techn Environ Policy 18:1849–1862. https://doi.org/10.1007/S10098-016-1159-2/METRICS

Torres-Tiji Y, Fields FJ, Mayfield SP (2020) Microalgae as a future food source. Biotechnol Adv 41:107536. https://doi.org/10.1016/J.BIOTECHADV.2020.107536

Tseng CC, Yeh HY, Liao ZH et al (2021) An in vitro study shows the potential of Nostoc commune (cyanobacteria) polysaccharides extract for wound-healing and anti-allergic use in the cosmetics industry. J Funct Foods 87:104754. https://doi.org/10.1016/j.jff.2021.104754

Úbeda B, Gálvez JÁ, Michel M, Bartual A (2017) Microalgae cultivation in urban wastewater: *Coelastrum* cf. *pseudomicroporum* as a novel carotenoid source and a potential microalga harvesting tool. Bioresour Technol 228:210–217. https://doi.org/10.1016/J.BIORTECH.2016.12.095

Van Iersel S, Flammini A (2010) Algae-based biofuels: applications and co-products

Varjani S, Rakholiya P, Shindhal T et al (2021) Trends in dye industry effluent treatment and recovery of value added products. J Water Process Eng 39:101734. https://doi.org/10.1016/J.JWPE.2020.101734

Vaz B d S, Costa JAV, Vaz B d S et al (2019a) Innovative nanofiber technology to improve carbon dioxide biofixation in microalgae cultivation. Bioresour Technol 273:592–598. https://doi.org/10.1016/j.biortech.2018.11.054

Vaz B d S, Vaz B d S, Costa JAV et al (2019b) Physical and biological fixation of CO_2 with polymeric nanofibers in outdoor cultivations of *Chlorella fusca* LEB 111. Int J Biol Macromol 151:1332–1339. https://doi.org/10.1016/j.ijbiomac.2019.10.179

Videau P, Videau P, Videau P et al (2019) *Anabaena* sp. strain PCC 7120: laboratory maintenance, cultivation, and heterocyst induction. Curr Protoc Microbiol 52(1):e71. https://doi.org/10.1002/cpmc.71

Volkmann M, Gorbushina AA, Volkmann M, Gorbushina AA (2006) A broadly applicable method for extraction and characterization of mycosporines and mycosporine-like amino acids of terrestrial, marine and freshwater origin. FEMS Microbiol Lett 255(2):286–295. https://doi.org/10.1111/j.1574-6968.2006.00088.x

Wang N, Xiang Y, Ma Y, Zhang P, Zhou X, Zhu H, Zhang Z, Li Z, Xiao X, He M, Mehmood MA, Zhu H (2024) Fermented *Spirulina platensis* alleviated DSS-induced ulcerative colitis by regulating gut microbiota and MyD88/TLR4 signaling pathway. Asian J Agric Biol 1(1):2023232. https://doi.org/10.35495/ajab.2023.232

Wiegand C, Pflugmacher S, Wiegand C, Pflugmacher S (2005) Ecotoxicological effects of selected cyanobacterial secondary metabolites: a short review. Toxicol Appl Pharmacol 203(3):201–218. https://doi.org/10.1016/j.taap.2004.11.002

Wolkers H, Barbosa MJ, Kleinegris DMM et al (2011) Microalgae: the green gold of the future?: large-scale sustainable cultivation of microalgae for the production of bulk commodities. Wageningen UR Food & Biobased Research

Yamaguchi Y, Yamaguchi Y, Koketsu M, Koketsu M (2016) Isolation and analysis of polysaccharide showing high hyaluronidase inhibitory activity in *Nostochopsis lobatus* MAC0804NAN. J Biosci Bioeng 121(3):345–348. https://doi.org/10.1016/j.jbiosc.2015.07.008

Zhang C, Show PL, Ho SH (2019) Progress and perspective on algal plastics – a critical review. Bioresour Technol 289:121700. https://doi.org/10.1016/J.BIORTECH.2019.121700

Zhuang D, Zhuang D, He N et al (2021) Application progress of bioactive compounds in microalgae on pharmaceutical and cosmetics. Chemosphere 291:132932. https://doi.org/10.1016/j.chemosphere.2021.132932

Chapter 3
Cyanobacterial Metabolic Pathways of Industrial Interests

Aqib Zafar Khan, Si Cai, Lingcheng Liu, Qingyue Cheng, Zhangli Hu, and Yihong Zheng

Abstract Cyanobacteria have substantial potential to utilize nitrogen and phosphorous for their growth alongside carbon dioxide and can transform these elements into both primary and secondary metabolites, making them a promising platform for the sustainable production of diverse biochemicals. These metabolites are not only pivotal in the realms of pharmaceuticals and nutraceuticals but also hold potential for a variety of industrial applications. These include but are not limited to bioplastic, anti-aging compounds, antibacterial, antifungal agents, and antioxidants. The market size of cyanobacteria-based bioproducts reached $348 million (CAGR:10.5) in 2020, while the algae-based bioproducts are estimated to be $19.85 billion (CAGR: 7.8) in 2028. Cyanobacteria biosynthesize these metabolites, which are regulated by different genes and cellular pathways, and produce hydrogen, terpenoids, ethanol, aliphatic alcohols, and more. Several methods for the uptake of nitrogen and phosphorus by cyanobacteria and the utilization of these nutrients in cellular metabolism are covered in this chapter. Cyanobacterial pathways responsible for producing hydrogen, terpenoids, and bioalcohols are also discussed. Besides, market and economic statistics of cyanobacterial biomass and its metabolites are also highlighted from an industrial point of view.

Keywords Cyanobacteria · Terpenoids · Hydrogen · Alcohols · Economics

---

A. Z. Khan · S. Cai · L. Liu · Q. Cheng · Z. Hu · Y. Zheng (✉)
Guangdong Engineering Research Centre for Marine Algal Biotechnology, Shenzhen Engineering Laboratory for Marine Algal Biotechnology, College of Life Sciences and Oceanography, Shenzhen University, Shenzhen, China
e-mail: zhengy@szu.edu.cn

M. A. Mehmood et al. (eds.), *Cyanobacteria Biotechnology*,
https://doi.org/10.1007/978-3-031-70698-1_3

3.1 Introduction

Cyanobacteria are ubiquitous in marine and freshwater ecosystems, where they are toxins that may harm both humans and other life; they also produce valuable primary and secondary metabolites via metabolic activities, including photosynthesis and carotenogenesis. Developmental activities of cyanobacteria, which include reproduction, growth, and cell division, are the exclusive domain of primary metabolites (Mazard et al. 2016). These primary metabolites, including carbohydrates, proteins, and lipids, have the potential to be transformed into valuable biotechnological products such as bioplastics, nutritional supplements, pigments, and biofertilizers (Agarwal et al. 2022; Chittora et al. 2020). Secondary metabolites, on the other hand, are not directly involved in the average growth and development processes of cyanobacteria. These metabolites, which include toxins, phenolic compounds, essential oils, alkaloids, steroids, and antibiotics, are unique to particular species and play a role in the stress response and defense mechanisms of cyanobacteria (Żymańczyk-Duda et al. 2022). An open-access database specifically for secondary metabolites produced by cyanobacteria, CyanoMetDB, contains the structural details and bioactivities of a large number of these chemicals (Jones et al. 2021). Recently, toxigenic cyanobacteria species have received more attention, with the goals of isolating and identifying toxins and developing techniques and systems for water purification (Massey et al. 2022; Burford et al. 2020). Interests in cyanobacteria have recently been shifted for their commercial application because they contain beneficial biologically active compounds, which are of interest for a variety of health, industrial, and environmental applications (Shahid et al. 2023, Jodlbauer et al. 2021; Wang et al. 2023).

Because cyanobacterial genomes are more straightforward when compared to eukaryotic genomes, they are considered the ideal microbes for screening for novel and medicinally significant compounds (Khan et al. 2019). Cyanobacteria can naturally manufacture various metabolites through diverse photosynthetic metabolism in response to environmental challenges such as changes in light intensity and nutrients. Our understanding of cyanobacterial regulatory processes may explain this disparity, the limitations of the genetic tools at our disposal, and the fact that various cyanobacterial strains exhibit varied levels of competence regarding defined genomic components (Rachedi et al. 2020). The availability of 923 cyanobacterial genome assemblies in GenBank (Sayers et al. 2021) represents a substantial increase in genomic sequencing data. This expansion of resources has dramatically facilitated the integration of studies in transcriptomics, proteomics, and metabolomics, thereby broadening the focus beyond model organisms and enabling the identification of novel species with industrial relevance (Khan et al. 2019).

Furthermore, diverse taxonomic cyanobacteria groups can produce various physiologically active compounds. The cyclic peptide anabaenopeptin, for instance, was initially isolated from the freshwater cyanobacterium *Dolichospermum flos-aquae* (*Anabaena flos-aquae*) (Harada et al. 1995). Around 260 distinct groups of cyanobacterial metabolites have been identified. Notably, organisms within the orders

Oscillatoriales (with 153 distinct metabolite groups) and *Nostocales* (with 98 distinct groups) produce the most diverse array of metabolites. Even in other genera, there were several metabolites with biological activity. As an example, cyanobacteria from the genus *Synechococcus* were discovered to have metabolites from more than 30 distinct families (Demay et al. 2019). In terms of biological activity, cyanobacterial metabolites are categorized into 14 classes, including neurotic, anti-inflammatory, antioxidant, antiviral, antibacterial, antifungal, antiprotozoal, and hepatotoxic effects (Demay et al. 2019). Numerous enterprises already produce such cyanobacteria-based bioproducts as whole biomass or specific extracted pigments (e.g., phycocyanin) for biotechnology applications, as shown in Table 3.2.

Many commercial uses have been discovered for bioactive metabolites produced by cyanobacteria, particularly in the nutraceutical and cosmetics sectors (Chakdar and Pabbi 2016). The global market for algal products is predicted to reach $6.09 billion by 2026, expanding at a compound annual growth rate (CAGR) of 6.7%. Forecasts indicate that the worldwide demand for cyanobacteria will more than two-fold rise between 2018 and 2025 (Amin et al. 2023). Among the various cyanobacterial products, pigments such as phycocyanin and phycoerythrin stand out as extensively used components in industrial applications, with a multitude of goods being marketed by multiple companies, as shown in Table 3.2 (Chakdar and Pabbi 2016).

This chapter discusses the cyanobacterial metabolic pathways for nitrate and phosphate metabolism, which will help us understand the metabolism of major inorganic nutrients. Some valuable products (such as hydrogen, terpenoids, polyketides, photosynthetic, and ethanolic pathways) obtained from the cyanobacteria are also discussed, which are the building blocks for the many biotechnological and industrial uses. Market data from the past, present, and future are also included.

3.2 Nutrient Uptake Pathway and Their Metabolism

Cyanobacteria are pivotal in the nitrogen cycle, particularly in removing nitrogen elements in aquatic environments. The nitrogen sources mainly used by cyanobacteria include nitrates and nitrites, ammonium salts, urea, dinitrogen (molecular nitrogen), cyanates, etc. Multiple genes are involved in the metabolism of each nitrogen source in cyanobacteria, which are given in Table 3.1.

3.2.1 Ammonium Uptake Pathways

Ammonium is often considered the preferred nitrogen source for cyanobacteria due to its higher reducibility than typical inorganic nitrogen compounds. Cyanobacteria usually prioritize using ammonium when presented with multiple nitrogen options. The metabolic process involves converting various nitrogen sources into

Table 3.1 List of genes encoded for nitrogen and phosphorous metabolism pathway in the cyanobacteria

Substrate	Gene/Protein	Functions	References
Ammonium	GS, GOGAT	Transform the ammonia into carbon skeleton	Muro-Pastor et al. (2005)
	Ammonium/methylammonium permease	Ammonium Transporter	
Nitrates and nitrites	Nitrate reductase and nitrite reductase	Convert substrate into ammonia	Maeda et al. (2014)
	[4Fe-4S] centers and a Mo-bis-molybdopterin guanine dinucleotide	Nitrite reduction	Flores et al. (2005)
Urea	*ureABCDEFG,*	Convert urea into ammonia	Veaudor et al. (2019)
	urtABCDE	Urease transporter	
Molecular nitrogen	*nifHDK*	Nitrogenase transformed nitrogen into ammonium	Bothe et al. (2010)
Cyanate	*cynS*	Cyanase converts cyanate into ammonia and CO_2	Kamennaya and Post (2011)
Orthophosphate	*Pit/PsT*	Phosphorous transporter	Xiao et al. (2022)
Organic phosphorus	*Alkaline phosphatase*	Transfer phosphomonoester into orthophosphate	

ammonium, which is then assimilated into biomass. After permease allows ammonium to enter cyanobacteria, glutamine synthetase (GS) and glutamate synthase (GOGAT) carry out a series of processes that integrate ammonium into the carbon skeleton. The transcription factor *NtcA* is crucial for the production of both *GS* and *IDH*, and 2-oxoglutarate is produced by *IDH*, which serves as the carbon skeleton substrate of the GS-GOGAT pathway (Muro-Pastor et al. 2005). Cyanobacteria can absorb ammonium natively due to the pull of GS and the diffusion of ammonia through their cytoplasmic membrane. Specific genera, such as *Chlorella* and *Anabaena*, have been reported to possess the ammonium/methylammonium permease, which allows them to absorb ammonium via an alternative transport mechanism. Methylammonium may be used to measure the activity of ammonium transport proteins (Muro-Pastor et al. 2005).

3.2.2 *Nitrates and Nitrites Uptake Pathways*

The active transport system is the primary pathway for nitrate uptake in cyanobacteria, with nitrate reductase and nitrite reductase enzymes catalyzing the conversion into ammonium inside the cells (Herrero et al. 2001). Cyanobacteria have an enzyme called nitrate reductase that breaks down nitrate into nitrite, and an enzyme called nitrite reductase further reduces nitrite to ammonium. Although the precise process

Table 3.2 List of companies cultivating cyanobacteria for commercial applications

Company and Origin	Description	Websites
Living Ink, USA, 2013	A biomaterials startup, uses algae pigments for ecofriendly ink production to replace petroleum-derived goods like ink	http://livingink.co/
CyanoCapture, UK, 2021	Genetically modified cyanobacteria are used as cellular factories for high-value bioproduction of biomass and products, using low-cost plug-and-play photobioreactors	https://www.cyanocapture.com/
Algenuity, UK, 2009	Developed the Algem® labscale photobioreactor for algae and cyanobacteria research and provides services to change microalgae for desired characteristics	https://www.algenuity.com/
Photanol, the Netherlands, 2008	Photanol's patents use genetically modified cyanobacteria to synthesize several biochemicals	https://photanol.com/
Solaga, Germany, 2015	Solaga uses cyanobacteria to produce biogas, by immobilizing them on a solar panel, using the energy they need to survive and creating intermediates that may be turned into biogas	https://www.solaga.de/
Growponics, UK	Develops automated hydroponic greenhouse factories using current agronomics and cutting-edge approaches to optimize food/plant production and profitability on annual crop yields	https://www.growponics.co.uk/
Algae Biotecnologia, Brazil, 2009	Pioneered microalgae and cyanobacteria cultivation systems in Brazil and Latin America for treating liquid and gaseous agro-industrial effluents, producing animal nutrition ingredients and additives, human food supplements, and biofuels	
BioFields, Mexico, 2008	Producing biofuels from cyanobacteria	https://grupoalego.com/
Bluetec, China Naturals, 1998	Produce Spirulina and phycocyanin	https://www.bestphycocyanin.com/
Fuqing King Dnarmsa, China, Spirulina, 1995	Produce Spirulina and phycocyanin	http://en.kingdnarmsa.cn/
Japan Algae, Japan, 1980	Produce Spirulina and natural pigments	http://www.sp100.com/
Taiwan Wilson Enterprise, Taiwan, 1978	Produce Spirulina tablets for human consumption	https://www.wilson-groups.com/

by which cyanobacteria adapt to low-nitrite environments is unclear, they can actively move nitrite via nitrite transport (Maeda et al. 2014). Both nitrate reductase and nitrite reductase are metalloenzymes that contain essential oxidized iron. Nitrate reductase features [4Fe-4S] centers and a Mo-bis-molybdopterin guanine dinucleotide, while nitrite reductase contains [4Fe-4S] and siroheme centers. Photosynthesis is the source of many cyanobacterial processes, including nitrate absorption, nitrite and ammonium conversion, and assimilation capacity reliant on ATP and iron-oxidizing proteins (Flores et al. 2005).

3.2.3 Urea Absorption Pathway

The absorption and assimilation of urea in cyanobacteria is facilitated by the enzyme urease, which can still meet the needs of some cyanobacteria at low concentrations and in the presence of alternative nitrogen substrates. Each molecule of urea can produce two molecules of ammonium through enzymatic reactions, and it also generates supplemental carbon sources (Erratt et al. 2020). This is because the decomposition products of urea after it enters the cyanobacteria cells through passive diffusion or active diffusion include ammonium and carbon dioxide. Urease is encoded by the gene *ureABCDEFG*, and urea transporter is encoded by the gene *urtABCDE*. Most cyanobacteria have both, but only a few cyanobacteria that grow with urea as the only nitrogen source systematically associate the two enzyme systems. Urease is also essential for the decomposition of endogenous urea into carbon and nitrogen (Veaudor et al. 2019).

3.2.4 Molecular Nitrogen Absorption Pathways

Cyanobacteria that can utilize molecular nitrogen are called diazotrophic cyanobacteria. In nitrogen-deficient environments, diazotrophic cyanobacteria, such as *Crocosphaera*, convert atmospheric dinitrogen (N_2) into ammonia to support primary production in low-nitrogen environments (Yang et al. 2022). This conversion is facilitated by the enzyme nitrogenase, which simultaneously consumes hydrogen ions to produce hydrogen gas (H_2) as a by-product. The hydrogen gas is recycled by hydrogenase, saving energy for the catalytic reaction of nitrogenase. The most common nitrogenase in cyanobacteria is molybdenum nitrogenase, encoded by the structural gene *nifHDK* and contains molybdenum in its cofactors. Other nitrogenases include iron nitrogenase and vanadium nitrogenase. Nitrogenase is very sensitive to oxygen and will be inactivated when exposed to oxygen, so cyanobacteria use heterocysts as nitrogen fixation sites to prevent the entry of oxygen from affecting the nitrogenase reaction (Bothe et al. 2010).

3.2.5 Cyanate Absorption Pathways

Some cyanobacteria, such as *Synechococcus* and *Prochlorococcus*, can use cyanate as a nitrogen source. These cyanobacteria use cyanase, encoded by the *cynS* gene, to convert cyanate into ammonia and carbon dioxide, a reaction that requires the participation of bicarbonate. Cyanobacteria growing with cyanate as the only nitrogen source must also express cyanate transport proteins. However, most cyanobacteria containing *cynS* cannot acquire cyanate, indicating that cyanase is used to detoxify endogenous cyanate (such as by-products of the urea cycle) (Kamennaya and Post 2011).

3.2.6 Phosphorus Uptake Pathways

Orthophosphate ions (Pi) are cyanobacteria's only directly usable phosphorus source. When inorganic phosphorus is used up, phosphatase will convert organic phosphorus into orthophosphate. When phosphorus supply is limited, cyanobacteria can continue to grow using stored phosphorus sources. The gene involved in phosphorus metabolism inside the cyanobacterial cells is highlighted in Table 3.1.

After orthophosphate diffuses to the outer membrane of cyanobacteria cells, it can enter the cells through two transporters (*Pit/PsT*). When the concentration outside the cell is higher than inside, a low-affinity Pi transport system is used, which is facilitated diffusion. When the external phosphorus concentration drops to a threshold, a high-affinity specific transport system is activated, driven by ATP. The high-affinity transport proteins can bind Pi more tightly than the low-affinity transport proteins (Xiao et al. 2022).

Dissolved organic phosphorus is mainly composed of low-molecular-weight esters and organic colloids, which cannot be directly absorbed. Cyanobacteria use phosphatase hydrolases of different origins, structures, and optimal pH ranges to obtain phosphates from organics. Among them, alkaline phosphatase is the most critical phosphatase hydrolase because it can convert the most abundant and unstable phosphomonoester in the dissolved organic phosphorus pool into orthophosphate. In addition, phosphonates are enzymes that some cyanobacteria use to utilize organic phosphorus. Phosphonate acquisition genes have also been detected in other cyanobacteria (Xiao et al. 2022).

In cyanobacteria, phosphorus is predominantly stored as polyphosphates (polyP). When there is an excess of phosphorus, cyanobacteria store it in the cell as polyP through luxury uptake. Phosphorus-starved cyanobacteria can rapidly accumulate even higher polyP levels when supplied with phosphorus (Jentzsch et al. 2023). Additionally, phosphorus is stored in organic forms like phosphoesters, nucleotides, phosphonates, and vitamins.

3.3 Metabolic Pathways of Cyanobacteria for Biosynthesis of High-Value Bioproducts

Cyanobacterial metabolism evolved to support a variety of complex functions and can be categorized into primary and secondary metabolism. Primary metabolism encompasses fundamental biochemical pathways essential for organismal survival and growth. In cyanobacteria, this involves the transformation of inorganic carbon dioxide and water into organic compounds such as carbohydrates, proteins, lipids, and nucleic acids, utilizing photosynthesis to generate energy (ATP) and reducing power (NADPH) (Schwarz et al. 2013).

Secondary metabolism, conversely, pertains to the synthesis of specialized metabolites. While not essential for basic cellular functions, these metabolites confer selective advantages for survival, growth, and defense (Forchhammer and Selim 2019). Cyanobacteria produce diverse secondary metabolites, including terpenoids, flavonoids, polyketides, cytokinin, and mycosporine. Notably, certain bloom-forming cyanobacteria produce multiple cyanotoxins in response to a combination of abiotic and biotic stresses induced by climate change, posing significant ecological and human health risks (Babele et al. 2023).

3.3.1 Hydrogen Biosynthesis Pathway

Cyanobacteria, categorized into filamentous nitrogen-fixing and non-nitrogen-fixing types, can produce hydrogen directly or indirectly through biophotolysis via nitrogenase or bidirectional hydrogenase (Zarei et al. 2024). Filamentous nitrogen-fixing cyanobacteria, such as *Nostoc*, *Anabaena azollae*, *Calothrix*, and *Oscillatoria*, can produce hydrogen under nitrogen limitation via nitrogenases located in heterocyst cells (Das and Veziroglu 2008). These cells provide an anaerobic environment for their oxygen-sensitive nitrogenase, separating oxygen and hydrogen production. Under aerobic conditions, these cyanobacteria fix N_2 and convert it to NH_3, producing H_2 as a by-product. The mechanism of hydrogen production by filamentous nitrogen-fixing cyanobacteria is described by Eq. (3.1) as follows:

$$N_2 + 8H + 8e^- + 16ATP \rightarrow 2NH_3 + H_2 + 16ADP + 16P_i \qquad (3.1)$$

Non-heterocystic cyanobacteria dominate hydrogen production under anaerobic conditions through the bidirectional hydrogenase pathway (Masukawa et al. 2001). They utilize the energy stored in endogenous carbohydrates (glycogen) to produce hydrogen through indirect biophotolysis. Commonly used genera include *Gloeobacter* PCC7421, *Synechococcus*, and *Aphanocapsa montana*. The mechanism of hydrogen production by non-nitrogen-fixing cyanobacteria is given in Eq. (3.2):

$$8H^+ + 8e^- + 16ATP \rightarrow 4H_2 + 16ADP + 16P_i \qquad (3.2)$$

Hydrogenases are located on the thylakoid membrane of filamentous cyanobacteria's heterocysts. They catalyze the oxyhydrogenation or Knallgas process, which involves reducing oxygen via the respiratory chain. Hydrogenases are made up of two subunits: the hupL-coded protein handles hydrogen absorption, whereas reduction is dealt with by the smaller subunit encoded by *hupS* (Tamagnini et al. 2007;

Aubert-Jousset et al. 2011). Under ambient circumstances, strains with uptake hydrogenases do not produce net hydrogen gas because the produced hydrogen is often re-oxidized by an uptake hydrogenase via a Knallgas reaction. The following is the structure of the process that the uptake hydrogenase catalyzes:

$$H_2 \rightarrow 2H^+ + 2e^- \quad (3.3)$$

Although its exact function in the cell remains indeterminate, bidirectional or reversible hydrogenase is believed to regulate intracellular ion concentrations. The cytoplasmic membrane is the location of reversible hydrogenase, which accepts electrons from NADH and H_2 (Boison et al. 1999). The number of subunits in the reversible hydrogenase seems to vary from species to species, with some containing four and others five (Boison et al. 1999; Schmitz et al. 1995). At its molecular level, it is a hydrogenase dimer encoded by the *hoxYH* gene that functions as a [Ni-Fe]-hydrogenase of the NAD(P) + reducing type. A group of auxiliary proteins designated as *hyp*—*hypF*, *hypC*, *hypD*, *hypE*, *hypA*, and *hypB*—must work together for maturation (Wünschiers et al. 2003). While the physiological role of bidirectional hydrogenase remains elusive, it is believed to contribute to hydrogen homeostasis. Unlike uptake hydrogenases, bidirectional enzymes are not universally present in cyanobacteria but are found in both nitrogen-fixing and non-nitrogen-fixing species. A gene called *HoxEFUYH* encodes the bidirectional Hox-hydrogenase. It contains a hydrogenase complex (*HoxY* and *HoxH*), a diaphorase complex (*HoxE, HoxF*, and *HoxU*), and a pentameric bidirectional hydrogenase. Every cyanobacterium has a unique enzyme called bidirectional Hox-hydrogenase, responsible for consuming and producing molecular hydrogen (Tamagnini et al. 2007; Aubert-Jousset et al. 2011).

3.3.2 Terpenoid Biosynthesis Pathway

Terpenoids, essential for all living organisms, exhibit great structural diversity. In cyanobacteria, they are synthesized via the methylerythritol 4-phosphate (MEP) pathway, using photosynthetically produced glyceraldehyde 3-phosphate and pyruvate as substrates to make the basic building blocks for all terpenes. Despite their importance, the extraction of terpenoids from plants or cyanobacteria is often costly due to their low abundance and complex structures. To overcome these challenges, metabolic engineering and synthetic biology techniques can be employed to increase terpenoid production (Pattanaik and Lindberg 2015).

Within cyanobacterial cells, two metabolic pathways are responsible for generating the precursors of terpenes, isopentenyl pyrophosphate (IPP) and dimethylallyl pyrophosphate (DMAPP), via the mevalonate pathway (MVA) and MEP. The MVA pathway was initially considered the only pathway for the synthesis of terpene precursors until the MEP pathway's existence was confirmed through C^{13} labeling (Disch et al. 1998). The MVA metabolic pathway mainly exists in the cytoplasm of eukaryotic organisms, such as plants, animals, and a small part of microorganisms, with acetyl-CoA as the first substrate of the MVA pathway. Plants typically possess both the MVA and MEP pathways. The MEP metabolic pathway is predominantly found in the plastids of plants, and it also plays a dominant role in most prokaryotes and green algae (Pattanaik and Lindberg 2015).

The starting substrates for the MEP pathway are pyruvate and glyceraldehyde 3-phosphate (GAP) (Fig. 3.1). Despite the MEP and MVA pathways culminating in the production of IPP and DMAPP, the primary raw materials for both pathways are different. In cyanobacteria, terpenes are synthesized via the MEP pathway using pyruvate and GAP, produced by photosynthesis, as substrates. These condense to form 1-deoxy-D-xylulose-5-phosphate (DXP), which undergoes reduction and isomerization to form MEP. MEP then forms methylerythritol cytidyl diphosphate (CDP-ME) due to *MCT*, which subsequently undergoes kinase (*MCK*), cyclization (*MDS*), and phosphorylation (*HDS*) to produce 4-hydroxy 3-methyl-butenyl 1-phosphate (HMBPP). HMBPP serves as a branching point in the pathway, where it can be converted into IPP or DMAPP through reductive dehydration. In the MEP pathway, isopentenyl diphosphate isomerase (IDI) only regulates the equilibrium between IPP and DMAPP (Pattanaik and Lindberg 2015).

Hemiterpenes (C5) are the simplest terpenes comprising a single terpene unit with five carbon atoms. The volatile hydrocarbon isoprene (2-methyl-1,3-butadiene) is a valuable precursor in the synthetic chemical industry and a potential biofuel. Most cyanobacteria lack the isoprene synthase gene for isoprene formation. However, isoprene was successfully produced from an isoprene synthase (*IspS*) gene heterologously expressed in *Synechocystis* from Kudzu plants, demonstrating its potential as a renewable biofuel or feedstock in the synthetic chemical industry (Lindberg et al. 2010).

Monoterpenes (C10), the most abundant secondary metabolites in plants, consist of 10 carbon atoms. They are derived from the condensation of IDP and DMADP in a head-to-tail, non-head-to-tail manner, or by the condensation of two DMADP monomers. The purpose of releasing monoterpenes remains to be determined, necessitating further physiological and biochemical evidence to understand this process. Cyanobacteria are capable of biotransformation, a set of reactions where cell cultures or endogenous enzymes can convert exogenous substrates into different products. Biotransformation can result in the production of novel or known compounds with higher yields and reduced costs (Giri et al. 2001).

Sesquiterpenes (C15), composed of 15 carbon atoms, are synthesized by the condensation of an isopentenyl diphosphate (IDP) monomer with the monoterpene geranyl diphosphate (GDP), leading to the formation of farnesyl diphosphate (FDP). The extended chain length introduces more significant variability, enabling the synthesis of numerous sesquiterpene compounds that contribute to diverse biological functions. Geosmin, a sesquiterpene-derived alcohol produced by various organisms, including cyanobacteria, is synthesized from the precursor FDP with the assistance of Mg^{2+}. This two-step conversion is catalyzed by the bifunctional geosmin synthase, encoded by the *geoA* gene (Bonsang et al. 2010).

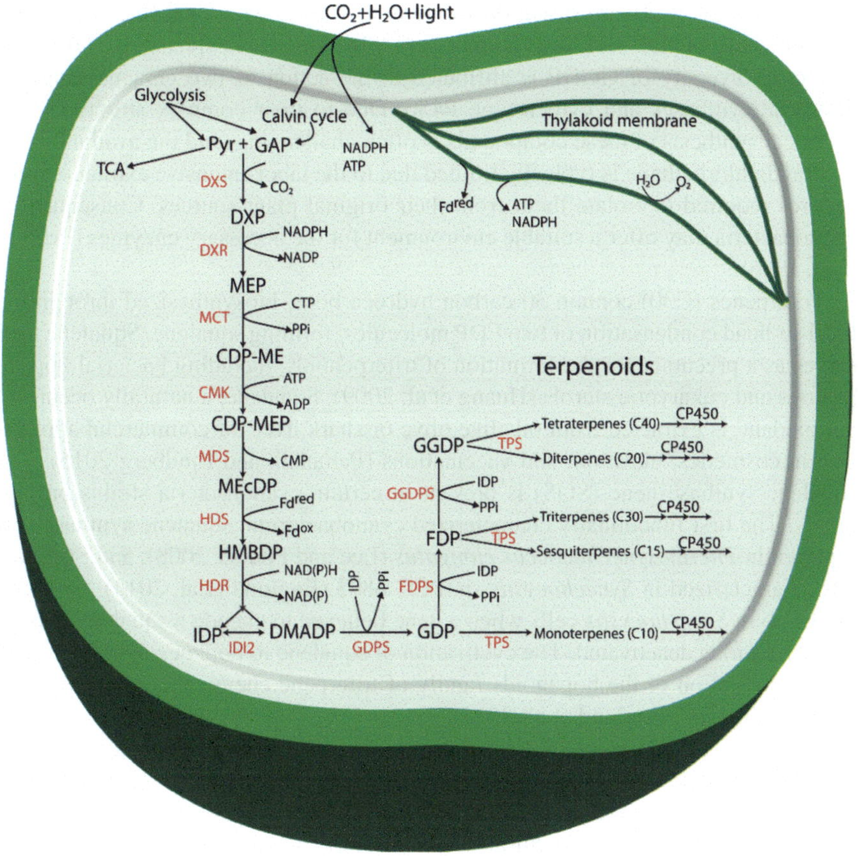

Fig. 3.1 Schematic representation of a proposed cyanobacteria-mediated methylerythritol-4-phosphate (MEP) route for terpenoid production. This figure has been adapted from (Pattanaik and Lindberg 2015). [Permission license is not required as per Open Access policy of the publisher]. Abbreviations: Pyr: Pyruvate; GAP: glyceraldehyde 3-phosphate; DXP: 1-deoxy-D-xylulose 5-phosphate; MEP: methylerythritol-4-phosphate; CDP-ME: 4-(cytidine 5-diphospho)-2-*C*-methyl-D-erythritol; CDP-MEP: 2-phospho-4-(cytidine 5-diphospho)-2-*C*-methyl-D-erythritol; MEcDP: 2-*C*-methyl-D-erythritol 2,4-cyclodiphosphate; HMBDP: 4-hydroxy-3-methylbut-2-enyldiphosphate synthase; IDP: isopentenyl diphosphate; DMADP: dimethylallyl diphosphate; GDP: geranyl diphosphate; FDP: farnesyl diphosphate; GGDP: geranylgeranyl diphosphate; DXS: 1-deoxy-D-xylulose 5-phosphate synthase; DXR: 1-deoxy-D-xylulose 5-phosphate reductoisomerase; MCT: 2-*C*-methyl-D-erythritol 4-phosphate cytidylyltransferase; CMK: 4-(cytidine 5-diphospho)-2-*C*-methyl-D-erythritol kinase; MDS: 2-*C*-methyl-D-erythritol 2,4-cyclodiphosphate synthase; HDS: 4-hydroxy-3-methylbut-2-enyl diphosphate synthase; HDR: 4-hydroxy-3-methylbut-2-enyldiphosphate reductase; IDI: isopentenyl diphosphate isomerase; GDPS: geranyl diphosphate synthase; FDPS: farnesyl diphosphate synthase; GGDPS: geranylgeranyl diphosphate synthase; TPS: terpene synthase; CP450: cytochrome P450 monooxygenase

Diterpenoids (C20) are structurally diverse C20 hydrocarbons formed by adding one molecule of IDP to FDP, yielding geranylgeranyl diphosphate (GGDP). The structural diversity of GGDP is attributed to the action of two enzyme families: diterpene synthases and cytochrome P450. Due to their complex structures, the chemical synthesis of these compounds is often challenging, and the availability of natural dichloroethane is typically limited due to the labor-intensive extraction procedures required to isolate them from their original plant sources. Consequently, cyanobacteria may offer a suitable environment for the necessary enzymes (Prinsep et al. 1996).

Triterpenes (C30) contain 30-carbon hydrocarbons, biosynthesized through the head-to-head condensation of two FDP molecules, forming squalene. Squalene then serves as a precursor for the formation of triterpenoids, including bacterial hopane analogs and eukaryotic sterols (Huang et al. 2009). Squalene, a naturally occurring antioxidant, is extracted from oils like olive or shark liver for commercial application in cosmetics, nutrition, and vaccinations (Pattanaik and Lindberg 2015). The squalene synthase gene (SQS) is present in certain cyanobacteria strains but not others. The first functionally characterized cyanobacterium squalene synthase was reported in *Thermosynechococcus elongatus* (Lee and Poulter 2008), and SQS was also characterized in *Synechocystis* sp. PCC 6803 (Englund et al. 2014). Squalene builds up in *Synechocystis* cells when a gene believed to encode a squalene hopene cyclase (*shc*) is deactivated. The cyclization of squalene to hopene, an intermediate in the production of the hopanoids family of triterpene chemicals, is catalyzed by the SHC enzyme (Englund et al. 2014).

Tetraterpenes (C40), on the other hand, 40-carbon compounds derived from octahydro-lycopene are formed in the head-to-head condensation reaction of two C20 GGDPs (Domonkos et al. 2013). A subset of tetraterpenes, carotenoid pigments, exhibit significant structural diversity. These pigments play crucial biological roles, including light harvesting, antioxidant activity, protection from free radicals, synthesis of phytohormones, and serving as structural components of membranes. Beyond their biological functions, carotenoids have a broad range of commercial applications, including food colorants, animal food supplements, nutraceuticals, and pharmaceuticals. While both photosynthetic and non-photosynthetic organisms can synthesize carotenoids, they are a vital component of the photosynthetic membrane in photosynthetic organisms, including cyanobacteria, where they stabilize photosynthetic reaction centers and regulate oxygen-producing photosynthesis in conjunction with chlorophyll.

3.3.3 Biofuels Biosynthesis Pathways

As aquatic photosynthetic microorganisms, cyanobacteria are considered a promising resource for biofuel production, specifically third-generation biofuels. They convert carbon dioxide and water into carbon-rich lipids, a substrate for biodiesel production, more efficiently than traditional oil-producing crops, positioning them as a superior alternative to conventional oilseed crops (Khan et al. 2019). Cyanobacteria biosynthesize ethanol, n-butanol, fatty alcohols, aliphatic alkanes, acetone, and others as described below. This metabolic diversity further enhances their potential as a renewable and sustainable source for biofuels and other industrial products.

Ethanol Pathway Ethanol is not a natural metabolite of cyanobacteria. The construction of an effective cyanobacterial cell factory for photosynthetic carbon fixation and ethanol synthesis relies on importing and optimizing exogenous ethanol synthesis pathways. At this stage, all cyanobacterial ethanol photosynthetic cell factory developments are based on the pyruvate decarboxylase-II-type ethanol dehydrogenase (Pdc-AdhII) pathway. Pyruvate decarboxylase uses the key node metabolite pyruvate in cyanobacteria as a substrate to catalyze the decarboxylation reaction to form acetaldehyde; ethanol dehydrogenase uses NADPH or NADH as a cofactor to reduce acetaldehyde to ethanol. Pyruvate decarboxylase (*Pdc*) and ethanol dehydrogenase (*Adh*) are key enzymes in ethanol production. *Pdc* converts pyruvate to acetaldehyde and carbon dioxide via nonoxidative decarboxylation, while *Adh* converts acetaldehyde to ethanol (Quintana et al. 2011).

Coenzyme-A Pathway-Dependent Alcohols Pathways The natural n-butanol biosynthetic pathway begins with the condensation of acetyl-CoA to acetoacetyl-CoA by acetyltransferase. β-Hydroxybutyl-coA dehydrogenase (*Hbd*) converts acetoacetyl-CoA to β-hydroxybutyl-coA, which is subsequently converted to crotonyl-CoA by 3-hydroxybutyl-coA dehydrogenase (*crotonase*). Butyl-coA-dehydrogenase (*Bcd*) in conjunction with electron transfer proteins A and B (*EtfAB*) to catalyze the generation of butyl-coA. NADH-dependent bifunctional aldehyde/alcohol dehydrogenase (*AdhE2*) converts butyl-coA to butanol. The BCD-EtfAB complex is thought to rely on oxygen-sensitive ferredoxin as an electron donor, making the conversion of the butanol synthesis pathway to other organisms challenging due to the additional source of reduction required (Mund et al. 2022). In *Synechococcus elongatus*, isopropanol is produced from acetyl-CoA by acetone in a four-step reaction (Kusakabe et al. 2013). Acetyl-CoA acetyltransferase (*Thl*) merges two acetyl-CoA into acetoacetyl-CoA. Then, catalyzed by acetoacetyl-CoA transferase (*CtfAB*), CoA is converted from acetoacetyl-CoA to acetate to produce acetoacetate, which is then decarboxylated by acetoacetate decarboxylase (*Adc*), yielding acetone and carbon dioxide. In the final step, acetone is reduced to isopropanol by the NADPH-dependent reaction of the primary-secondary alcohol dehydrogenase (*Adh*) (Mund et al. 2022). It is also known that numerous species of cyanobacteria can generate acetate from stored glycogen under oxygen and light-limited conditions.

Alcohols Dependent Fatty Acid Pathways Fatty acid (FA) synthesis is a fundamental pathway for constructing membrane lipids and synthesizing various signaling molecules. This pathway is categorized into two main types: FAS-I and FAS-II. FAS-I functions in animals and fungi, while FAS-II operates in prokaryotes and phytoplasma. In *E. coli*, the FA synthesis pathway requires an acetyl-coA carboxylase (*AccABCD*) complex, a multi-enzyme complex that catalyzes the first commitment step, converting acetyl-coA to malonyl-coA. Malonyl-coA is converted to acetoacetylacp via β-ketoacyl-ACP synthase III (*FabH*), and subsequently to acyl-acyl carrier protein (acyl-acp) in the presence of β-ketoacyl-acp reductase (*FabG*), β-hydroxylacyl-acp dehydrogenase (*FabA* or *FabZ*), and enoyl-acp reductase (*FabI*) (two carbon extensions per cycle) (Janßen and Steinbüchel 2014). The final product of bacterial FA synthesis, Acyl-ACP, is a substrate for membrane synthesis. While FA synthesis genes are similar in cyanobacteria and *E. coli*, the *fabB* and *fabA* genes are notably absent in cyanobacteria.

Fatty Alcohols Pathways The production of fatty alcohols involves different pathways in prokaryotic and eukaryotic systems. In prokaryotes, fatty acyl-coA is first reduced to a fatty aldehyde by fatty acyl-coA reductase (*Acr*), and the aldehyde is subsequently reduced to an alcohol by fatty aldehyde reductase (*Ahr*) as depicted in Fig. 3.2 (Mendez-Perez et al. 2011). In contrast, in eukaryotes, individual enzymes convert acyl-coA or acyl-acp to fatty alcohols without forming intermediate fatty aldehydes. Fatty alcohol synthesis in cyanobacteria shares the aliphatic alkane syn-

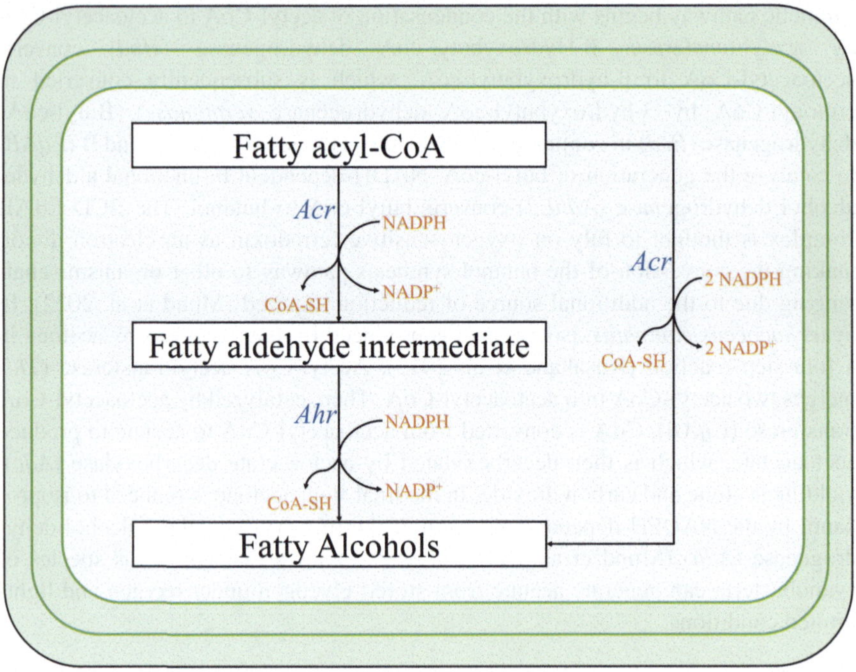

Fig. 3.2 The processes in cyanobacteria that produce fatty alcohols are shown schematically. Fatty alcohol biosynthesis has two distinct reaction pathways. A fatty acyl-CoA reductase (*Acr*) and a fatty aldehyde reductase (*Ahr*) might undergo sequential reduction processes to produce fatty alcohol, or *Acr* could be reduced in a single step

thesis pathway, in which aldehyde intermediates serve as substrates for the corresponding alcohols (Rowland and Domergue 2012).

Aliphatic Alkanes Pathways Cyanobacteria are among the few organisms capable of producing alkanes, with the highest reported production level being 0.12% of dry biomass. The alkane synthesis pathway involves two steps: ACP is first reduced to an aldehyde by *Aar*, and the resulting aldehyde is further oxidized to the corresponding Cn-1-terminal alkane or olefin, depending on the structure of the original ACP substrate. Contrary to the AR-ADO pathway, which generates Cn-1, the α-olefin synthesis pathway involves olefin synthase (OLS), converting fatty acyl-ACP to odd-chained α-olefins (+1 carbon). Hexadecanes with unique terminal double bonds are the most abundant hydrocarbons in cyanobacteria with an AR-ADO pathway. In cyanobacteria with the OLS pathway, two hydrocarbons, 1-nonadiene and 1,14-nonadiene, have been identified.

In membrane fluidization and related gene expression investigations, hexane-1-ol is an invaluable primary aliphatic alcohol. Transcriptome analysis provides a reference that might help find the best strategies to increase cyanobacterial resistance to alcohols (Mironov et al. 2021), and it was shown that alcohol treatment suppresses the expression of fatty acid desaturase genes (Mironov et al. 2018). One potential approach to enhance resistance to alcohol effect is stabilizing desaturases that are inhibited by alcohol. The negative regulation of Hik34, a heat shock-induced suppressor of chaperone expression, might strengthen cellular resistance to alcohol stress. Increasing cellular resilience to alcohol stress may be best achieved, in the end, by modifying specific antisense RNA interactions (Mironov et al. 2021).

3.4 Economic Layout of Cyanobacteria-Based Products

The algae-based product market is estimated to reach $19.85 billion by 2028, up from $13–14 billion in 2023. This represents a CAGR of 7.8, and over the forecast period, the market is expected to capture a CAGR of 9.1%. Product sales in the aquaculture, pharmaceutical, and nutraceutical industries are expected to be boosted by the increased use of cyanobacteria components and proteins (Table 3.2). A significant driver of the expansion might be the growing number of government programs aimed at standardizing the cyanobacterial farming process. *Spirulina* spp. continue to command a premium price tag due to their high nutritional profile. Currently, Spirulina is the most protein-rich meal on the market due to the presence of 60% protein content (TBRC 2024). The Spirulina market was worth $348 million in 2020, as reported by Allied Market Research (Research 2021). The projected growth rate from 2019 to 2026 is 10.5%, which is anticipated to reach $779 million by that year. The North American market dominated the worldwide Spirulina market in 2018 and is expected to continue its leadership from 2019 to 2026, driven by the rising demand for all-natural food colorants. Market participants may anticipate lucrative prospects in the Asia-Pacific region, which is projected to see a CAGR of 11.2% from 2019 to 2026 (Research 2021).

Research from Future Market Insights estimated that the worldwide market for Spirulina pigment, phycocyanin, was worth $112.3 million in 2018. Assuming a CAGR of 7.6% from now until the end of 2025, its predicted value would exceed $232.9 million (Insights 2019). The newly authorized European project SPIRALG (Horizon 2020) (www.spiralg.eu) demonstrates the increasing interest in phycocyanin pigment. The project aims to establish a demonstration plant for phycocyanin manufacturing, to support sustainability via the valorization of the produced co-products. In addition, a new extraction line will allow Naturex (www.naturex.com), a Givaudan company that once specialized in natural specialty ingredients for food, health, and cosmetics, to increase phycocyanin output by a factor of three. The food and cosmetics industries already use this spirulina blue pigment in drinks like Bloo Tonic® and B-blue Spirulina, as well as in candies like M&Ms.®. Value Market Research estimates that the worldwide market for chlorophyll extract will grow from an initial valuation of $279.5 million in 2018 to $463.7 million in 2025, a CAGR of 7.5%.

Phycocyanin is also considered a food additive and a colorant. Phycocyanin is manufactured by the French firm AlgoSource (www.algosource.com) and offered in phials (10 mL) as a liquid extract under the brand name Spirulysat®. Spira (www.spirain.com) is another business that makes powdered phycocyanin (Electric Sky®) for use as a dietary supplement and for coloration, among other functional health benefits. The scientific literature also backs up these tendencies, with some research reporting that phycocyanin is added to food as a bioactive natural colorant. In study on phycocyanin-enriched yogurt, color stability was stable for 14 days and it was found to have a higher pH than the plain control and its viscosity rose in tandem with the concentration of phycocyanin (Mohammadi-Gouraji et al. 2019). A gelatin-based film was created that included phycocyanin from *Spirulina* sp., which showed promise as a colored material for outer or inner package (Chentir et al. 2019).

3.5 Conclusion and Prospects

Cyanobacteria can biosynthesize a broad variety of products, including hydrogen and terpenoids, showcasing their potential as cellular factories in biotechnology and their vast array of commercial uses. The production of such bioproducts is regulated by the metabolic pathways of cyanobacteria, which are controlled by genes, regulatory factors, and some abiotic factors, such as nutrient deficiency. The yield of these bioproducts in cyanobacterial cells can be low, but improvements can be achieved through pathway engineering, enzyme engineering, evolutionary engineering, and synthetic biology. New cutting-edge approaches also need to be developed to engineer cyanobacterial strains to get high yields of these compounds. Most cyanobacteria resources still need to be explored, so finding new indigenous cyanobacterial strains that produce high amounts of metabolites for industrial interest is essential.

Omics-based and mass spectrometry-based studies can be employed to identify new pathways and explore new metabolites. The widespread commercial use of these potent phototrophic cell factories still needs to be improved, mainly due to the lack of a solid grasp of intracellular processes and a summary of process design abilities. Developing novel nutraceutical and pharmaceutical businesses based on cyanobacteria and their use in bioactive compound synthesis needs further exploration.

References

Agarwal P, Soni R, Kaur P et al (2022) Cyanobacteria as a promising alternative for sustainable environment: synthesis of biofuel and biodegradable plastics. Front Microbiol 13. https://doi.org/10.3389/fmicb.2022.939347

Amin M, Haq MAU, Xu JR, Li H, Wang N (2023) A bibliometric analysis of the growing market of *Spirulina*-based algal products and emerging trends. Adv Green Chem 1:23002

Aubert-Jousset E, Cano M, Guedeney G et al (2011) Role of HoxE subunit in *Synechocystis* PCC6803 hydrogenase. FEBS J 278:4035–4043. https://doi.org/10.1111/j.1742-4658.2011.08308.x

Babele PK, Srivastava A, Young JD (2023) Metabolic flux phenotyping of secondary metabolism in cyanobacteria. Trends Microbiol 31:1118–1130. https://doi.org/10.1016/j.tim.2023.05.005

Boison G, Bothe H, Hansel A et al (1999) Evidence against a common use of the diaphorase subunits by the bidirectional hydrogenase and by the respiratory complex I in cyanobacteria. FEMS Microbiol Lett 174:159–165. https://doi.org/10.1016/S0378-1097(99)00136-6

Bonsang B, Gros V, Peeken I et al (2010) Isoprene emission from phytoplankton monocultures: the relationship with chlorophyll-*a*, cell volume and carbon content. Environ Chem 7:554–563. https://doi.org/10.1071/EN09156

Bothe H, Schmitz O, Yates MG et al (2010) Nitrogen fixation and hydrogen metabolism in cyanobacteria. Microbiol Mol Biol Rev 74:529–551. https://doi.org/10.1128/mmbr.00033-10

Burford MA, Carey CC, Hamilton DP et al (2020) Perspective: advancing the research agenda for improving understanding of cyanobacteria in a future of global change. Harmful Algae 91:101601. https://doi.org/10.1016/j.hal.2019.04.004

Chakdar H, Pabbi S (2016) Cyanobacterial phycobilins: production, purification, and regulation. In: Shukla P (ed) Frontier discoveries and innovations in interdisciplinary microbiology. Springer India, New Delhi. https://doi.org/10.1007/978-81-322-2610-9_4

Chentir I, Kchaou H, Hamdi M et al (2019) Biofunctional gelatin-based films incorporated with food grade phycocyanin extracted from the Saharian cyanobacterium *Arthrospira* sp. Food Hydrocoll 89:715–725. https://doi.org/10.1016/j.foodhyd.2018.11.034

Chittora D, Meena M, Barupal T et al (2020) Cyanobacteria as a source of biofertilizers for sustainable agriculture. Biochem Biophys Rep 22:100737. https://doi.org/10.1016/j.bbrep.2020.100737

Das D, Veziroglu TN (2008) Advances in biological hydrogen production processes. Int J Hydrog Energy 33:6046–6057. https://doi.org/10.1016/j.ijhydene.2008.07.098

Demay J, Bernard C, Reinhardt A et al (2019) Natural products from cyanobacteria: focus on beneficial activities. Mar Drugs 17:320. https://doi.org/10.3390/md17060320

Disch A, Schwender J, Müller C et al (1998) Distribution of the mevalonate and glyceraldehyde phosphate/pyruvate pathways for isoprenoid biosynthesis in unicellular algae and the cyanobacterium *Synechocystis* PCC 6714. Biochem J 333:381–388. https://doi.org/10.1042/bj3330381

Domonkos I, Kis M, Gombos Z et al (2013) Carotenoids, versatile components of oxygenic photosynthesis. Prog Lipid Res 52:539–561. https://doi.org/10.1016/j.plipres.2013.07.001

Englund E, Pattanaik B, Ubhayasekera SJK et al (2014) Production of squalene in *Synechocystis* sp. PCC 6803. PLoS One 9:e90270. https://doi.org/10.1371/journal.pone.0090270

Erratt KJ, Creed IF, Trick CG (2020) Differential drawdown of ammonium, nitrate, and urea by freshwater chlorophytes and cyanobacteria. J Phycol 56:458–468. https://doi.org/10.1111/jpy.12960

Flores E, Frías JE, Rubio LM et al (2005) Photosynthetic nitrate assimilation in cyanobacteria. Photosynth Res 83:117–133. https://doi.org/10.1007/s11120-004-5830-9

Forchhammer K, Selim KA (2019) Carbon/nitrogen homeostasis control in cyanobacteria. FEMS Microbiol Rev 44:33–53. https://doi.org/10.1093/femsre/fuz025

Giri A, Dhingra V, Giri CC et al (2001) Biotransformations using plant cells, organ cultures and enzyme systems: current trends and future prospects. Biotechnol Adv 19:175–199. https://doi.org/10.1016/S0734-9750(01)00054-4

Harada K-I, Fujii K, Shimada T et al (1995) Two cyclic peptides, anabaenopeptins, a third group of bioactive compounds from the cyanobacterium *Anabaena flos-aquae* NRC 525-17. Tetrahedron Lett 36:1511–1514. https://doi.org/10.1016/0040-4039(95)00073-L

Herrero A, Muro-Pastor AM, Flores E (2001) Nitrogen control in cyanobacteria. J Bacteriol 183:411–425. https://doi.org/10.1128/jb.183.2.411-425.2001

Huang Z-R, Lin Y-K, Fang J-Y (2009) Biological and pharmacological activities of squalene and related compounds: potential uses in cosmetic dermatology. Molecules 14:540–554. https://doi.org/10.3390/molecules14010540

Insights FM (2019) Phycocyanin market to grow at a CAGR of 7.6% by 2028 future market insights

Janßen HJ, Steinbüchel A (2014) Fatty acid synthesis in *Escherichia coli* and its applications towards the production of fatty acid based biofuels. Biotechnol Biofuels 7:7. https://doi.org/10.1186/1754-6834-7-7

Jentzsch L, Grossart H-P, Plewe S et al (2023) Response of cyanobacterial mats to ambient phosphate fluctuations: phosphorus cycling, polyphosphate accumulation and stoichiometric flexibility. ISME Commun 3:6. https://doi.org/10.1038/s43705-023-00215-x

Jodlbauer J, Rohr T, Spadiut O et al (2021) Biocatalysis in green and blue: cyanobacteria. Trends Biotechnol 39:875–889. https://doi.org/10.1016/j.tibtech.2020.12.009

Jones MR, Pinto E, Torres MA et al (2021) CyanoMetDB, a comprehensive public database of secondary metabolites from cyanobacteria. Water Res 196:117017. https://doi.org/10.1016/j.watres.2021.117017

Kamennaya NA, Post AF (2011) Characterization of cyanate metabolism in marine *Synechococcus* and *Prochlorococcus* spp. Appl Environ Microbiol 77:291–301. https://doi.org/10.1128/AEM.01272-10

Khan AZ, Bilal M, Mehmood S et al (2019) State-of-the-art genetic modalities to engineer cyanobacteria for sustainable biosynthesis of biofuel and fine-chemicals to meet bio–economy challenges. Life 9:54. https://doi.org/10.3390/life9030054

Kusakabe T, Tatsuke T, Tsuruno K et al (2013) Engineering a synthetic pathway in cyanobacteria for isopropanol production directly from carbon dioxide and light. Metab Eng 20:101–108. https://doi.org/10.1016/j.ymben.2013.09.007

Lee S, Poulter CD (2008) Cloning, solubilization, and characterization of squalene synthase from *Thermosynechococcus elongatus* BP-1. J Bacteriol 190:3808–3816. https://doi.org/10.1128/jb.01939-07

Lindberg P, Park S, Melis A (2010) Engineering a platform for photosynthetic isoprene production in cyanobacteria, using *Synechocystis* as the model organism. Metab Eng 12:70–79. https://doi.org/10.1016/j.ymben.2009.10.001

Maeda S-I, Konishi M, Yanagisawa S et al (2014) Nitrite transport activity of a novel HPP family protein conserved in cyanobacteria and chloroplasts. Plant Cell Physiol 55:1311–1324. https://doi.org/10.1093/pcp/pcu075

Massey IY, Al Osman M, Yang F (2022) An overview on cyanobacterial blooms and toxins production: their occurrence and influencing factors. Toxin Rev 41:326–346. https://doi.org/10.1080/15569543.2020.1843060

Masukawa H, Nakamura K, Mochimaru M et al (2001) Photobiological hydrogen production and nitrogenase activity in some heterocystous cyanobacteria. In: Miyake J, Matsunaga T, Pietro AS (eds) Biohydrogen II. Pergamon, Oxford. https://doi.org/10.1016/B978-008043947-1/50005-3

Mazard S, Penesyan A, Ostrowski M et al (2016) Tiny microbes with a big impact: the role of cyanobacteria and their metabolites in shaping our future. Mar Drugs 14:97. https://doi.org/10.3390/md14050097

Mendez-Perez D, Begemann MB, Pfleger BF (2011) Modular synthase-encoding gene involved in α-olefin biosynthesis in *Synechococcus* sp. strain PCC 7002. Appl Environ Microbiol 77:4264–4267. https://doi.org/10.1128/AEM.00467-11

Mironov KS, Shumskaya M, Sidorov RA et al (2018) Membrane physical state and stress regulation in *Synechocystis*: fluidizing alcohols repress fatty acid desaturation. Plant J 96:1007–1017. https://doi.org/10.1111/tpj.14086

Mironov KS, Kupriyanova EV, Shumskaya M et al (2021) Alcohol stress on cyanobacterial membranes: new insights revealed by transcriptomics. Gene 764:145055. https://doi.org/10.1016/j.gene.2020.145055

Mohammadi-Gouraji E, Soleimanian-Zad S, Ghiaci M (2019) Phycocyanin-enriched yogurt and its antibacterial and physicochemical properties during 21 days of storage. LWT 102:230–236. https://doi.org/10.1016/j.lwt.2018.09.057

Mund NK, Liu Y, Chen S (2022) Advances in metabolic engineering of cyanobacteria for production of biofuels. Fuel 322:124117. https://doi.org/10.1016/j.fuel.2022.124117

Muro-Pastor MI, Reyes JC, Florencio FJ (2005) Ammonium assimilation in cyanobacteria. Photosynth Res 83:135–150. https://doi.org/10.1007/s11120-004-2082-7

Pattanaik B, Lindberg P (2015) Terpenoids and their biosynthesis in cyanobacteria. Life 5:269–293. https://doi.org/10.3390/life5010269

Prinsep MR, Thomson RA, West ML et al (1996) Tolypodiol, an antiinflammatory diterpenoid from the cyanobacterium *Tolypothrix nodosa*. J Nat Prod 59:786–788. https://doi.org/10.1021/np9602574

Quintana N, Van Der Kooy F, Van De Rhee MD et al (2011) Renewable energy from cyanobacteria: energy production optimization by metabolic pathway engineering. Appl Microbiol Biotechnol 91:471–490. https://doi.org/10.1007/s00253-011-3394-0

Rachedi R, Foglino M, Latifi A (2020) Stress signaling in cyanobacteria: a mechanistic overview. Life 10:312. https://doi.org/10.3390/life10120312

Research AM (2021) Spirulina market size, share, competitive landscape and trend analysis report by type, application and formulation: global opportunity analysis and industry forecast, 2020–2027. Allied Market Research

Rowland O, Domergue F (2012) Plant fatty acyl reductases: enzymes generating fatty alcohols for protective layers with potential for industrial applications. Plant Sci 193-194:28–38. https://doi.org/10.1016/j.plantsci.2012.05.002

Sayers EW, Bolton EE, Brister JR et al (2021) Database resources of the national center for biotechnology information. Nucleic Acids Res 50:D20–D26. https://doi.org/10.1093/nar/gkab1112

Schmitz O, Boison G, Hilscher R et al (1995) Molecular biological analysis of a bidirectional hydrogenase from cyanobacteria. Eur J Biochem 233:266–276. https://doi.org/10.1111/j.1432-1033.1995.266_1.x

Schwarz D, Orf I, Kopka J et al (2013) Recent applications of metabolomics toward cyanobacteria. Meta 3:72–100. https://doi.org/10.3390/metabo3010072

Shahid A, Khan AZ, Jabeen F et al (2023) Cyanobacteria-based biorefineries for a sustainable future of bioindustry. In: Oncel SS (ed) A sustainable green future: perspectives on energy, economy, industry, cities and environment. Springer International Publishing, Cham. https://doi.org/10.1007/978-3-031-24942-6_24

Tamagnini P, Leitão E, Oliveira P et al (2007) Cyanobacterial hydrogenases: diversity, regulation and applications. FEMS Microbiol Rev 31:692–720. https://doi.org/10.1111/j.1574-6976.2007.00085.x

Tbrc (2024) Algae products global market report 2024- by type algae products global market report 2024- by type (lipids, carrageenan, carotenoids, alginate, algal protein), By Form (Liquid, Solid), BySource (Blue-Green Algae, Brown Algae, Green Algae, Red Algae), By Distribution Channel (Online, Offline), By Application (Food and Beverage, Nutraceutical and Diet Supplement, AnimalFeed, Personal Care, Pharmaceutical, Other Applications)- Market Size, Trends, And GlobalForecast 2024-2033. The Business Research Company

Veaudor T, Cassier-Chauvat C, Chauvat F (2019) Genomics of urea transport and catabolism in cyanobacteria: biotechnological implications. Front Microbiol 10. https://doi.org/10.3389/fmicb.2019.02052

Wang N, Xiang Y, Ma Y, Zhang P, Zhou X, Zhu H, Zhang Z, Li Z, Xiao X, He M, Mehmood MA, Zhu H (2023) Fermented *Spirulina platensis* alleviated DSS-induced ulcerative colitis by regulating gut microbiota and MyD88/TLR4 signaling pathway. Asian J Agric Biol 1(1):2023232. https://doi.org/10.35495/ajab.2023.232

Wünschiers R, Batur M, Lindblad P (2003) Presence and expression of hydrogenase specific C-terminal endopeptidases in cyanobacteria. BMC Microbiol 3:8. https://doi.org/10.1186/1471-2180-3-8

Xiao M, Burford MA, Wood SA et al (2022) Schindler's legacy: from eutrophic lakes to the phosphorus utilization strategies of cyanobacteria. FEMS Microbiol Rev 46. https://doi.org/10.1093/femsre/fuac029

Yang N, Lin Y-A, Merkel CA et al (2022) Molecular mechanisms underlying iron and phosphorus co-limitation responses in the nitrogen-fixing cyanobacterium *Crocosphaera*. ISME J 16:2702–2711. https://doi.org/10.1038/s41396-022-01307-7

Zarei Z, Malekshahi P, Morowvat MH et al (2024) A review of bioreactor configurations for hydrogen production by cyanobacteria and microalgae. Int J Hydrog Energy 49:472–495. https://doi.org/10.1016/j.ijhydene.2023.09.108

Żymańczyk-Duda E, Samson SO, Brzezińska-Rodak M et al (2022) Versatile applications of cyanobacteria in biotechnology. Microorganisms 10:2318. https://doi.org/10.3390/microorganisms10122318

Chapter 4
Metabolic Pathway Engineering in Cyanobacteria for Biohydrogen Production

Ashutosh Kumar, Sanchita Paul, Divya Das, and Meenakshi Singh

Abstract Switching to cleaner and greener energy sources is now imperative to meet the ever-increasing global demands while keeping the carbon footprint low to minimize the impacts of climate change. Cyanobacteria, having higher photosynthetic efficiency and biomass productivity among all terrestrial plants, are robust cellular factories with extraordinary potential to produce biohydrogen. Commercial biohydrogen production using cyanobacteria is still at the nascent stage due to several challenges, including (1) cyanobacteria's hydrogenase sensitivity to oxygen, (2) low hydrogenase efficiency, (3) competing pathways and metabolism flux, (4) photobioreactor design and scale-up, (5) strain stability, (6) multi-parameters optimization, (5) regulatory and ethical issues with genetic transformation, and (6) techno-economic challenges. However, metabolic pathway engineering techniques could improve the low biohydrogen yield in cyanobacteria. In cyanobacteria farming, along with genetic modifications, optimizing the various cultivation parameters (nutrients, reactor design, strain selection, air flow, temperature, and light intensity) could help enhance biomass productivity, leading to higher biohydrogen production. This chapter discusses various molecular approaches and genetic engineering techniques to overcome the challenges mentioned above. In the future, Artificial Intelligence and mathematical models can help optimize multiple parameters to enhance the growth of genetically modified cyanobacteria that can pave the path to commercial scale biohydrogen production using cyanobacteria.

A. Kumar
Department of Biotechnology, Smt. Chandibai Himathmal Mansukhani College, Mumbai University, Ulhasnagar, Maharashtra, India

S. Paul
Department of Botany, West Bengal State University, Kolkata, India

D. Das
Department of Botany, Faculty of Science, The Maharaja Sayajirao University of Baroda, Vadodara, Gujarat, India

M. Singh (✉)
Department of Ecology and Biodiversity, Sustaina Greens LLP, Vadodara, Gujarat, India

M. A. Mehmood et al. (eds.), *Cyanobacteria Biotechnology*,
https://doi.org/10.1007/978-3-031-70698-1_4

Keywords Cyanobacteria · Biohydrogen production · Metabolic pathway engineering · CRISPR-Cas · Multi-parameter optimization · Photobioreactors

4.1 Introduction

The scientific community is constantly searching for alternative sustainable biological resources to avoid any harm to the environment (Satpati et al. 2022). Globally, the world consumed about 100.87 million barrels of oil per day in 2019 and about 92.2 million barrels in 2020; still, the demand appears to be increasing exponentially (Josephs et al. 2022). Biohydrogen, a biomass product, is an attractive alternative to clean, green, environment-friendly, inexhaustible, sustainable, and economical energy resources in the wake of natural resource depletion and global warming (Parambath et al. 2022).

There are several methods of hydrogen production, such as biological, direct thermal, electrochemical, and thermochemical [example: electrolysis, stream reforming] but among all, the biological method seems to be the most promising for producing sustainable and cost-effective hydrogen (Kumar et al. 2024). Because other methods not only deplete fossil fuels but also consume lots of energy while polluting the environment. In this scenario, cyanobacteria/blue-green algae/cyanophyta seem to be a potential candidate to work as a fossil fuel alternative because of their immense ability to convert solar energy into hydrogen (H_2) just by feeding on waste material, unlike Enterobacteriaceae, which require the expensive substrate to feed on (Kumar et al. 2024). Besides cost, the other issue linked to Enterobacteriaceae-based hydrogen production via dark fermentation reaction is its inability to decompose maximum substrate, i.e., only 20% of the stored substrate decomposes in total, and resulting contamination together proves to be a barrier in upscaling the hydrogen production despite their fast reaction rate (Kumar et al. 2024). The H_2 possesses the highest energy per unit weight (141.65 MJ kg^{-1}) compared to any fuel; therefore, cyanobacteria's ability to produce hydrogen is a fundamental process (Vu et al. 2020). These oxygenic microbes synthesize value-based products in a carbon-neutral manner as they fix carbon dioxide using water and nutrients through photosynthesis and the Calvin–Benson cycle (CBC). It seems like plants also contribute to photosynthesis. However, there is a physiological difference; the cyanobacteria lack membrane-bound organelles such as chloroplast, nucleus, and Golgi apparatus, and the chlorophyll pigment located in the thylakoid membrane of cytoplasm comes into the function for light absorption and assisting photosynthesis (Govindjee 2011).

Interestingly, the thylakoid membrane is in the same location where cyanobacteria hold significant amounts of valuable lipids. In contrast to photosynthesis, the light-independent cyanobacteria, for example, Nostoc and Oscillatoria, also exist and can grow in dark and anaerobic conditions; they are also called diazotrophic cyanobacteria, particularly for their role of nitrogen-fixation (Kollmen and Strieth 2022). Despite the physiological differences in the way they photosynthesize, there is still a phylogenetic similarity between the cyanobacterial genome and chloroplast, which led researchers to propose a theory claiming cyanobacteria symbiont to be the origin of chloroplasts of the plant as well as algae (Sato 2021). These gram-negative cyanobacteria have a peptidoglycan layer along with an outer membrane/plasma membrane which is composed of carotenoids, unusual lipids (β-hydroxypalmitic acid), and proteins that originated around 2.4 billion years back and laid the foundation of formation, evolution, and advancement of multicellular life by producing tons of oxygen as atmospheric constituents and the organic substances which serves as a food supply for eubacteria [the first life forms] on earth (Rahmatpour et al. 2021; Hoiczyk and Hansel 2000). There are numerous ways in which the cyanobacteria contributed and are still contributing. The cyanobacterial rate to convert solar energy into biomass is whooping 10% higher when compared to first and second generation biofuels, which are focused on exploiting crops such as sugarcane, wheat, peanuts, soybean, etc., without disrupting the food supply chain (Zahra et al. 2020). All credits go to the superior photosynthetic capabilities, higher oil contents, and shorter growth duration of cyanobacteria. Nevertheless, cyanobacteria lack a fully developed pathway that is as sophisticated. However, now, large-scale commercial production of biofuel, especially biohydrogen, seems possible and quite exciting with modern scientific technologies like metabolic engineering and synthetic biology.

Employing genetic engineering techniques and understanding the molecular basis of hydrogen (H_2) production is crucial to improve and upscale biohydrogen production (Das et al. 2014). Therefore, this chapter focuses on the plethora of genetic tools and strategies associated with genetic modifications aiming to produce economically feasible hydrogen (H_2) production. Day-to-day advancements in genetic manipulation tools lead to significant changes in cyanobacterial systems; right from (1) manipulating the central carbon metabolism skeleton, (2) improving the enzymatic activities, (3) redirecting the metabolic pathways, and (4) introducing the foreign genes into cyanobacteria, for all such roles, the genetic tools come into employment. Additionally, this chapter delves into the cyanobacterial strain selection strategies, various multi-parameter optimization methods, and several photobioreactors [PBRs] models, as well as the associated challenges of their design and the ability of light attenuation. Finally, the chapter winds up by acquainting us about the current global scenario of the cyanobacterial hydrogen market and significant prospects and widespread commercialization.

4.2 Molecular Approaches for Cyanobacterial Hydrogen Production

Cyanobacteria are excellent candidates for producing hydrogen due to their inherent hydrogen metabolic routes and capacity to utilize solar energy for photosynthesis (Kumar et al. 2024). They mainly deploy two groups of enzymes for hydrogen production: Hydrogenase (H_2ase) and Nitrogenase (N_2ase). At the same time, the second one also contributes to fixing environmental nitrogen and is found in filamentous or unicellular cyanobacteria. The development of a wide range of molecular tools has greatly facilitated our ability to introduce foreign DNA into different cyanobacteria strains and modify the genetic sequences (Kamshybayeva et al. 2024). Furthermore, the emergence of novel sequencing technologies has enabled the sequencing of the genomes of over 200 cyanobacterial species, enhancing our understanding of their genetic composition and potential biotechnological applications (Kamshybayeva et al. 2024). Genetic engineering is typically employed as a molecular tool for improving biological hydrogen. Now, researchers can significantly increase the efficiency and sustainability of cyanobacterial-based hydrogen production systems by focusing on genes associated with the enzymes hydrogenase and nitrogenase, as well as by optimizing the electron transport and adaptability to environmental stress (Kamshybayeva et al. 2024).

4.2.1 Improvement of Hydrogen-Producing Enzymes

Hydrogenase is the primary catalyst for producing hydrogen; however, the synthesis of H_2 in diazotrophic bacteria and diazotrophs cyanobacteria is primarily associated with nitrogen fixation, facilitated by the enzyme nitrogenase (Fig. 4.1) (Kumar et al. 2024). To gain a deeper understanding of the properties and constraints of enzymes, it is essential to examine phylogenetically distinct enzymes along with their related genes, maturation processes, catalytic domains, and mechanisms of action. This helps researchers perk up hydrogen production by modifying things molecularly (Das et al. 2014).

4.2.2 Improvement of Hydrogenase

Hydrogenase is ubiquitous in bacteria, archaea, and algae. It has the potential to catalyze both hydrogen oxidation and evolution. Though photosynthetic and fermentative bacteria are used to be anoxygenic during hydrogen production, cyanobacteria develop oxygen due to their PSII system. This oxygen inhibits hydrogenase activity. The hydrogen-producing enzymes catalyze an elementary redox reaction by accepting two electrons: $2H^+ + 2e \rightarrow H_2$.

These enzymes can be classified into two groups according to their metal content: nickel-iron [Ni-Fe] and di-iron [Fe-Fe] (Fig. 4.2). Di-iron hydrogenase is the most biased and sensitive to oxygen, whereas Ni-Fe is oxygen tolerant. It works 15-fold better without the help of ATP and nitrogenase, reducing NADPH and uptaking hydrogenase to utilize the electrons from hydrogen.

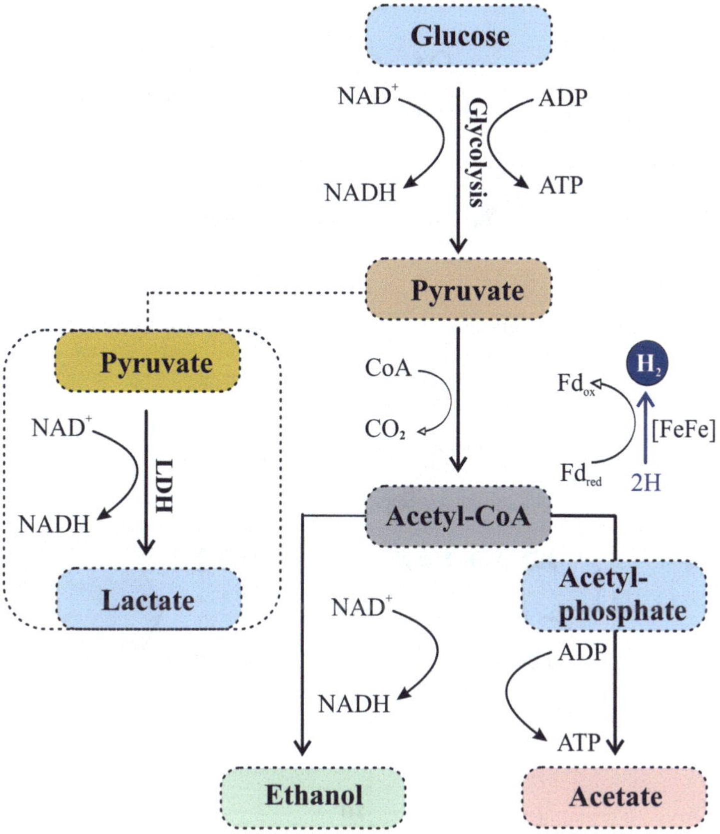

Fig. 4.1 Metabolic pathway of hydrogenase inhibiting the acid and alcohol production to increase hydrogen production

The uptake hydrogenase of Ni-Fe is the potential obstacle to high-level production. The group A (membrane-bound) cytoplasmic hydrogenase Ni-Fe (hupSL) is situated in the heterocyst of the cyanobacteria, where it couples the hydrogen oxidation with an aerobic chain to uphold intercellular anoxygenic condition. The group provides both oxygen inactivation by nitrogenase protection and recycling of the obligate side product (H_2) of nitrogen fixation. Group B (loosely membrane-bound) Ni-Fe is a bidirectional hydrogenase (hoxEFUYHW) and soluble, found in both nitrogen-fixing and non-nitrogen-fixing bacteria. It is different from group A from both catalytic and physical perspectives (Fig. 4.3). Biohydrogen production has been improved by redirecting electron flux towards group B to maintain a distance from the electron-entailing metabolic network. It can be restrained by hydrogen partial pressure build-up. Hence, constant sparging can amplify overall production. A set of genes like *hypABCDEF* that encodes chaperon proteins implements Ni incorporation and CO/CN synthesis, which helps in the maturation of cyanobacterial Ni-Fe hydrogenases. However, it can be problematic to express Fe-Fe

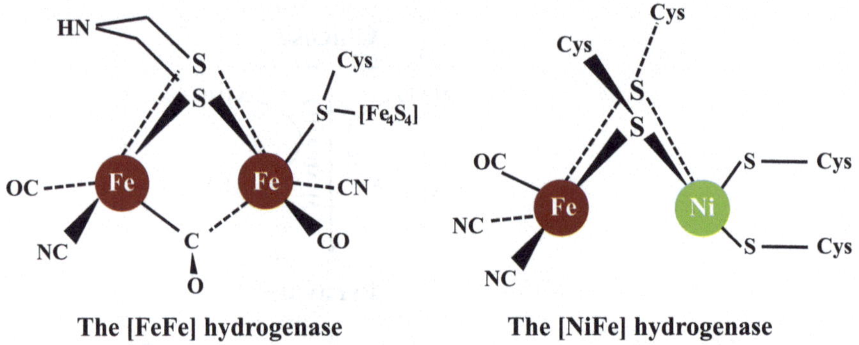

Fig. 4.2 Two different enzyme formations according to their metal content

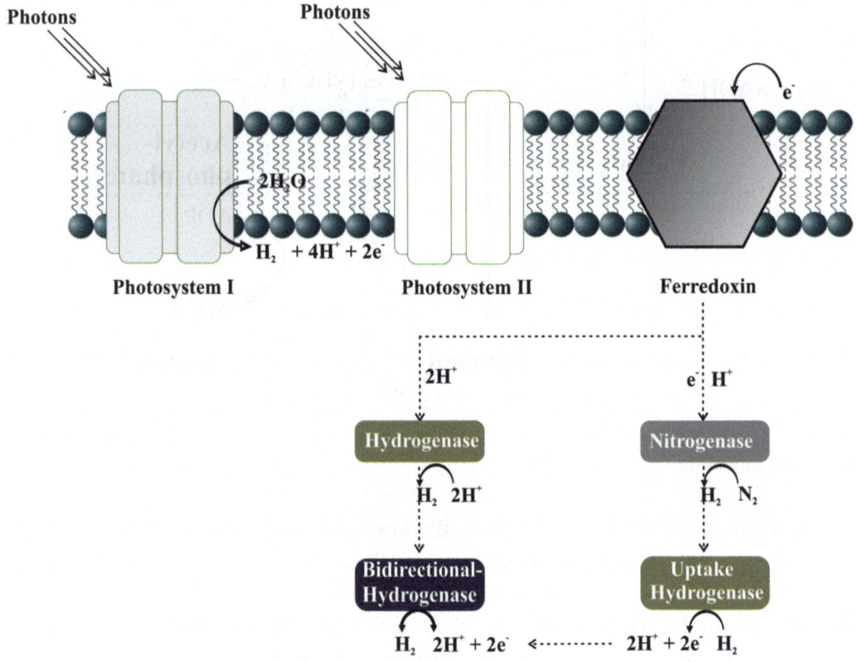

Fig. 4.3 Hydrogen production methods in cyanobacteria via three main evolving enzymes

hydrogenase. Fe-Fe can be matured by systems like HydE, HydF, and HydG, which are absent here. Heterologous expression of Fe-Fe, along with the maturation system proteins under the command of the heterocyst-specific promoter, develops active hydrogenase. The overexpression of hetR, pats, and hetN inactivation can increase the heterocyst frequency. The combined approaches can be a way to increase the potential production.

Algal hydrogenase consists of a simple H-cluster that is oxygen inactivation sensitive. In an investigation using *Ralstonia eutropha,* crystal oxygen-tolerant Ni-Fe from the aerobic hydrogen oxidizer was found to be heterodimeric. It contains a sizeable catalytic center docking in a hydrogen-reduced state and a small electron relaying 3Fe-S cluster. The cofactors operating the electronic switch function as electron acceptors for hydrogen oxidation and as delivering agents upon oxygen attacks in the active site. In *Chlamydomonas,* 10-fold more oxygen tolerance has been acquired by intruding random and site-directed mutagenesis.

4.2.3 Improvement of Nitrogenase

Nitrogenase is a complex of proteins containing two distinct enzymes, which are (1) Dinitrogenase, also called Mo-Fe protein, and (2) Dinitrogenase reductase, also called Fe protein, both of which participate in two important roles as follows:

During nitrogen fixation, nitrogenase converts atmospheric nitrogen (N_2) into ammonia (NH_3) by using electrons from ferredoxin and energy from adenosine triphosphate (ATP), thereby converting nitrogen into the form, which is vital for plant growth, survival, and ultimately the whole food chain.

During hydrogen production, nitrogenase favors hydrogen (H_2) production instead of the conversion of nitrogen (N_2) into ammonia (NH_3), which typically happens in the scenario of nitrogen starvation conditions or the presence of excess protons (H^+) when they redirect the flow of electrons to reduce protons (H_+) into hydrogen (H_2). The hydrogen evolution reaction allows bacteria to release excess reducing equivalents to maintain redox balance despite being energetically less advantageous than nitrogen fixation.

Because of several characteristics, nitrogenase-mediated H_2 production seems far more promising and exciting than hydrogenase-mediated H_2 production. Firstly, nitrogenase is less sensitive to oxygen than hydrogen and ammonium ions, and not O_2 primarily inhibits oxygen. Cyanobacteria employ the following two mechanisms to protect nitrogenase efficiently from oxygen inhibition. (1) Unique separation (heterocysts): cyanobacteria develop specialized cells called heterocysts with thick walls that restrict oxygen diffusion into the cell during nitrogen starvation conditions. This is how nitrogenase contained within heterocysts lowers the oxygen concentration, creating an anaerobic environment, thereby assisting in fixing nitrogen into ammonia without oxygen. (2) Temporal separation (light/dark cycle): another approach cyanobacteria employ to deal with oxygen sensitivity is to initiate nitrogen fixation only in the dark phase and not during the light cycle when they perform photosynthesis at the time when oxygen is released. Secondly, the ability of enzymes to produce hydrogen under the same partial pressure is another reason nitrogenase is particularly interesting to us. Studies have proved that cyanobacteria of diazotrophs category/heterocystous cyanobacteria show more excellent H_2 production than non-diazotrophic cyanobacteria (Martínez-Merino et al. 2013). Some examples of diazotrophic cyanobacterial species include *Nostoc*, marine *Aphanocapsa montana*, nonmarine *Anabaena* species, marine *Anabaena* species (*A. variabilis* PK84, *Anabaena* AMC 414, *Anabaena cylindrica*, *Anabaena variabilis*) and various members belong to genera: *Calothrix, Entophysalis, Gloeobacter, Lyngbya, Nodularia, Oscillatoria,* and *Synechococcus* (Titlyanov et al. 2017).

4.2.4 Overexpression of Enzymes

Overexpression of enzymes can be defined as the technique of artificially increasing the synthesis of a specific enzyme inside a cell or organism. This is typically achieved by genetic engineering techniques, in which the gene encoding the target enzyme is inserted into the organism's genome to stimulate its expression at higher levels than under normal conditions. The idea is to get more copies of the gene of interest to increase the production of enzymes. In particular, the strategy of homologous overexpression is based on the introduction and amplification of genes similar, homologous, or closely related to the target genes, thus leading to overall higher expression. For the first time in history, [FeFe]-hydrogenase was isolated, characterized, and homologous overexpressed from *Clostridium perfringens* strain intending to boost hydrogen productivity (Morra et al. 2016). Results were promising; along with enhancing the hydrogenase activity, there was an increase in hydrogen productivity of about 1.7-fold and 1.5-fold (Morra et al. 2016). *Enterobacter cloacae* IIT-BT 08 is another example of hydrogenase activity enhancement up to 1.3-fold through overexpression of enzymes (Khanna et al. 2011).

4.3 Genetic Engineering Techniques and Tools

4.3.1 CRISPR-Cas9 Technology

All credit to CRISPR-Cas12 technologies, metabolic engineers can now effectively manipulate or modify the cyanobacterial genomes in several ways through gene substitutions, marker-free point mutations, and gene knock-ins and knockouts (Xue and Greene 2021). Cas nucleases play a pivotal role in accurate and hassle-free genome editing. Cas nucleases work by cleaving the double strand of DNA at desired locations within a genome. Later, the cleaved cellular double strand undergoes repair machinery assisting desired mutations, gene deletions, or sequence additions. There are primarily four types of DNA repair pathway mechanisms that take place in a cell. They are (1) homologous recombination, (2) non-homologous end joining, (3) single-stranded annealing, and (4) microhomology-mediated end joining. In particular, homologous recombination, also called the homology-directed repair mechanism, is correlated to enhance the effectiveness of genome editing.

This increase in efficiency was demonstrated in cyanobacteria using CRISPR-Cas9 technology for the first time (Pooja et al. 2023). CRISPR-dCas9 and Cas12a systems have shown promising results in several model cyanobacteria, including *Anabaena*, *Synechocystis* 6803, and *Synechococcus* 7942 (Pooja et al. 2023). CRISPRi has effectively been used in these applications to dynamically up or downregulate the target genes in several synthetic pathways for increased output of biofuels, be it fatty alcohols or fatty acids or metabolites, for example, amino acids, lactate, pyruvate, and succinate (Pooja et al. 2023). Apart from gene editing efficiency, CRISPR-Cas9 is also appreciated for some of the other features in cyanobacteria, such as shorter homology arms ranging from 400 bp to 700 bp, which can facilitate effective homologous recombination (Pooja et al. 2023). This opens the door to enabling gene integration into small genomic target sites, thereby avoiding unwanted recombination, which can manipulate phenotype and host behavior. Not only this but CRISPR-Cas also helps reduce template plasmid amount for homologous recombination (Pooja et al. 2023). However, CRISPR-Cas9 technology also possesses certain demerits, including the fact that Cas9 nuclease is toxic. However, Cpf1, an innovative and revolutionary RNA-directed dsDNA nuclease harmless to cyanobacteria, has demonstrated a promising role in making numerous marker-less alterations in cyanobacteria. Moreover, the Cpf system, compared to Cas9, not only prevents cyanobacteria from toxicity but is also quite economical and comes in low cost. This is reflected by the fact that the Cpf system only requires a 42 nt RNA component, while Cas9 would require more than 100 nt gRNA (Pooja et al. 2023).

4.3.2 Plasmid Design and Transformation

The cyanobacterial ability to transform (a procedure in which foreign DNA is introduced in the cell) is continuously propelling the advancement in the molecular biology of cyanobacteria, which is becoming the base for several potential biotechnological applications like biofuels, value-based agricultural, environmental, cosmeceutical, nutraceutical, pharmaceutical, and medicinal products. Some cyanobacterial strains undergo genetic manipulations through (1) natural transformations, while several other cyanobacterial strains are made to transform in the laboratory either through (2) conjugation or (3) electroporation. Only a few unicellular cyanobacterial strains such as *Synechocystis* sp. PCC6803 and *Synechococcus* sp. PCC7942 have high DNA uptake efficiency and are capable of natural transformations, which has driven scientists to learn more about them. However, the underlying mechanisms behind natural transformations are not yet fully understood (Al-Haj et al. 2016). Heterocyst-forming filamentous cyanobacteria contain extracellular nucleases that cause incoming DNA to be degraded, leading to low transformation efficiency. Apart from strain type, the other factors affecting efficiency and frequency of transformation worth mentioning include length, concentration, form, and the transformation method (Al-Haj et al. 2016). The high efficiency of DNA in natural transformation has simplified both processes: (1) to develop the mutants using insertional mutagenesis and (2) to clone using genomic libraries. The second method, electroporation, has been employed for both filamentous and unicellular strains; however, the electroporation process might be mutagenic. Thirdly, conjugation, the method initially devised by "Wolk and coworkers," is the most used and often preferred for cyanobacterial transformation because this method can be applied amongst several species' varieties with maximum efficiency (Satta et al. 2023). In this method of DNA transfer to cyanobacteria, two types of plasmids are utilized along with the two types of *E. coli* strains in triparental mating (Satta et al. 2023). The first type, called "conjugative plasmid," contains the *mob* genes carried by the helper strain of *E. coli* and helps transfer small fragments of genetic elements. The second type is the "cargo plasmid," which is the vector used to clone the DNA, i.e., the gene of interest contains the *bom* (oriT-origin of transfer) site called *nic*. This cargo plasmid is carried by another type of *E. coli* strain called donor strain in combination with (1) an origin of replication [oriT], (2) a gene of interest to be transferred, and (3) the sequence required for conjugation transferred, to the recipient cyanobacterial cells. In addition to these two discussed vectors, some cyanobacterial species employ additional types of vectors to maximize conjugation efficiency. For example, Ana7120 requires helper plasmid pRL623, having the ability to encode three methylases, which helps prevent the mobilizer plasmid from being digested in conjugation transformation (Satta et al. 2023).

4.3.3 Selection and Screening Strategies

Different cyanobacterial species are classified based on their families, such as *Bacillariophyceae, Chlorophyceae, Rhodophyceae*, and morphological features. For example, some cyanobacterial species including *Chroococcus giganteus, Cyanothece* sp., *Gloeothece* sp., and *Synechocystis* sp., are classified under the unicellular group, other examples such as *Arthrospira platensis* (also popularly known as *Spirulina platensis*), *Leptolyngbya* sp., J40 (*Aphanizomenon issatschenkoi*) are classified under filamentous group. Lastly, examples including *Nodularia harveyana, Nostoc carneum, Nostoc muscorum,* and *Nostoc punctiforme* are classified under the heterocystous group (Catarina Guedes et al. 2011). However, strain identification and classification into various categories according to their characteristics is an ongoing process. For instance, a recently identified cyanobacterium called *Anthocerotibacter panamensis*, belonging to the Gloeobacteria group is believed to unfold a mystery about the evolution of photosynthesis on Earth (Rahmatpour et al. 2021). There are few cyanobacterial species or strains that are of great importance in the context of large-scale biohydrogen production. *Anabaena* spp. has been found to have the highest metabolic ability to synthesize hydrogen, with an estimated 68 μmol mg^{-1} chl a h^{-1}, and has drawn the attention of scientists in various species identification investigations (Salleh et al. 2016). The yield was significantly high when compared to the other 13 cyanobacterial species, such as *Anabaenopsis, Aphanocapsa, Calothrix, Chroococcidiopsis, Cyanothece, Gloeocapsa,*

Gloeobacter, Microcoleus, Macrocystis, Nostoc, Oscillatoria, Synechococcus, and *Synechocystis* as reported in the study (Quintana et al. 2011). Another recently discovered cyanobacterium worth mentioning for its significant commercial importance is *Anabaena siamensis* TISTR 8012, which demonstrated increased H_2 productivity of 0.057 µmolH_2 mg chl a h^{-1} when cultivated in BG11 medium supplemented with 4 µM nickel ion and was isolated from the rice field in Thailand (Taikhao and Phunpruch 2017). The fact that these recently discovered strains have both nitrogenase and bidirectional hydrogenase accounts for their remarkable success in producing biohydrogen (Taikhao and Phunpruch 2017).

4.4 Metabolic Pathway Engineering Approaches for Improved Hydrogen Production

Among the various ways to produce hydrogen, cyanobacteria stand out due to their unique ability to utilize solar energy and produce hydrogen through separate molecular mechanisms. The complex metabolic pathways involved in hydrogen production in cyanobacteria provide a fascinating landscape involving diverse groups of microorganisms, including both heterocystic and non-heterocystic cyanobacteria. However, challenges remain, especially about nitrogen-dependent hydrogen evolution in heterocystic cyanobacteria. Issues such as carbon sequestration limitations and lower-than-theoretical conversion efficiencies highlight the need to investigate metabolic pathways thoroughly. Non-heterocystic cyanobacteria, on the other hand, offer alternative hydrogen production reactions that pave the way to innovative solutions. In the area of metabolic pathways, the emphasis is on metabolic engineering, which involves redirecting a production pathway to increase production. Computational methods such as OptKnock, OptStrain, and OptReg are crucial in optimizing modulation and detuning strategies. Understanding regulatory mechanisms, alternative metabolic pathways and the effect of nutrient limitations are emphasized, which requires proteome analysis, electron diversion to hydrogenase enzymes, and optimization of carbon fixation pathways (Tiwari and Pandey 2012). Such a multifaceted approach has become central, involving proteomic insights, computational strategies, and genetic interventions. It promises not only to meet the challenges of the process but also to provide a sustainable alternative to traditional energy sources that meet the global drive for a cleaner and greener energy future.

4.4.1 Proteomics Analyses

Proteomics, the comprehensive study of proteins, provides an invaluable perspective for understanding the cellular processes of hydrogen production in cyanobacteria. In the field of metabolic pathways, proteomics is becoming a versatile toolbox.

This emerging field focuses on identifying and characterizing proteins and their various forms, including isoforms and post-translational modifications. Quantitative proteomic methods provide snapshots of cells and proteins at different stages of bioproduction and provide insight into optimization and process monitoring mechanisms (Landels 2016). The development of laboratory methods aimed to enhance proteomic techniques by improving extraction techniques and discovering new ways to measure proteins. The use of data-driven techniques tackles the issues of tag-based proteomic studies, missing tags, lower cutoffs, and pooling datasets. Furthermore, using broader data-driven approaches will lead to improvements in proteome research, as it will address false discovery rates, identify post-translational modification frequencies, and interpret cluster-based data with unspecified assumptions (Landels 2016). Modern proteomic tools such as Matrix-Assisted Laser Desorption/Ionization Time-Of-Flight Mass Spectrometry (MALDI-TOF-MS) and Accurate Mass Temporal (AMT) analysis using Liquid Chromatography-Mass Spectrometry (LC-MS) are indispensable for the interpretation of metabolic pathways. These methods enable precise identification of proteins and provide deep insights into cyanobacterial responses to complex substrates (Sinha et al. 2016). Table 4.1 shows the recent research on proteomic approaches, offering insights into the latest advancements in enhancing cyanobacterial hydrogen production.

4.4.2 Redirecting the Electrons Toward Hydrogenase Enzymes

Redirecting electron flow to hydrogenase enzymes is a strategic way of increasing hydrogen generation in cyanobacteria. The revolutionary research by Khetkorn et al. (2013) reported that site-directed mutagenesis or targeted knockouts were used to inactivate hydrogenase uptake successfully. Compared to their wild-type counterparts, mutants, such as Δ*hupSL* in *Anabaena variabilis* AVM13 and ΔhupS in *A. siamensis* TISTR 8012, demonstrate much increased and persistent hydrogen production (Happe et al. 2000; Khetkorn et al. 2012). It is important to remember that a bidirectional Hox hydrogenase can still be involved in hydrogen consumption, which is a problem with that diversion strategy. The *ldhA* mutation in the *Synechococcus* sp. strain PCC 7002 study is a promising method to increase hydrogen production. This requires removing the enzyme that converts pyruvate to D-lactate in an NADH-dependent manner, increasing the NAD(P)H/NAD(P) + ratio during auto-fermentation. As a potential electron source for bidirectional [NiFe] hydrogenase, the increased ratio significantly increases hydrogen production compared to the wild type (McNeely et al. 2010).

Additionally, the study of the *Cyanothece* sp. strain ATCC 51142 provides insights into its adaptive metabolism, which allows it to thrive in different environments. The strain's ability to produce hydrogen was demonstrated by its bidirectional hydrogenase, uptake hydrogenase, and nitrogenase (Min and Sherman 2010). Nitrogen-fixation conditions lead to the highest rates of hydrogen formation, underscoring the importance of nitrogen incorporation. According to the study, electron

Table 4.1 Proteomic approaches for cyanobacterial hydrogen production. The table showcases the recent advancements in research in the context of hydrogen production in various species and strains of cyanobacteria with the help of proteomics

Species	Experimental approach	Key findings	References
Cyanothece sp. ATCC 51142 and PCC 7822	Potential replacement of *Cyanothece* 7822's peroxiredoxin promoter with that of *Cyanothece* 51142	*Cyanothece* 51142 exhibits robust uptake hydrogenase activity, managing reactive oxygen species (ROS) effectively, leading to sustained nitrogenase activity and potential for increased H_2 production compared to *Cyanothece* 7822. Genetic modification in *Cyanothece* 7822's peroxiredoxin promoter with that of *Cyanothece* 51142 could enhance H_2 production	Aryal et al. (2013)
Anabaena sp. PCC 7120	Label-Free Quantitative Proteomics: Investigation of Δ*hupL* and Δ*hoxH* mutants	Proteomic analysis identifies 2728 proteins, with 79 differently expressed in Δ*hupL* and 117 in Δ*hoxH* mutants. Sheds light on Anabaena's basic metabolism. Uncharacterized proteins are also identified as potential targets for maximizing H_2 generation	Kourpa et al. (2019)
Synechocystis sp. PCC6803 wild type (WT)	Transcriptome Analysis Pan-genomic DNA microarray and SVGMapping R package	Identification of *AbrB2* as a master regulator affecting genes involved in oxidative stress, metal transport, and H_2 production. Emphasizes the complexity of regulating hydrogen generation and the significance of considering plasmids in metabolic pathway analysis	Leplat et al. (2013)
Cyanothece sp. ATCC 51142	Chemoproteomic Method: System-level understanding of protein thiol redox during photosynthesis and hydrogen generation	Dinitrogenase NifHDK acts as an emergency electron valve responding to intracellular redox circumstances. Proteomic insights offer a plan for scalable and effective cyanobacterial hydrogen synthesis.	Sadler et al. (2016)
Nostoc punctiforme ATCC 29133	Uptake Hydrogenase Mutant (NHM5): Comprehensive Proteomic Analysis	Inactivation of uptake hydrogenase in NHM5 leads to redirected metabolism, influencing oxygen reduction and processes supplying reducing equivalents. Despite reduced growth, NHM5 compensates for electron loss, providing insights for enhancing nitrogenase-mediated hydrogen photogeneration	Ekman et al. (2011)

transport through photosystem II (PS II) is a prerequisite for hydrogenase activity (Min and Sherman 2010). Ducat et al. (2011) successfully redirected electrons toward hydrogenase enzymes by introducing [FeFe] hydrogenase (HydA) from *Clostridium acetobutylicum* in a study utilizing *Synechococcus elongatus* sp. strain 7942. This modified strain exhibited a much higher light-dependent hydrogenase efficiency than the native hydrogenation enzyme. Internal energy reserves are dark fermented to hydrogen gas by regulating electron flow through external ferredoxins, increasing yield. A comprehensive analysis of the problems associated with low hydrogen yields in cyanobacteria has been described previously (Khanna and Lindblad 2015). It is proposed that ferredoxin and flavodoxin can act as electron donors, reversing the traditional function of NAD(P)H and enabling the bidirectional Hox hydrogenase to operate. The review highlights the need to engineer [NiFe] hydrogenases to express optimal catalytic efficiency or high turnover rate [FeFe] hydrogenases in cyanobacteria to achieve biohydrogen targets, suggesting that successful hydrogen-producing strains may require the integration of enzymatic properties from different species.

Hence, electron transfer to hydrogenase enzymes seems an effective method to increase hydrogen production in cyanobacteria. The viability of this strategy is demonstrated by the efficient use of mutants such as Δ*hupSL*, the *ldhA* mutant, and modified strains with added hydrogenases. Nevertheless, challenges still need to be addressed, such as the bidirectional hydrogenase activity and the necessity for a comprehensive understanding of metabolism. A deeper understanding of electron flow regulation and optimizing the capacity of modified strains presents a significant opportunity to increase the efficiency and scalability of hydrogen production in cyanobacteria. By advancing the science of biohydrogen production, this line of research highlights the potential of using metabolic engineering to harness these photosynthetic microorganisms.

4.4.3 Carbon Fixation Pathways Optimization

Carbon fixation pathways in cyanobacteria are essential for continuous energy production through photosynthesis, providing essential reducing equivalents for hydrogenase enzymes in hydrogen production. The efficiency of cyanobacterial hydrogen production is affected by the pathways that regulate cellular redox balance, nitrogenase activity, and fermentation ATP production. Optimizing carbon fixation is necessary to ensure a constant and sufficient supply of electrons, which in turn improves hydrogen supply.

The importance of inhibitors that target the Calvin–Benson cycle, like dl-glyceraldehyde, which has shown a significant increase in hydrogen production under light conditions, is highlighted by the study of Khetkorn et al. (2013). In addition, the use of specific inhibitors such as DCMU to inhibit PSII activity, thereby increasing nitrogenase activity and protecting against oxygen inhibition, has been investigated. According to Khetkorn et al. (2013), this deliberate redirection of electron flow increases the number of electrons available for hydrogen metabolism. The work of Hallenbeck et al. (1979) highlights the need for a concentrated effort to optimize the carbon fixation pathway to improve cyanobacterial hydrogen production, emphasizing the energetic effect of H_2 evolution in glycolysis and substrate catabolism. Ananyev et al. (2008) provide an essential overview of nutritional and environmental factors that significantly influence H_2 evolution during fermentation processes. The results show nitrate removal during fermentative H_2 evolution significantly increases H_2 formation. This finding highlights the importance of maintaining the intracellular reduction potential (NADH/NAD ratio), optimizing conditions for anaerobic ATP generation, and pre-accumulating glycogen to promote H_2 evolution. The study of Ananyev et al. (2008) strongly supports a comprehensive optimization strategy to increase cyanobacterial hydrogen production through efficient carbon fixation pathways. Moreover, in *Cyanothece* sp., the metabolic adaptability of strain ATCC 51142 sheds more light on the mechanisms behind carbon fixation. According to Min and Sherman (2010), this strain can flourish with or without fixed nitrogen, as it has variable growth mode. The research emphasizes the significance of light levels and the presence or absence of glycerol in achieving maximum hydrogen production rates. These results suggest that optimizing carbon fixation pathways is essential for efficient hydrogen synthesis, particularly when nitrogen is fixed. *Cyanothece* sp. highlights the need to thoroughly understand and refine carbon fixation pathways to improve hydrogen production in cyanobacteria (Min and Sherman 2010).

In summary, the intentional optimization of carbon fixation pathways—as evidenced by the targeted use of inhibitors and the understanding of environmental factors—remains a crucial avenue for cyanobacterial hydrogen production. Based on a range of studies, these findings demonstrate how intricate cyanobacteria's metabolic engineering is. By optimizing carbon fixation pathways, researchers may make the most of these microorganisms and lay the groundwork for a sustainable and effective hydrogen production process.

4.5 Optimization Strategies

Cyanobacterial farming requires multi-parameter optimization to be successful, especially given the high demand from industry. Due to this commercial necessity, research into using various scientific techniques in cyanobacterial cultivation to increase biomass output, titer, and yield of high-quality goods while lowering production and maintenance costs has increased. Furthermore, cyanobacterial species differ not just in terms of the composition of their biomass but also in terms of their capacity to tolerate different environmental factors, such as pH, temperature, salinity, availability of nutrients, CO_2, contaminants, grazers, light quality, and duration of exposure, etc. (Pathak et al. 2018). In this regard, response surface methodology (RSM) is a solid and effective mathematical technique that is frequently used as a multi-parameter optimization method because it can simultaneously provide multiple responses, information about the interaction between variables, and information required for process and design optimization (Ojha et al. 2020) Additionally, another well-known technique for figuring out the combination of critical variables and ideal circumstances that lead to high biomass production is the Taguchi model (Bentahar and Deschênes 2022).

4.5.1 Light Intensity and Photobioreactor Design

The amount and quality of light significantly impact the efficiency of cyanobacterial photosynthesis, which in turn determines biomass composition and output. One of the main problems is photoinhibition, which is brought on by exposure to light at an intensity higher than ideal and lowers photosynthetic efficiency and biomass growth (Grama et al. 2022). Notably, the relationship between high light intensity and high lipid synthesis or other biochemical composition varies among species (Nzayisenga et al. 2020)—*Synechocystis* sp. PCC6803 can exhibit an increase in lipid content when exposed to high-light conditions (Wang et al. 2024). Even though too much light damages cells oxidatively, cells attempt to store the light energy as lipids. Therefore, strains that can withstand high light are ideal for outdoor, large-scale production systems. Recently, in genetically modified cyanobacteria, the operon CPC was suppressed to reduce the size of the phycobilisome (PBS) antennae complex, which enabled them to prevent excessive solar absorption. This means that the cyanobacteria can only absorb incident sunlight to the extent necessary for their survival and metabolic processes (Grama et al. 2022). In addition, the increased light intensity creates a problem of overheating outdoor PBRs. Two methods are being tried to manage the temperature: water bath immersion and shading. To prevent overheating, it has also been recommended to choose an algae production site where the local temperature does not change significantly at night and the environment is moderate, meaning it is neither too hot nor too cold. According to a study, PBR orientation and position are essential as well. Outdoor PBRs positioned vertically in the east–west direction can gather 5% more solar energy than PBRs oriented horizontally in the north–south direction (Prabha et al. 2022). Also, choosing the appropriate LED light can lower electric power costs and improve the composition of proteins, lipids, pigments, and value-added chemicals. Open-raceway systems present several issues, such as excessive light exposure, overheating, oxygen build-up, water loss from evaporation, bacterial invasion, low CO_2 supply, salinity, cultivation vessels, contamination, and uncontrollable mixing rate that can cause shear stress-cell destruction. Closed PBR systems, which offer greater process parameter regulation with little to no external environmental impact, can overcome the drawbacks of open ponds (Fig. 4.4). However, compared to open-raceway ponds, the design, manufacture, operation, surface sterilization, biofouling, and cooling of PBRs are more costly and challenging, which results in lower biomass output and higher harvesting costs. A few pieces of data already suggest that PBR improvement is progressing. For instance, it was found that using gold nanoparticles (AuNPs) in PBRs can save costs since the suspended AuNPs are recyclable and do not readily pollute, allowing for at least five reuses (Sitther et al. 2020). Applying silver nanoparticles (AgNPs) coating, which kills microorganisms upon contact while protecting viable algal cells in bulk media, is an example of nanotechnology that can stop the biofouling problem in PBRs. After a given amount of time, AgNP coatings begin to degrade or lose their bacterial activity (Hamida et al. 2020).

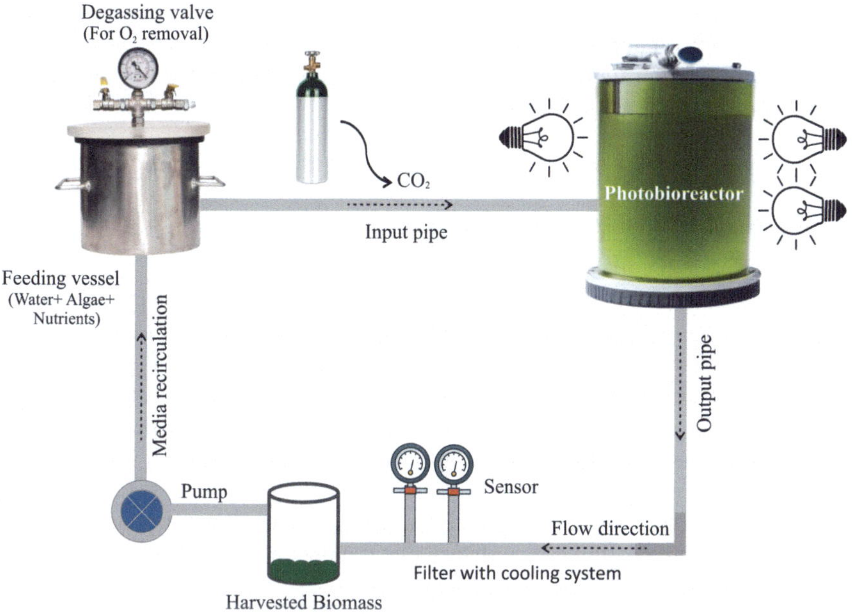

Fig. 4.4 Schematic representation of a photobioreactor for biohydrogen production

Transparency, cleanliness, convenience of use, and a high surface-to-volume ratio for intensive mixing are essential for photobioreactor design.

4.5.2 *Nutrient Management*

Several critical components regarding nutrients must be managed to produce biohydrogen at optimal yields. Among the essential elements needed for cyanobacterial enzyme cofactors, transport, and dehydrogenase activities are iron, sodium, zinc, nickel, phosphate, nitrogen, and magnesium. In a comprehensive study, Das et al. (2014) described that dark hydrogen synthesis is aided by ferredoxin intermediates, which require iron. Magnesium is essential for glycolytic activities and maintains the integrity of cell walls and membranes by activating enzymes. The production of hydrogen is impacted variably by heavy metals like chromium, copper, and zinc, depending on their concentration levels. In certain circumstances, copper shows severe toxicity. For anaerobic digestion and phosphate buffering to function properly, the ratios of carbon to nitrogen and carbon to phosphate must be maintained. While high quantities of phosphate can increase production rates, too much of it can have the reverse effect and reduce yields overall (Das et al. 2014). Ammonia may have been a cost-effective alternative, but it causes problems by releasing hydronium (H) ions, which causes pH levels to fluctuate. In contrast, agro-industrial wastewater, rich in nitrogen and phosphorus, is an attractive solution since it can reduce the cost of nutritional media (Kamravamanesh et al. 2019). Thus, nutrient management strategies are much required and must be researched further for efficient H_2 production.

4.5.3 pH Control and Gas Exchange

In cyanobacteria, pH plays an important role in optimizing hydrogen production. H_2 is produced during the exponential growth phase, but pH shifts toward solvent production in the stationary stage. The maintenance of a particular pH may prevent this phase shift. The concentration of volatile fatty acids (VFAs), like acetate, butyrate, propionate, ethanol, etc., produced simultaneously with H_2, is pH dependent. Due to the increased ionic strength, the accumulation of VFAs decreases pH, which hinders the production of H_2 because of the shift from acidogenesis to solventogenesis (Harirchi et al. 2022). The high extracellular pH facilitates VFA entry into lower pH cells, causing an intracellular disturbance that requires ATP to drive a proton gradient and prevents hydrogenase denaturation. To produce H_2 efficiently, optimal substrate concentration, minimum operating pH, and ion balance are essential. Sometimes, hydrogen accumulates in the upper space of the bioreactor, which can prevent production by raising the partial pressure. Optimum agitation (>2.5 rpm) facilitates hydrogen transition to the gas phase, reducing accumulation without mechanical damage. Table 4.2 showcases the recent research studies on different cyanobacteria species by optimizing various parameters for enhanced hydrogen production.

4.6 Conclusion and Prospects

Various metabolic pathways are essential to enhance biohydrogen production by following various strategies. Biomass composition and environmental tolerance variables underscore the importance of optimizing factors like pH, temperature, nutrients, and light exposure. Mathematical techniques like Response Surface Methodology (RSM) and the Taguchi model play pivotal roles in multi-parameter optimization, offering insights into variable interactions and ideal conditions for maximizing biomass production. Genetic modulation by altering genes involved in NH_4 transport can also stabilize nitrogenase activity, making mutants NH_4-insensitive and thus boosting hydrogen generation. Double mutation in homocitrate synthase and hydrogenase genes in *Nostoc* PCC 7120 can improve long-term hydrogen production capability by disrupting non-essential pathways and enhancing hydrogenase activity. Therefore, cyanobacterial farming necessitates meticulous optimization across various parameters to meet industrial demands, driving increased research into cultivation techniques for enhanced biomass output, quality, and cost-effectiveness.

Table 4.2 Optimization strategies for enhancing cyanobacterial hydrogen production

Cyanobacterial species	Optimization of parameters	Key findings	References
Anabaena cylindrica B629 *Tolypothrix* sp. B379 *Nostoc punctiforme* PCC 73102 *Anabaena* sp. PCC 7120 *Fischerella muscicola* PCC 7414 *Nostoc commune* MFG-1 *Spirirestis rafaelensis* LQ-10 *Scytonema hylanium* NCC-4B *Calothrix* sp. MCC-3	O_2 and N_2 Concentration, Light intensity, Glucose	In every culture examined, nitrogen (N_2) significantly reduces the synthesis of hydrogen (H_2); at the air ratio (4:1), N_2 is a more powerful inhibitor than oxygen (O_2). The addition of glucose in most species significantly increased H_2 generation.	Yeager et al. (2011)
Arthrospira sp. PCC 8005	Fe^{2+}, β-mercaptoethanol, Glucose, pH	Adding Fe^{2+} and β-mercaptoethanol to nitrogen- and sulfur-starved cells significantly increases H_2 production; it reaches a maximum of 5.91 ± 0.14 µmol H_2 mg $Chla^{-1}$ h^{-1}. Slight osmolality increase and glucose promote the synthesis of H_2. High production rates are attained in a medium lacking N_2 and with a pH of neutral to alkaline; in an acidic environment, these rates decrease	Raksajit et al. (2012)
Spirulina platensis	Photobioreactor design, Light intensity, pH	By optimizing illumination and reducing shading and photolysis effect in the dense cell cultures, a conical photobioreactor with an excavated base facilitated enhanced hydrogen evolution in photobiological production by *S. platensis*	Ainas et al. (2017)
Geitlerinema sp. RMK-SH10	$NaNO_3$-free ASN III medium, NaCl, glucose, Ni^{2+}	This marine species generated a maximum H_2 production rate of 0.271 ± 0.013 µmol H_2 mg^{-1} dry weight h^{-1} when cultivated in $NaNO_3$—Free ASN III medium containing 0.2 M NaCl, 18.9 mmol C-atom of glucose L^{-1}, and 0.1 µM Ni^{2+}	Tinpranee et al. (2018)

References

Ainas M, Hasnaoui S, Bouarab R, Abdi N, Drouiche N, Mameri N (2017) Hydrogen production with the cyanobacterium Spirulina platensis. Int J Hydrog Energy 42(8):4902–4907. https://doi.org/10.1016/j.ijhydene.2016.12.056

Al-Haj L, Lui Y, Abed R, Gomaa M, Purton S (2016) Cyanobacteria as chassis for industrial biotechnology: progress and prospects. Life 6(4):42. https://doi.org/10.3390/life6040042

Ananyev G, Carrieri D, Dismukes GC (2008) Optimization of metabolic capacity and flux through environmental cues to maximize hydrogen production by the Cyanobacterium "Arthrospira (Spirulina) maxima". Appl Environ Microbiol 74(19):6102–6113. https://doi.org/10.1128/AEM.01078-08

Aryal UK, Callister SJ, Mishra S, Zhang X, Shutthanandan JI, Angel TE, Shukla AK, Monroe ME, Moore RJ, Koppenaal DW, Smith RD, Sherman L (2013) Proteome analyses of strains ATCC 51142 and PCC 7822 of the diazotrophic cyanobacterium Cyanothece sp. under culture conditions enhanced H2 production. Appl Environ Microbiol 79(4):1070–1077. https://doi.org/10.1128/AEM.02864-12

Bentahar J, Deschênes J-S (2022) Media optimization design towards maximizing biomass production of Tetradesmus obliquus under mixotrophic conditions. Bioresourc Technol Rep 17:100885. https://doi.org/10.1016/j.biteb.2021.100885

Catarina Guedes A, Barbosa CR, Amaro HM, Pereira CI, Xavier Malcata F (2011) Microalgal and cyanobacterial cell extracts for use as natural antibacterial additives against food pathogens. Int J Food Sci Technol 46(4):862–870. https://doi.org/10.1111/j.1365-2621.2011.02567.x

Das D, Khanna N, Dasgupta CN (2014) Biohydrogen production. CRC Press. https://doi.org/10.1201/b16574

Ducat DC, Sachdeva G, Silver PA (2011) Rewiring hydrogenase-dependent redox circuits in cyanobacteria. Proc Natl Acad Sci USA 108(10):3941–3946. https://doi.org/10.1073/pnas.1016026108

Ekman M, Ow SY, Holmqvist M, Zhang X, Van Wagenen J, Wright PC, Stensjö K (2011) Metabolic adaptations in a H2 producing heterocyst-forming cyanobacterium: potentials and implications for biological engineering. J Proteome Res 10(4):1772–1784. https://doi.org/10.1021/pr101055v

Govindjee (2011) Adventures with cyanobacteria: a personal perspective. Front Plant Sci 2. https://doi.org/10.3389/fpls.2011.00028

Grama SB, Liu Z, Li J (2022) Emerging trends in genetic engineering of microalgae for commercial applications. Mar Drugs 20(5):285. https://doi.org/10.3390/md20050285

Hallenbeck PC, Kostel PJ, Benemann JR (1979) Purification and properties of nitrogenase from the cyanobacterium, Anabaena cylindrica. Eur J Biochem 98(1):275–284. https://doi.org/10.1111/j.1432-1033.1979.tb13186.x

Hamida RS, Ali MA, Redhwan AMO, Bin-Meferij MM (2020) Cyanobacteria—a promising platform in green nanotechnology: a review on nanoparticles fabrication and their prospective applications. Int J Nanomed 15:6033–6066. https://doi.org/10.2147/IJN.S256134

Happe T, Schutz K, Schutz S, Bohme H, Bohme B (2000) Transcriptional and mutational analysis of the uptake hydrogenase of the filamentous cyanobacterium Anabaena variabilis ATCC 29413. J Bacteriol 182(6). https://journals.asm.org/journal/jb

Harirchi S, Wainaina S, Sar T, Nojoumi SA, Parchami M, Parchami M, Varjani S, Khanal SK, Wong J, Awasthi MK, Taherzadeh MJ (2022) Microbiological insights into anaerobic digestion for biogas, hydrogen or volatile fatty acids (VFAs): a review. Bioengineered 13(3):6521–6557. https://doi.org/10.1080/21655979.2022.2035986

Hoiczyk E, Hansel A (2000) Cyanobacterial cell walls: news from an unusual prokaryotic envelope. J Bacteriol 182(5):1191–1199. https://doi.org/10.1128/JB.182.5.1191-1199.2000

Josephs RE, Onuh CY, Orodu OD, Olabode OA, Nchila YT, Dinga CN (2022) Nigerian marginal oilfield development program: PIA and current issues, pp 83–95. https://doi.org/10.1007/978-3-030-96721-5_8

Kamravamanesh D, Kiesenhofer D, Fluch S, Lackner M, Herwig C (2019) Scale-up challenges and requirement of technology-transfer for cyanobacterial poly (3-hydroxybutyrate) production in industrial scale. Int J Biobased Plast 1(1):60–71. https://doi.org/10.1080/24759651.2019.1688604

Kamshybayeva GK, Kossalbayev BD, Sadvakasova AK, Kakimova AB, Bauenova MO, Zayadan BK, Lan C-W, Alwasel S, Tomo T, Chang J-S, Allakhverdiev SI (2024) Genetic engineering contribution to developing cyanobacteria-based hydrogen energy to reduce carbon emissions and establish a hydrogen economy. Int J Hydrog Energy 54:491–511. https://doi.org/10.1016/j.ijhydene.2022.12.342

Khanna N, Lindblad P (2015) Cyanobacterial hydrogenases and hydrogen metabolism revisited: recent progress and future prospects. Int J Mol Sci 16(5):10537–10561. MDPI AG. https://doi.org/10.3390/ijms160510537

Khanna N, Dasgupta CN, Mishra P, Das D (2011) Homologous overexpression of [FeFe] hydrogenase in Enterobacter cloacae IIT-BT 08 to enhance hydrogen gas production from cheese whey. Int J Hydrog Energy 36(24):15573–15582. https://doi.org/10.1016/j.ijhydene.2011.09.020

Khetkorn W, Lindblad P, Incharoensakdi A (2012) Inactivation of uptake hydrogenase leads to enhanced and sustained hydrogen production with high nitrogenase activity under high light exposure in the cyanobacterium Anabaena siamensis TISTR 8012. J Biol Eng 6. https://doi.org/10.1186/1754-1611-6-19

Khetkorn W, Khanna N, Incharoensakdi A, Lindblad P (2013) Metabolic and genetic engineering of cyanobacteria for enhanced hydrogen production. In: Biofuels, vol 4, issue 5, pp 535–561. https://doi.org/10.4155/bfs.13.41

Kollmen J, Strieth D (2022) The beneficial effects of cyanobacterial co-culture on plant growth. Life 12(2):223. https://doi.org/10.3390/life12020223

Kourpa K, Manarolaki E, Lyratzakis A, Strataki V, Rupprecht F, Langer JD, Tsiotis G (2019) Proteome analysis of enriched heterocysts from two hydrogenase mutants from anabaena sp. PCC 7120. Proteomics 19(19). https://doi.org/10.1002/pmic.201800332

Kumar A, Mishra B, Singh M (2024) Threats, challenges and issues of large-scale cyanobacterial cultivation. In: Pharmaceutical and nutraceutical potential of cyanobacteria. Springer International Publishing, pp 245–275. https://doi.org/10.1007/978-3-031-45523-0_10

Landels AR (2016) Improving proteomic methods and investigating H 2 production in Synechocystis sp. PCC6803

Leplat C, Champeimont R, Saenkham P, Cassier-Chauvat C, Jean-Christophe A, Chauvat F (2013) Genome-wide transcriptome analysis of hydrogen production in the cyanobacterium synechocystis: towards the identification of new players. Int J Hydrog Energy 38(4):1866–1872. https://doi.org/10.1016/j.ijhydene.2012.11.118

Martínez-Merino V, Gil MJ, Cornejo A (2013) Biological hydrogen production. In: Renewable hydrogen technologies. Elsevier, pp 171–199. https://doi.org/10.1016/B978-0-444-56352-1.00008-8

McNeely K, Xu Y, Bennette N, Bryant DA, Dismukes GC (2010) Redirecting reductant flux into hydrogen production via metabolic engineering of fermentative carbon metabolism in a cyanobacterium. Appl Environ Microbiol 76(15):5032–5038. https://doi.org/10.1128/AEM.00862-10

Min H, Sherman LA (2010) Hydrogen production by the unicellular, diazotrophic cyanobacterium cyanothece sp. strain ATCC 51142 under conditions of continuous light. Appl Environ Microbiol 76(13):4293–4301. https://doi.org/10.1128/AEM.00146-10

Morra S, Mongili B, Maurelli S, Gilardi G, Valetti F (2016) Isolation and characterization of a new [FeFe]-hydrogenase from *Clostridium perfringens*. Biotechnol Appl Biochem 63(3):305–311. https://doi.org/10.1002/bab.1382

Nzayisenga JC, Farge X, Groll SL, Sellstedt A (2020) Effects of light intensity on growth and lipid production in microalgae grown in wastewater. Biotechnol Biofuels 13(1):4. https://doi.org/10.1186/s13068-019-1646-x

Ojha SK, Singh PK, Mishra S, Pattnaik R, Dixit S, Verma SK (2020) Response surface methodology based optimization and scale-up production of amylase from a novel bacterial strain, Bacillus aryabhattai KIIT BE-1. Biotechnology Reports 27:e00506. https://doi.org/10.1016/j.btre.2020.e00506

Parambath SD, Bhombal Y, Kumar A, Singh M, Chavali M, Chandrasekhar K (2022) The multifaceted microalgal approach to wastewater treatment to generate energy and essential chemicals, pp 139–176. https://doi.org/10.1007/978-981-19-0793-7_6

Pathak J, Rajneesh, Maurya PK, Singh SP, Häder D-P, Sinha RP (2018) Cyanobacterial farming for environment friendly sustainable agriculture practices: innovations and perspectives. Front Environ Sci 6. https://doi.org/10.3389/fenvs.2018.00007

Pooja P, Lekshmi KE, NS P (2023) Biofuels from cyanobacteria – a metabolic engineering approach. Plant Sci Today. https://doi.org/10.14719/pst.2505

Prabha S, Vijay AK, Paul RR, George B (2022) Cyanobacterial biorefinery: towards economic feasibility through the maximum valorization of biomass. Sci Total Environ 814:152795. https://doi.org/10.1016/j.scitotenv.2021.152795

Quintana N, van der Kooy F, van de Rhee MD, Voshol GP, Verpoorte R (2011) Renewable energy from cyanobacteria: energy production optimization by metabolic pathway engineering. Appl Microbiol Biotechnol 91(3):471–490. https://doi.org/10.1007/s00253-011-3394-0

Rahmatpour N, Hauser DA, Nelson JM, Chen PY, Villarreal A, Ho M-Y, Li F-W (2021) A novel thylakoid-less isolate fills a billion-year gap in the evolution of cyanobacteria. Curr Biol 31(13):2857–2867.e4. https://doi.org/10.1016/j.cub.2021.04.042

Raksajit W, Satchasataporn K, Lehto K, Mäenpää P, Incharoensakdi A (2012) Enhancement of hydrogen production by the filamentous non-heterocystous cyanobacterium Arthrospira sp. PCC 8005. Int J Hydrog Energy 37(24):18791–18797. https://doi.org/10.1016/j.ijhydene.2012.10.011

Sadler NC, Bernstein HC, Melnicki MR, Charania MA, Hill EA, Anderson LN, Monroe ME, Smith RD, Beliaev AS, Wright AT (2016) Dinitrogenase-driven photobiological hydrogen production combats oxidative stress in Cyanothece sp. strain ATCC 51142. Appl Environ Microbiol 82(24):7227–7235. https://doi.org/10.1128/AEM.02098-16

Salleh SF, Kamaruddin A, Uzir MH, Karim KA, Mohamed AR (2016) Investigation of the links between heterocyst and biohydrogen production by diazotrophic cyanobacterium A. variabilis ATCC 29413. Arch Microbiol 198(2):101–113. https://doi.org/10.1007/s00203-015-1164-6

Sato N (2021) Are cyanobacteria an ancestor of chloroplasts or just one of the gene donors for plants and algae? Genes 12(6):823. https://doi.org/10.3390/genes12060823

Satpati GG, Kumar A, Parambath SD, Bagchi T, Singh M, Chavali M, Chandrasekhar K (2022) Recent development in microalgal cultivation and biomass harvesting methods. In: Algal biorefineries and the circular bioeconomy. CRC Press, pp 61–91. https://doi.org/10.1201/9781003195429-2

Satta A, Esquirol L, Ebert BE (2023) Current metabolic engineering strategies for photosynthetic bioproduction in cyanobacteria. Microorganisms 11(2):455. https://doi.org/10.3390/microorganisms11020455

Sinha P, Roy S, Das D (2016) Genomic and proteomic approaches for dark fermentative biohydrogen production. In Renewable and sustainable energy reviews. Elsevier, vol 56, pp 1308–1321. https://doi.org/10.1016/j.rser.2015.12.035

Sitther V, Tabatabai B, Fathabad SG, Gichuki S, Chen H, Arumanayagam ACS (2020) Cyanobacteria as a biofuel source: advances and applications. In: Advances in cyanobacterial biology. Elsevier, pp 269–289. https://doi.org/10.1016/B978-0-12-819311-2.00018-8

Taikhao S, Phunpruch S (2017) Effect of metal cofactors of key enzymes on biohydrogen production by nitrogen fixing cyanobacterium Anabaena siamensis TISIR 8012. Energy Procedia 138:360–365. https://doi.org/10.1016/j.egypro.2017.10.166

Tinpranee N, Incharoensakdi A, Phunpruch S (2018) Screening cyanobacteria from marine coastal waters of Thailand for biohydrogen production. J Appl Phycol 30(6):3471–3481. https://doi.org/10.1007/s10811-018-1490-6

Titlyanov EA, Titlyanova TV, Li X, Huang H (2017) Marine plants of coral reefs. In Coral reef marine plants of Hainan Island. Elsevier, pp 5–39. https://doi.org/10.1016/B978-0-12-811963-1.00002-0

Tiwari A, Pandey A (2012) Cyanobacterial hydrogen production - a step towards clean environment. Int J Hydrog Energy 37(1):139–150. https://doi.org/10.1016/j.ijhydene.2011.09.100

Vu HP, Nguyen LN, Zdarta J, Nga TT, Nghiem LD (2020) Blue-green algae in surface water: problems and opportunities. Curr Pollut Rep 6(2):105–122. https://doi.org/10.1007/s40726-020-00140-w

Wang M, Ye X, Bi H, Shen Z (2024) Microalgae biofuels: illuminating the path to a sustainable future amidst challenges and opportunities. Biotechnol Biofuels Bioprod 17(1):10. https://doi.org/10.1186/s13068-024-02461-0

Xue C, Greene EC (2021) DNA repair pathway choices in CRISPR-Cas9-mediated genome editing. Trends Genet 37(7):639–656. https://doi.org/10.1016/j.tig.2021.02.008

Yeager CM, Milliken CE, Bagwell CE, Staples L, Berseth PA, Sessions HT (2011) Evaluation of experimental conditions that influence hydrogen production among heterocystous cyanobacteria. Int J Hydrog Energy 36(13):7487–7499. https://doi.org/10.1016/j.ijhydene.2011.03.078

Zahra Z, Choo DH, Lee H, Parveen A (2020) Cyanobacteria: review of current potentials and applications. Environments 7(2):13. https://doi.org/10.3390/environments7020013

Chapter 5
Metabolic Pathway Engineering in Cyanobacteria for Environmental Applications

Aqib Zafar Khan, Azeem Asghar, Muhammad Nabeel Haider, Asadullah Khan, Archishman Bose, and Linda O'Higgins

Abstract Cyanobacteria have an efficient and well-controlled metabolic architecture. Their abilities to adapt to harsh environments and unpredictable stresses have led to the development of sophisticated molecular editing tools to fully harness their machinery for diverse applications. Improved knowledge of cellular metabolism under heavy metal stress, pollutant removal, and salt tolerance utilizing omics-based techniques is crucial for manipulating cyanobacteria's genome for environmental applications. This chapter covers the elucidation and engineering of the molecular pathways in cyanobacteria for removal of heavy metals, phosphorus, and excessive salts. Promoter systems, ribosome-binding sites (RBSs), CRISPR interference (CRISPRi), and synthetic short regulatory RNA (sRNA) are among the well-established regulatory tools in cyanobacteria. A promising strategy for creating multi-knock-in marker-free mutants in *Escherichia coli*, serine integrase recombinational engineering may soon be investigated in cyanobacteria. Synthetic genetic circuits that operate as Boolean logic functions have been built in model strains, while these circuits allow switch-like control over gene expression in response to

A. Z. Khan (✉) · A. Khan
Guangdong Provincial Key Laboratory for Plant Epigenetics, Guangdong Engineering Research Centre for Marine Algal Biotechnology, College of Life Sciences and Oceanography, Shenzhen University, Shenzhen, China
e-mail: aqib.z.khan@szu.edu.cn

A. Asghar
Department of Bioinformatics and Biotechnology, Government College University Faisalabad, Faisalabad, Pakistan

M. N. Haider · A. Bose
MaREI Centre, Environmental Research Institute, University College Cork, Cork, Ireland

Process and Chemical Engineering, School of Engineering and Architecture, University College Cork, Cork, Ireland

L. O'Higgins (✉)
Process and Chemical Engineering, School of Engineering and Architecture, University College Cork, Cork, Ireland
e-mail: linda.ohiggins@ucc.ie

M. A. Mehmood et al. (eds.), *Cyanobacteria Biotechnology*,
https://doi.org/10.1007/978-3-031-70698-1_5

environmental inputs. The possibility of elucidating and manipulating the resistance mechanisms has also been discussed as a tool to improve cyanobacterial bioremediation capabilities. Lastly, possible obstacles and opportunities for enhancing cyanobacteria-based bioremediation of heavy metals, phosphorus, and excessive salts are also highlighted.

Keywords Cyanobacteria · Heavy metals · Salt-stress · Bioremediation · Wastewater

5.1 Introduction

Over the previous decades, environment has been contaminated due to the overuse of synthetic fertilizers and pesticides to improve the agricultural yields. Various toxic compounds, heavy metals, and other undesirable by-products have been adversely impacting the environment (Cui et al. 2020b; Dong et al. 2023; Chakdar et al. 2022; Debnath et al. 2024). It happened because emerging nations quickly underwent through industrialization to meet the global market demands without implementing strong regulatory framework to curb the pollution that followed, unless these problems became much more apparent. The majority of pollution in the environment comes from man-made chemicals, while the situation is even worst in the developing and underdeveloped nations (Francová et al. 2017). There is a strong correlation between pollution and advances in water resource engineering, and around 80% of the world's population is in high danger of water insecurity (Vörösmarty et al. 2010).

Some cases of pollutant remediation have been accomplished through the use of long-established physicochemical methods, such as filtration, solidification, incineration, evaporation, oxidation and reduction, chemical precipitation, electrochemical treatment, and ion-exchange approaches (Shahid et al. 2021). Unfortunately, these traditional methods have their limitations, such as high processing costs, reagent needs, volume, and the creation of secondary environmental contaminants. Wastewaters from industries such as mining, contamination of groundwater sources, and others highlight the limitations of these procedures (Bhalara et al. 2014; Dasgupta et al. 2015).

The utilization of oxygen-evolving photosynthetic cyanobacteria is one environmentally benign and cost-effective remediation strategy. Heavy metals, antibiotics, phenanthrene, pesticides, nitrogen, detergents, phosphorus, naphthalene, phenol, catechol, and carbon dioxide are just a few of the many xenobiotic and naturally occurring/synthetic recalcitrant contaminants that can be effectively bioremediated using cyanobacteria (Singh et al. 2016). In addition to this bioremediation capability, cyanobacteria fix 25 Gt/annum CO_2 into biomass through photosynthesis (Pisciotta et al. 2010). It is more acceptable to use cyanobacteria for bioremediation since they participate in metabolic pathways that rely mostly on the enzymes they generate. These cyanobacteria can either totally break down the contaminants or can transform them into less dangerous forms, depending upon the nature and concentration of the contaminants. Besides, these organisms are gaining attention because of easier amenability of their metabolic engineering when compared to plants or eukaryotic algae. Studying cyanobacteria's biological systems and the networks that comprise them at the molecular, cellular, population, community, and ecosystem levels is made possible by genomics and proteomics, two complementary types of genomic data (Chakdar et al. 2022; Debnath et al. 2024; Behler et al. 2018; Sun et al. 2018; Xia et al. 2019; Bradley et al. 2016; Higo et al. 2017; Hendry et al. 2020; Gu et al. 2019). These methods yield comprehensive and useful information on biological processes that may be modified to optimize the removal of a specific pollutant or pollutants at once (Ng et al. 2020; Sanz-Luque et al. 2020; Sun et al. 2018).

Improved knowledge of cellular metabolism under heavy metal stress, EPS synthesis, phosphate removal, and salt tolerance utilizing omics-based techniques is crucial for manipulating cyanobacteria for environmental applications, in addition to developing better genetic tools. Cyanobacteria such as *S. elongatus* 6803, *S. elongatus* 7942, and *S. elongatus* 2973 are among the few genome-scale established models (Hendry et al. 2020). They require marker-free genome editing to use as HM-bioremediation tool. Genomes of various cyanobacteria have been edited using genome editing techniques based on CRISPR (clustered regularly interspaced short palindromic repeats) (Behler et al. 2018; Niu et al. 2019). A promising strategy for creating multi-knock-in marker-free mutants in *Escherichia coli*, serine integrase recombinational engineering may soon be investigated in cyanobacteria (Snoeck et al. 2019). Promoter systems, ribosome-binding sites (RBSs), CRISPR interference (CRISPRi), and synthetic short regulatory RNA (sRNA) are among the well-established regulatory tools in cyanobacteria (Sun et al. 2018; Xia et al. 2019). Synthetic genetic circuits that operate as Boolean logic functions have been built in model strains targeting the regulatory tools. These circuits allow digital-like control over gene expression in response to environmental inputs (Bradley et al. 2016). Also, using DNA-binding proteins and antisense RNAs in *Anabaena* 7120, a positive feedback loop was created to improve reporter protein expression (Higo et al. 2017). This chapter presents an analysis of genomic and proteomic data derived from cyanobacteria which provides insights into the underlying processes involved and identifies prospective targets for enhancing their bioremediation capabilities. Current engineered cyanobacterial strains that have shown significant capacity to tolerate high concentrations of heavy metals, producing extracellular polymeric substances (EPS) for bioremediation purposes, remove phosphorus, and tolerate high salt environments are also discussed.

5.2 Exploring Cyanobacteria's Molecular Mechanisms for Bioremediation

5.2.1 Heavy Metals Sequestration, Chelation, and Detoxification by Cyanobacteria

Cyanobacteria have a wide variety of strategies to deal with cellular accumulation of heavy metals (HM) that may be quite harmful (Fig. 5.1). Phytochelatins (PCs) and high-affinity metal-binding metallothionein (MT) proteins are crucial for metal homeostasis and resistance to toxicity (Balaji et al. 2016). Cyanobacteria can endure HM stress because of their enzymatic (catalase, peroxidase, superoxide dismutase) and non-enzymatic (carotenoids, glutathione) redox machinery (Leong and Chang 2020). For instance, a freshwater cyanobacterium *Synechococcus* sp. showed differential expression of 278 genes in response to zinc stress ($ZnCl_2$, 25 mgL^{-1}) which included metallothionein (*smtA*), cation efflux (*catT*, *pacS*), and ABC transporters (Newby et al. 2017). Notably, ZiaA efflux pumps and other mechanisms have

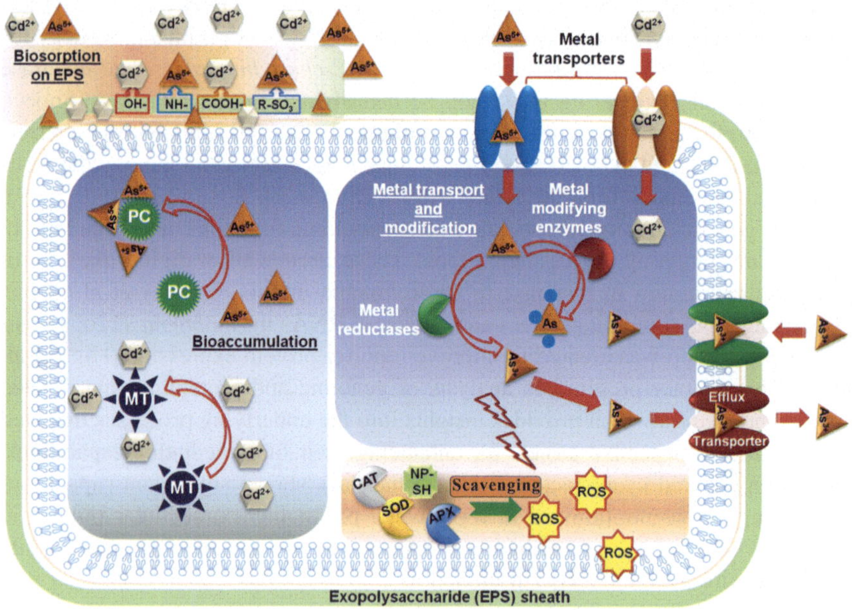

Fig. 5.1 Diagram showing extra- and intra-cellular mechanisms by which cyanobacteria limit HM. Metal ions may be entrapped by EPS's negatively charged functional groups. After entering cells, metal ions may undergo valence conversion with enzymes and be expelled by efflux transporters. Many HMs like As can be converted into cell-safe organoarsenicals. Phytochelatins (PC) and Metallothioneins (MT) bind to HMs to maintain metal homeostasis. Antioxidant enzymes like SOD, CAT, APX, or NP-SH may scavenge reactive oxygen species (ROS) produced by HMs to reduce their toxicity. This figure is adapted from Chakdar et al. (2022) under the licence number (5750320473254)

evolved in the unicellular model cyanobacterium *Synechocystis* sp. to eliminate excess $ZnCl_2$ from cells (Badarau and Dennison 2011). Additionally, it has been claimed that the protein known as thioredoxin (*Trx*), which provides protection to organisms against "reducing stress" may chelate HMs, perhaps by a process similar to that of metallothioneins (Lemaire et al. 1999). *Trx* expression rises in cyanobacteria in response to Cd stress ($CdSO_4$: 9.0 μM), although the exact method by which it chelates is yet unknown; however, it was predicted that cyanobacterial *Trx* binds to Cd ions and functions as a HM sink to reduce Cd stress (Xu et al. 2018). A variety of conditions place significant demands on ferredoxins (Fed2-Fed9), a group of iron-sulfur proteins encoded by various stress response genes in cyanobacteria (Cassier-Chauvat et al. 2021). One example is the ferredoxin-glutaredoxin-thioredoxin crosstalk pathway, which is involved in protecting cells against oxidative and metal stressors and includes the ferredoxin proteins Fed1, Fed7, and Fed9. Chaperones are also thought of as a crucial protein for survival in very stressful environments (Voth and Jakob 2017). *Synechococcus* sp. contains 6–9 *Fed* genes which facilitate protein refolding and inhibit denaturing stress-induced protein aggregation (Cassier-Chauvat et al. 2021). For example, upon exposure to Cd, a *Nostoc entophytum* strain, collected from oil-contaminated areas in southern Iran, exhibited an increase in the expression of the GroEL and HptG chaperones (High temperature protein G). The strong antioxidant system of this strain, including catalase, peroxidase, and chaperons, may be responsible for its remarkable Cd-uptake capabilities (Alidoust et al. 2016, 2019).

Multiple ion transporters and specific enzymes help cyanobacteria to ingest, biotransform, and ultimately externalize HM. Cyanobacteria like *Synechocystis* sp. PCC 6803, for example, contain *arsBHC* operons to biotransform hazardous metals (López-Maury et al. 2003). The *arsB* gene produces an arsenite or As(III)-specific efflux pump that removes As(III) from the cell following its conversion from arsenate or As(V) (Zhao et al. 2015). The thioredoxin reductase-like enzyme *ArsC* is crucial to this process because it accelerates reduction via the glutathione/glutaredoxin pathway (Sánchez-Riego et al. 2014). The operon's other gene, *arsH*, is thought to encode an FMN-quinone reductase (Hervás et al. 2012). In *Synechocystis* sp. PCC 6803, *arsI*, an arsenate reductase gene, also protects against Arsenic (As) stress (Sánchez-Riego et al. 2014) while ArsR repressor protein controls the *arsBHC* operon and de-repression of ArsR increases As(III) and As(V) resistance (López-Maury et al. 2003). Cyanobacteria oxidize or methylate As(III) to reduce its toxicity because As(V) and organic arsenicals are less poisonous. Cyanobacteria like *Nostoc*, *Microcystis*, *Synechocystis*, and *Prochlorococcus* sp. have widespread homologs of modified enzymes like arsenite S-adenosylmethionine methyltransferases (*ArsM*), making them attractive targets for engineering HM resistance. Cyanobacteria may oxidize arsenite extracellularly to As(V) in vacuoles after being exposed to As(III) (Yin et al. 2011). These arsenates are reduced to arsenite and pumped out of cells via efflux transporters or methylated to form non-toxic, volatile dimethyl arsenates. Cyanobacteria may also produce organoarsenicals, arsenosugars, arsenolipids, and volatile arsenicals and detoxify hazardous materials by converting heavy metal ions to non-toxic sulfides. Similarly, *Synechococcus leopoliensis* has been shown to biotransform Cd to CdS when supplemented with sulfite, cysteine, and sulfate. High cysteine desulfhydrase activities were seen under settings that led to a rise in cadmium sulfide formation, providing compelling evidence that cysteine is the sulfur-rich building block for CdS synthesis (Edwards et al. 2013).

Meanwhile, cyanobacteria have developed efficient transporters to remove HM from their cells. Like metal ions, heavy metal nanoparticles (HMNP) accumulate intracellularly, likely via transporter proteins. Cyanobacteria may have HMNP transporter proteins due to manufactured nanoparticles' charged nature (Mahana et al. 2021). A novel von Willebrand A (vWA) domain-containing protein in *Microcystis aeruginosa* controls Ag (Silver) absorption from AgNPs during heterologous expression in *E. coli* (Chen et al. 2016). The genome of Cd-resistant cyanobacterial strain JJU (Hapalosiphonaceae family) contains Cd-tolerant genes including *czcA*, *czcB*, and *cadA* (Victoria Angelo et al. 2018) while membrane-bound CzcA and CzcB proteins provide a cation efflux system to protect microorganisms against Co, Zn, and Cd stress (Kim et al. 2011; Nesler et al. 2017). ATP hydrolysis and HM efflux are coordinated by the CadA protein, a transporting ATPase. Cyanobacteria significantly control the expression of these transporter proteins, but their mechanisms are unknown. A Cd-stress-related response regulator (Sll0649) regulates *Synechocystis* sp. PCC 6803's MntC (Sll1598) and ZiaA (Slr0798) (Chen et al. 2016). The *slr0946* was reported to be a new Sll0649 protein target, joining *mntC* and *ziaA* genes (Sun et al. 2017). It was also found that overexpressing *sll0649* gene improved Cd stress tolerance. *Slr0946*'s genomic position offers potential for gene control. The *arsBHC* operon cluster in *Synechocystis* sp. PCC 6803 contains *slr0944-slr0945-slr0946* (López-Maury et al. 2003) while Slr1738, a regulatory protein in *Synechocystis* sp. PCC 6803, also activates *slr0946* in Cd absorption (Houot et al. 2007). HM-resistance genes therefore have a complicated regulation network, and omics analysis may assist in revealing their signaling cascade patterns in cyanobacteria.

5.2.2 *Proteomic Exploration in Cyanobacteria Against Heavy Metals*

Gene expression has been confirmed in proteomic investigations while other unknown proteins in the cyanobacterial proteome affected by HM treatment also show significant changes (Pandey et al. 2012) (Fig. 5.2). This section discusses HM-associated proteome amendments in cyanobacteria and highlights significant proteins that might be genetically altered to improve bioremediation.

Proteomics investigations on *Synechocystis* sp. PCC 6803 showed that certain proteins (Ni, Co, Cd) were expressed in response to most HMs, whereas others (e.g., chaperons) were particularly induced by specific HMs (Mehta et al. 2014). Exposure to Ni, Co, and Cd (40 μM $NiSO_4$, 20 μM $CoCl_2$, and 15 μM $CdCl_2$) often down-regulates the periplasmic iron-binding protein (FutA2/Slr0513), among other

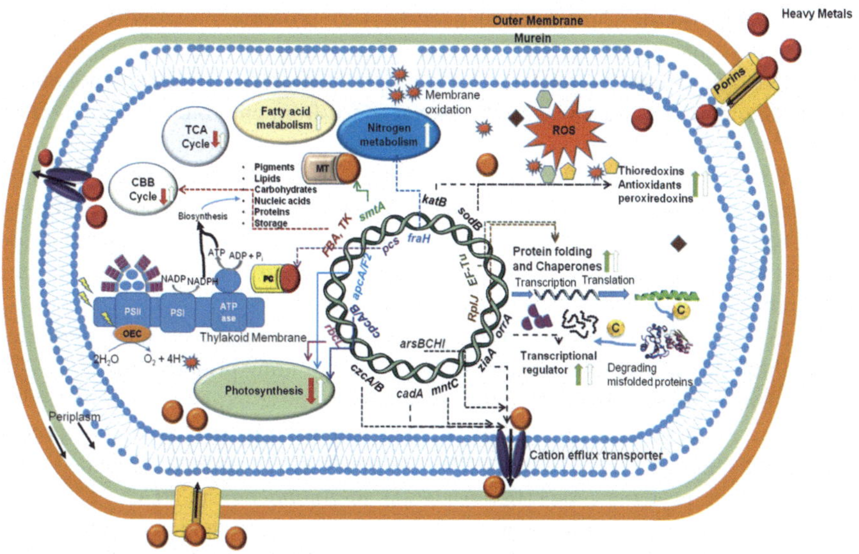

Fig. 5.2 Mechanism of heavy metal stress affects cyanobacterial proteomes. Regulation is going up when the green arrows are pointing up and down when the red arrows are pointing down. Bold arrows indicate the reaction after 24 h of exposure, whereas dotted arrows reflect the response after more than 10 days of exposure. The presence of porin molecules in the cell membrane enables HMs to penetrate which initially targets proteins that are essential for respiration, photosynthesis, and nitrogen fixation which resulted in improvement of antioxidant defense, chaperone, and transport proteins. Conversely, several activity-related proteins have their levels increased by extended exposure. It was found that genes involved in biochemical processes that have been markedly up- or down-regulated, in addition to their target process, their respective proteins are involved in energy metabolism: *FBA* and *TK*, *rbcL* in photosynthesis, *orrA* in signal transduction, *fraH* in heterocyst differentiation, *cpcA*, *cpcB*, *apcA*, and *apcF2* in water-soluble phycobiliproteins (PBPs), *smt* OEC, ROS, phytochelatins, and metallothioneins. This figure is adapted from Chakdar et al. (2022) under the licence number (5750320473254)

unique features. This protein's involvement in Ni, Co, and Cd homeostasis is unknown; however, it has been shown to absorb Fe and Cu. Ni stress uniquely upregulated production of the thermotolerance-related HspA chaperon. HspA stabilizes and protects photosynthetic apparatuses from oxidative stress, although its involvement in Ni tolerance is unknown (Sakthivel et al. 2009). In addition, Co selectively stimulated upregulation of the chaperonin GroES (Slr2075). In contrast to Ni and Co stress, prolonged exposure (i.e., ≥48-h) to Cd stress did not selectively express chaperones (Sakthivel et al. 2009) suggesting engineering pathways that use chaperones and photosynthetic apparatus-stabilizing proteins to genetically create bioremediating cyanobacterial strains. The proteomes of three *Anabaena* species—*Anabaena* L31, *Anabaena* sp. PCC 7120, and *Anabaena doliolum*—varied across species, with L31 being the most Cd-resistant. Cd toxicity among *Anabaena* strains may explain their early upregulation of transporters and signaling proteins and considerable fall of nitrogen absorption proteins. The study proposed the use of

GroEL (chaperonin), ClpP (ATP-dependent Clp protease proteolytic subunit), Hsp70 (chaperone protein), NusG (transcription termination/antitermination protein), GST (glutathione S-transferase), Cph2 (phytochrome-like protein) and others as targets to enhance cyanobacterial removal of Cd from environment (Singh et al. 2015). Cyanobacterial proteome responses differ by element essentiality. *Cyanothece* sp. CCY 0110 was investigated for proteome responses to Cd and Cu stress by time and metal concentration and found that Cu, an important electron-transporting metal, helps cells recuperate from prolonged stress by slowing their metabolic rate to invest energy to eliminate the Cu toxic effects, whereas Cd toxicity accumulates and inhibits the cell growth (Mota et al. 2015).

Arsenic (As) HM-exposure causes cyanobacteria to generate catalase, peroxiredoxin, thioredoxin, and oxidoreductase and protein folding proteins (Ef-Ts, Ef-Tu, GroEL, DnaJ), it may also induce these proteins to cause oxidative stress (Pandey et al. 2012; Prajapati et al. 2020). For example, Zhou and Wolk et al. found upregulation/production of phytochelatins (PC synthase), GST, arsenate reductase, and arsenite efflux proteins (Asr1102, Alr1097) in *Anabaena* sp. PCC 7120 were induced to likely sequester intracellular As and resulted in formation of akinete-like structures. Alr4050 protein, which was 98% identical to the akinete marker protein of the related cyanobacterium *Anabaena variabilis*, was thought to induce this structure (Zhou and Wolk 2002). Cyanobacterial signals to develop dormant spores when cells were exposed to HM stress and endure extended stress may cause this phenotype. The *Anabaena* sp. PCC7120 also increased cysteine synthase (CS) and L-asparaginase induction (Pandey et al. 2012). CS maintains cysteine levels for the manufacture of cysteine-rich polypeptides such as metallothioneins, phytochelatins, and GSH for As detoxification (Pickering et al. 2000). Proteins and carbon-rich storage polymer cyanophycin are degraded by asparaginase (Hejazi et al. 2002) which may help stressed cyanobacterial cells reduce photosynthesis and maintain metabolic activity by upregulating the stored carbon. All4894, a fasciclin superfamily protein involved in cell adhesion and extracellular matrix interaction (Kim et al. 2000). This protein may aggregate cyanobacterial filaments to prevent As exposure. Again, these proteins are promising targets for the engineering of more tolerant cyanobacterial strains capable of bioremediation for commercial applications.

5.2.3 Engineering Cyanobacteria for Metal Bioremediation

Cyanobacteria store heavy metals using metal-binding entities. Previous research found that the marine cyanobacterium *Synechococcus* sp. NKBG 15041c, expressing the MT protein from Syn7942, could withstand 4 μM of $CdCl_2$, which killed the wild-type (Sode et al. 1998). Later, it was found that the K45C mutation of MT protein SmtA in *Synechococcus* 7942 increased Cd(II) absorption (Saffar et al. 2015). NmtA, a new MT, was found in *Anabaena* sp. PCC 7120 which under 10 μM Cd(II) concentration in media grew 40% faster than the control strain with overexpressed NmtA (T V et al. 2018). In *Anabaena* sp. PCC 7120, the gene *alr0975* encodes a PC synthase-like protein (Harada et al. 2004). Similar research found that overexpressing *alr0975* in modified *Anabaena* sp. PCC 7120 led to a 22.3% increase in growth under $CdCl_2$ conditions (Chaurasia et al. 2017).

Metal-binding entities are connected to hazardous heavy metals in cyanobacteria, however, are poorly understood. The use of integrated and multi-omics may be useful in the discovery of innovative metal-binding entities and processes. In addition, heterologous production of highly efficient MBPs and phytochelatins may improve HM storage (Kumari and Das 2019; Wei et al. 2014). Finally, protein engineering methods such as error-prone PCR-based mutation and phage-assisted continuous evolution (PACE) might speed up screening of new and novel metal-binding proteins (Morrison et al. 2020).

Two-component systems play a crucial role in signal transduction, enabling microbes to detect and react to changes in their environment and stress levels. It was shown that Sll0649 is an important positive regulator that changes the expression of several genes involved in the transport and use of various metal ions, including Cu(II), Fe(II), Mn(II), and Zn(II) (Chen et al. 2014). It was also found that Syn6803 has a novel target slr0946 of the response regulator (RR) Sll0649 that is involved in Cd(II) tolerance (Sun et al. 2017). It should be noted that under a 5 μM Cd(II) environment, the development of *Synechocystis* sp. 6803 was enhanced about twofold when the genes *slr0946*, *sll1598*, or *slr0798* were separately overexpressed (Sun et al. 2017). In order to modify tolerance, various techniques such as overexpression/silencing of stress responsive gene(s), Adaptive Laboratory Evolution, CRISPR, mutagenesis, and synthetic circuits designing using synthetic biology can be exploited. To improve bioremediation in the not-too-distant future, it would be worthwhile to do more research to discover new two-component systems using multi-omics methods.

For the purpose of increasing microbial tolerance, Adaptive Laboratory Evolution (ALE) has been used more often than any other engineering strategy (Gao et al. 2023). For example, during an ALE procedure that lasted about 802 days, the tolerance of *Synechocystis* sp. 6803 to Cd(II) was enhanced from 4.6 to 9.0 μM. By comparing genomic sequences, the essential altered genes were identified which are linked to Cd(II) tolerance in the evolved strain. For example, in *Synechocystis* sp. 6803, *slr0454* was shown to have a direct correlation with Cd(II) tolerance. In particular, it was found that the developed *Synechocystis* strain could withstand high levels of light stress, zinc, and copper (Xu et al. 2018). Results showed that ALE may be used to modify heavy metal tolerance in cyanobacteria, and the research also offered helpful information for cyanobacterial resilience. Nevertheless, cells would need a considerable amount of time in an ALE continuous culture system to amass random genetic mutations. Another possible solution may be to use atmospheric and room temperature plasma, which can cause more DNA damage and a greater mutation rate (Zhang et al. 2014). Similarly, *Synechococcus* elongatus PCC 7942 outgrew its ancestor by 6 times after 1200 generations after exposure to constant, intense light (1313 μmol photons m^{-2} s^{1}). Rapid development required a mutation in *SasA*, an important regulator of the circadian rhythm. The circadian clock is an adaptable system that optimizes development across varied ecological settings, in addition to acting as a timekeeping device (Mendaña et al. 2024).

5.3 Exploring the Role of Cyanobacteria's Exopolysaccharides (EPS)

5.3.1 Metabolic Pathways of EPS Production

In cyanobacteria, EPS are essential because they produce a hydrated biofilm matrix that includes proteins, lipids, nucleic acids, and polysaccharides. This matrix increases the survival and ecological success of cyanobacteria by mediating adherence to surfaces, facilitating cell-to-cell contacts, and providing protection against environmental challenges (Debnath et al. 2024). There is a lack of understanding about the complex control of EPS biosynthesis in cyanobacteria. It is still crucial to fully understand the metabolic pathways, rate-limiting processes, and genetic controls to enhance cyanobacteria's EPS secretion. Even though research into the metabolic pathways involved in cyanobacterial EPS synthesis has only just begun, there has been substantial evidence of process conservation across different bacterial strains. Cyanobacteria, like bacteria, have a three-stage process for making extracellular polysaccharides (EPS): first, production of sugar nucleotides and monosaccharides; second, production of polysaccharides; and third, exporting the polysaccharides to an external location. Monosaccharide and sugar-nucleotide biosynthesis typically takes place inside cells using carbon that is received by photosynthesis and central carbon metabolism (Maeda et al. 2021; Nishanth et al. 2021). Figure 5.3 shows that four pathways have been proposed to explain the formation and export of polysaccharides: the Wzx-Wzy dependent pathway, the ABC transporter-dependent pathway (Wzm/Wzt), the Synthase-dependent pathway, and the sucrase-mediated extracellular biosynthesis (Potnis et al. 2021). Cyanobacterial strains differ greatly in these pathways, which causes distinct polymerization methods and EPS types to be produced (Nishanth et al. 2021). To exemplify the difference, heteropolysaccharide synthesis is supported by the Wzx-Wzy and ABC transporter pathways, while homopolysaccharide synthesis is carried out by the synthase-dependent and sucrase-mediated routes.

The sugar-nucleotide biosynthesis precursor monosaccharides fructose-6-phosphate (Fru6P), glucose-6-phosphate (Glc6P), and glucose-1-phosphate (Glc1P) are produced via gluconeogenesis from 3PG (3-phosphoglycerate) that is produced during photosynthesis and the Calvin-Benson pathways. *Synechocystis* PCC6803's sugar-nucleotide biosynthesis enzymatic route is shown in Fig. 5.3 (Schmid 2018). *Synechocystis* sp. synthesizes a wide range of nucleotide sugars, including ADP-glucose, GDP-mannose, GDP-fucose, CDP-glucose, GDP-rhamnose, and a few more types. Among these sugar nucleotides are muramic acid, n-acetyl glucosamine, and other UDP sugars. The second phase involves the addition of sugar

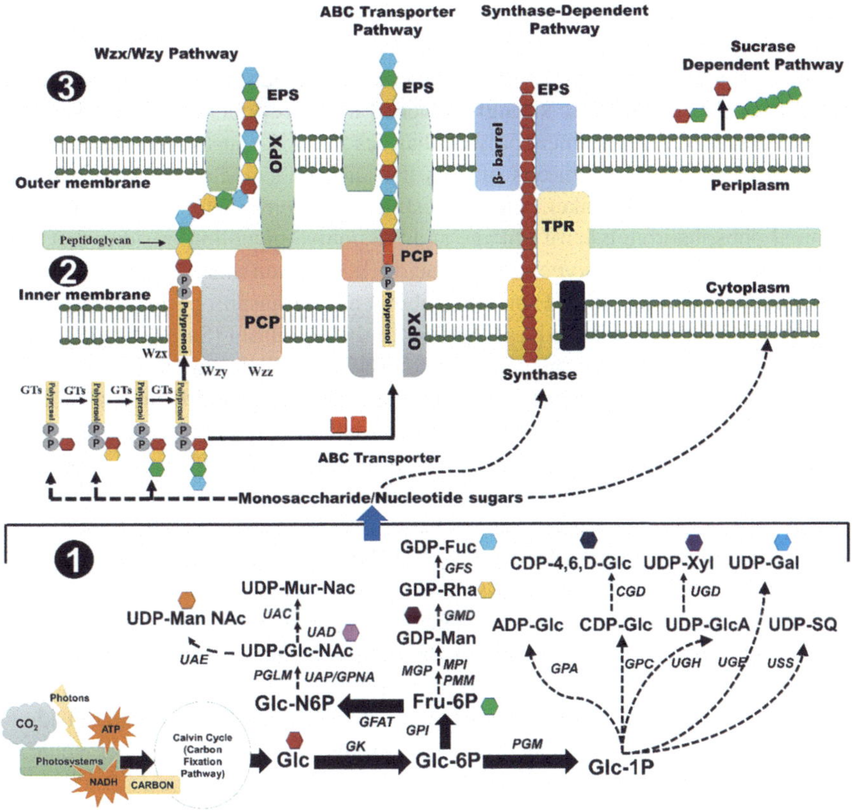

Fig. 5.3 A diagram indicating the processes involved in cyanobacterial EPS synthesis. This figure is adapted from Debnath et al. (2024) under the licence number (5751441091634)

moieties to the oligosaccharide chain by means of glycosyl-transferases (GTs) in the cytoplasm, which connect the sugar nucleotides and monosaccharides to particular linkers/acceptors. This is the point at which the Wzx, ABC, synthase-dependent, and sucrase-dependent pathways diverge greatly from one another.

Depending on the substrate specificity of the GTs and the expressing operons, the GTs enzymes in the Wzx-Wzy pathway produce a hetero-oligosaccharide repeating unit containing various monomeric nucleotide sugar molecules (Schmid 2018; Cuthbertson et al. 2009; Islam and Lam 2014). A cytosolic anchoring molecule that aids in oligosaccharide elongation, undecaprenyl diphosphate is linked to the nucleotide sugars via the GTs as well. Moreover, the next phase involves the synthesis of polysaccharide employing the polymerizing enzyme provided by the *wzy* gene and the flippase enzyme represented by the *wzx* gene for transporting the oligosaccharide-linker complex from the cytoplasm to the periplasm. Enzymes wzy and wzz code for polymerizing and protein-polysaccharide co-polymerase, respectively, which determine the length and termination of the expanding polysaccharide chain. An outer membrane non-specific OPX exporter then transports the terminated polysaccharide to the extracellular phase.

The ATP-binding cassette (ABC) transporter is crucial in the synthesis of exterior polysaccharides that are strongly attached to the cell membrane (CPS) (Rana and Upadhyay 2020). The linear polysaccharides for ABC transporters are produced in the cytosolic phase by means of GTs that are connected to the Und-P linker. Aside from the Und-P linker, another distinctive feature of CPS is the reducing terminal phosphatidylglycerol linkage and the unique glycolipid residue that consists of poly-2-keto-3-deoxyoctulosonic acid (PKD). Even more specifically, a protein complex including an ATP-binding subunit and a transport permease facilitates the export of the produced polysaccharides to the periplasm (Santos et al. 2021b). Lastly, in a manner comparable to the Wzx-Wzy route, PCP and OPX transport the expanding polysaccharide to the extracellular phase. In a linked mechanism, the length of the chain is determined by the rivalry between components of the assembly complex and those that bind ATP, which in turn dictates when the developing polysaccharide stops and when it is exported. Another possibility is an uncoupled system in which the production and release of polysaccharides are separate processes. The alteration of the Und-P linker at the non-reducing terminal causes the end of chain length expansion in this scenario (Wang et al. 2008).

Thirdly, a synthase-dependent route integrates polymerization and translocation, two activities that are made possible by the same protein. This route also facilitates the production of cellulose and other homopolymers such as alginates (Schmid 2018). The fourth kind of route is dependent on an enzyme that is found outside of cells called sucrase. This enzyme forms glucans and fructans from monosaccharide units by cleaving sucrose molecules.

5.3.2 Engineering Cyanobacteria for EPS Production

Analyzing the genetic makeup of several cyanobacterial strains revealed functionally redundant genes spread out over the genomic DNA, clustered operons, and dispersed genes. Mutagenesis was used on the unicellular cyanobacterium *Synechocystis* sp. PCC6803 (Cassier-Chauvat et al. 2021; Zuniga et al. 2020) to uncover a high copy number of genes with redundant functions. The presence of the *Wzx-Wzy-Wzz* and *Wza-Wzb-Wzc* genes in the freshwater unicellular cyanobacterium *Aphanothece sacral* suggests a WzY-dependent pathway for EPS generation (Potnis et al. 2021). Wzx-Wzy-Wzz proteins were shown to polymerize EPS in unique ways across species, as shown by the variation in copy numbers and degree of similarity across these proteins. *Synechocystis* sp. exhibited no phenotypic changes in response to selective truncation or deletion of the *wzy* and *wzx*. There seems to be significant functional redundancy and route cross-talks in *Synechocystis* sp. (Pereira et al. 2019). Nevertheless, EPS production was impacted by mutations in the *WzC* and *WzB*. Specifically, a mutant strain without the WzC protein impaired

RPS (discharged polysaccharides outside of cells) and CPS synthesis, whereas a mutant strain lacking the WzB protein solely affected RPS synthesis (Li et al. 2021).

It was shown that *kpsM* gene is essential for the phosphorylation and dephosphorylation of tyrosine residues. It was shown that the *kpsM* mutant strain of *Synechocystis* sp., PCC6803 significantly reduced RPS concentration, decreased CPS somewhat, and increased poly-hydroxybutyrate accumulation compared to wild-type *Synechocystis* sp., PCC6803 (Santos et al. 2021b). A possible role for *exoD* in EPS biosynthesis has been suggested by the observation that inducing its expression in *Nostoc* 7120 improved EPS production. There was also a lack of the *kpsC* and *kpsS*, which are responsible for the formation of capsular polysaccharides in other microbes (Karlyshev et al. 2000).

To regulate EPS production and add desirable properties, genetic engineering might change the composition or chain length. To reroute the flow toward the manufacturing of monosaccharide sugars and nucleotide sugars—essential components of EPS—the genes encoding the core carbon metabolic pathway should be the principal targets of genetic engineering approach. Furthermore, GTs are the most important targets because they open the door to optimizing the structure and composition of EPS by allowing alternative monosaccharide sugar compositions in the polymeric chain. After EPS undergoes pyruvylation, methylation, acetylation, and sulfation, it undergoes further modifications by additional enzymes. Cyanobacterial defense against heavy metal stress can be enhanced via EPS's genetic and metabolic engineering approach by optimizing its chemical and rheological properties, such as increasing functional group density, negative charge, viscosity, elasticity, and network formation (Gongi et al. 2022; Can et al. 2022). These changes increase the ability of EPS to bind, sequester, and immobilize heavy metals, lowering their toxicity and safeguarding cyanobacterial cells in the process. Lastly, the rates of EPS release into the extracellular phase may be directly influenced by targeting the polymerizing enzymes and exporters. To regulate these target enzymes and exporters, a comprehensive study of the genetic composition of EPS synthesis in diverse strains is still required. The synthesis of valuable compounds such as astaxanthin, lutein, beta-carotene, and others has been made easier with the recent advances in synthetic toolboxes, multi-omics data, and comprehensive genomic databases (Show et al. 2022; Eltanahy and Torky 2022). In addition, metabolic models that include the whole genome, *in-silico* genomics, and proteomics all work together to help find the reasonable targets that may be modified to increase EPS production (Schmid 2018). Creation of insertional mutants, which are essential for the advancement of genetic technology in microalgae and cyanobacteria, may confirm the genome-scale models. The genomes of model microalgae species (*Chlamydomonas reinhardtii, Chlorella sorokiniana, Chlorella vulgaris, Phaeodactylum tricomutum, Nannochloropsis oceanica, Tetraselmis* sp.,) have been edited using new genomic editing technologies such as (Trovão et al. 2022; Liu et al. 2022; Chu 2022) identified zinc-finger nuclease (ZFN), transcription activator-like effector nucleases (TALENs), and clustered regularly interspaced short palindromic repeats with associated protein (CRISPR/Cas) as these tools. Genome editing tools are an additional crucial component of genetic engineering in microalgae and cyanobacteria (Young et al. 2021).

Advances in genome engineering for microalgae and cyanobacteria have been made possible by the CRISPR/Cas system, which allows for multiple target-editing and has a simpler target design. For versatile gene expression regulation in photosynthetic microbes, researchers have developed CRISPR interference systems that bind target DNA with a nuclease-deficient dead Cas9 (dCas9) (Lu et al. 2022). *Anabaena* sp. PCC 7120, *Synechococcus elongatus* PCC7942, *Synechococcus* sp. PCC7002, and *Synechococcus* sp. PCC 6803 are among the microorganisms that have had their metabolic pathways engineered using CRISPR (Ng et al. 2020; Santos et al. 2021a).

5.3.3 Role of EPS in Bioremediation

Cyanobacterial endophytic symbionts such as *Anabaena* sp., *Spirulina* sp., *Lyngbya* sp., *Nostoc* sp., and *Microcystis* sp. have the potential to detoxify soil contaminated with the herbicide glyphosate (Lipok et al. 2009) because of bio-adhesive characteristics, making EPS the best colonizers of bare soil types. When mixed with other microbes like bacteria, lichens, mosses, fungus, and algae, they can form biological soil crusts that can halt desertification in semi-arid and arid regions across the world. Therefore, cyanobacterial EPSs have potential for usage in soil and water purification (Rossi et al. 2017). In study on the function of cyanobacterial inoculation in soil structure maintenance, it was found that primary clay aggregation and soil structure maintenance were caused by EPS produced by the *Nostoc* strains AfS49 and KaS35 (Rossi et al. 2022). The EPS characterization in *Cyanothece* sp. showed that sulfates were also present on saccharides and uronic acids which enhance the anionic ability for metal binding (Ohki et al. 2014). It is also possible to increase EPS production and tailor EPS to certain needs (e.g., heavy metal binding) by modifying its composition or structure such as to stop soil erosion in farming (Bhuyan et al. 2023).

5.4 Exploring the Role of Cyanobacteria for Phosphate Removal

The functional component of numerous cell molecules, such as nucleic acids, phospholipids, phosphoproteins, and metabolites in most catabolic and anabolic pathways and signaling molecules, phosphorus (P), primarily in the form of inorganic or orthophosphate (PO_4^{3-}), is essential to metabolic processes. Polyphosphate (polyP) chains, which can be formed by various organisms through polymerization (Sanz-Luque et al. 2020), consist of three to hundreds of PO_4^{3-} groups connected by high-energy phosphoanhydride bonds. These bonds enable ATP to function as the cell's energy currency (Sanz-Luque et al. 2020).

In order to reduce the harmful effects of adding too much PO_4^{3-} to soils, researchers have looked at techniques to redirect polyP metabolism toward bioremediation, water treatment, and biofertilizer synthesis utilizing cyanobacteria and algae (Siebers et al. 2019; Solovchenko et al. 2016). Different species of cyanobacteria such as *Aphanothece* sp., *Spirulina* sp., *Arthrospira* sp., *Lyngbya* sp., *Anabaena* sp., and *Phormidium* sp. have been utilized to extract nutrients from wastewater (Ray et al. 2013). A significant portion of the excess PO_4^{3-} absorbed by these organisms is converted into polyP. The large quantities of polyP that cyanobacteria produce have the potential to be used in methods to economically remove PO_4^{3-} from wastewaters on a commercial scale, as well as to produce biomass rich in PO_4^{3-} that might be used as fertilizer in agricultural products. The decomposition or hydrolysis of these biofertilizers by microorganisms would determine the pace of PO_4^{3-} mobilization from the soil. A delayed or moderate release might promote sustained plant growth with lower PO_4^{3-} runoff.

Engineered cyanobacterial strains that collect excess polyP might be useful for developing biological systems for bioremediation and producing biofertilizers. Both photosynthetic and heterotrophic bacteria have shown that such strains may be developed. The *ppk* gene, which normally helps *Pseudomonas putida* remove PO_4^{3-} from its surroundings, was transferred to it from the cyanobacterium *Microcystis aeruginosa* NIES-843 to develop a cyanobacterium that accumulates a significant amount of polyP. This particular strain outperformed the control strain (which did not have the *ppk* gene added) in terms of its ability to remove PO_4^{3-} from the media when cultured in a sequencing batch biofilm reactor (Du et al. 2012). Another strategy took use of the observation that low amounts of polyP were also seen in mutants of the *PhoU* gene, which encodes a negative regulatory element of the Pho regulon. The capacity of the cells to remove PO_4^{3-} from the media was four times higher in *Synechococcus* sp. PCC6803 cells that had their phoU gene rendered inactive compared to the wild-type cells, and there was an increased build-up of intracellular polyP (Morohoshi et al. 2002).

Overexpression of PipY or RelQ in *S. elongatus* PCC7942 led to similar phenotypic responses as *PipX* overexpression, including growth arrest, loss of photosynthetic activity and viability, increased cell size, and accumulation of large polyphosphate granules. This study aimed to gain insights into the cellular functions of *PipY*, which is involved in signaling the intracellular energy status and carbon-to-nitrogen balance. On the other hand, cells whose *PipX* expression was increased had shorter lifespans, suggesting that *PipX* and *PipY* had opposing effects on cell development. While the wild-type strain of *Synechocystis* sp. PCC6803 grew more quickly and produced more ATP under high-carbon circumstances, a *ppk* knock-out strain showed a growth deficiency in response to many stress factors, including low light, acidity, and glucose. These findings provide credence to the idea that polyphosphate synthesis and degradation play a part in energy control (Sebesta et al. 2024). They also indicate that these processes might be good targets for biocontainment strategies. *Synechocystis* sp. PCC6803 (ΔSphU), which is deficient in the Pho regulon's SphU regulator, has the potential to reduce the concentration of Pi in recirculating aquaculture wastewater to below the detection limit of 0.01 mg PL^{-1} at a rate of 2.07 ± 0.33 mg P h^{-1} g^{-1} DW. It was suggested that the flocculated cells might be recycled for three more cycles of effective P removal (Burut-Archanai et al. 2013).

5.5 Exploring the Role of Cyanobacteria for Salt Tolerance

To adapt to harsh environments, cyanobacteria have evolved special mechanisms that have served them well throughout their long history of evolution. Some cyanobacteria are the primary autotrophs in many hypersaline habitats, which places them in a pivotal role in ecological niches (Waditee-Sirisattha and Kageyama 2023) while some cyanobacterial strains had been engineered to adapt salt-stress by manipulations of genes (Table 5.1). Such environments often host a diverse array of cyanobacterial species, including those with a single cell, like *Halothece* and *Euhalothece*, and those without a cell wall, such as *Halospirulina* and *Halomicronema* (Nübel et al. 2000; Li et al. 2014).

In salt-stress conditions (5 M NaCl or 3 M KCl), engineered *S. elongatus* 2973 with overexpressed *clscB* secreted over 94% sucrose from cells, reaching 8.7 g L^{-1} in 21 days through semi-continuous culture (Song et al. 2016) and 35.5 mg L^{-1} h^{-1} through a column bioreactor (Tan et al. 2018). Additionally, in order to improve salt tolerance, Lin et al. overexpressed *cscB* together with *sps*, which encodes sucrose-phosphate synthase in *S. elongatus* 2973 which resulted in a sucrose productivity of 1.1 g L^{-1} D^{-1} (Lin et al. 2020). *S. elongatus* 2973 showed a 24% improvement in salt tolerance based on cell growth after having the glucosylglycerol biosynthetic pathway introduced (Cui et al. 2021). The synthesis of this compound reached 51 mg g^{-1} under 0.4 M NaCl after 60 h. Transforming *S. elongatus* 2973 into *S. elongatus* 7002 by overexpressing eight genes and producing an *Mrp* antiporter that increased its growth in a 0.4 M NaCl environment by around 57.7% (Cui et al. 2020b). In addition, the results of the targeted metabolomics study demonstrated that increasing 7002-Mrp expression decreased total carboxylic acid (TCA), increased metabolic flux into sucrose and glycogen biosynthesis, and increased NADPH and ATP content; these changes may enhance growth in salt-stress conditions (Cui et al. 2020a). Transporters, suitable solutes, stress-induced enzymes and proteins, and ALE are the four areas that have recently seen advancements summarized here. One suitable solute that may impart salt tolerance is ectoine, which is present in halophilic microbes. Here, the ectoine biosynthetic pathway (*actA, actB, actC*) was used to enhance *S. elongatus*'s salt tolerance; the result was a final OD_{750} that was 20% higher under 300 mM NaCl and 80% higher under 400 mM NaCl compared to the wild-type (WT) strain. Not only could the altered strain decrease the requirement for intense sulfur metabolism, but it could also live in environments with 500 mM NaCl, which was deadly to WT (Cui et al. 2020b).

Table 5.1 A summary of salt tolerance engineering in cyanobacteria

Strain name	Approach	Genetic manipulation	Phenotype
S. elongatus 7942	Ion transporters	Expression of *ApNhaP* gene	Growth in 0.5 M NaCl
		Expression of *NapA*-type Na^+/H^+ antiporter gene	
		Expression of the *ApNa⁺-atp* operon	
	Compatible solutes	Expression of *ApGSMT-DMT* genes	Growth improved by 70% in 0.12 M NaCl
		Expression of serine hydroxymethyltransferase gene	Growth improved by onefold in 0.3 M NaCl
		Expression of *codA* gene	Growth in 0.4 M NaCl
	Salt-induced proteins and enzymes	Overexpression of *katE* and *bet* genes	Growth improved by twofold in 0.38 M NaCl
		Overexpression of heat shock and heat shock genes	Growth improved by 24% in 0.5 M NaCl
		Expression of H2846 gene	Growth improved by twofold in 0.3 M NaCl
S. elongatus 2973	Ion transporters	Expression of Mrp-system	Growth improved by 57.7% in 0.4 M NaCl
S. elongatus 6803	Salt-induced proteins and enzymes	Overexpression of drought-resistant gene *drnf1*	Growth improved by threefold in 1 M NaCl
	Adaptive laboratory evolution	ALE process with 3% NaCl	Growth improved by onefold in 3% NaCl
A. doliolum	Compatible solutes	Expression of *ApGSMT-DMT* genes	Growth in 0.5 M NaCl
Anabaena sp. *PCC 7120*	Salt-induced proteins and enzymes	Overexpression of *groESL* operon	Growth improved by 50% in 0.15 M NaCl
		Overexpression of *all3940* gene	Growth improved by twofold in 0.15 M NaCl
A. fertilissima	Adaptive laboratory evolution	ALE process with 0.1–0.5 M NaCl	Growth in 0.5 M NaCl
M. aeruginosa	Adaptive laboratory evolution	Salt-stress, the fluctuation analysis	Growth in 0.24 M NaCl

This table is adapted from Cui et al. (2020a) under the licence number (5754310044730)

5.6 Conclusion and Prospects

Cyanobacteria have an efficient and well-controlled metabolic architecture and their ability to adapt to harsh environments and unpredictable stresses has led to the development of sophisticated molecular editing tools to harness this machinery. Their remarkable adaptability and dynamic metabolism have allowed them to flourish in almost every habitat and region of the world making them the most suitable microbes for environmental decontamination. With their primary removal processes, cyanobacteria also outperform traditional bioremediation approaches in terms of both environmental friendliness and economic feasibility. To promote cyanobacteria growth and reduce growth inhibitory factors in wastewater, the cyanobacterial strains may be engineered for the HM stress, high concentration of pollutants, higher growth rate, and improved tolerance against salt-stress. Besides, employing cyanobacteria for environmental mitigation offers low carbon footprint of the process for waste remediation and HM detoxification, nutrient reuse, and the value-added benefits offered.

References

Alidoust L, Soltani N, Modiri S, Haghighi O, Azarivand A, Khajeh K, Shahbani Zahiri H, Vali H, Akbari Noghabi K (2016) Cadmium uptake capacity of an indigenous cyanobacterial strain, *Nostoc entophytum* ISC32: new insight into metal uptake in microgravity-simulating conditions. Microbiology 162:246–255

Alidoust L, Zahiri HS, Maleki H, Soltani N, Vali H, Noghabi KA (2019) *Nostoc entophytum* cell response to cadmium exposure: a possible role of chaperon proteins GroEl and HtpG in cadmium-induced stress. Ecotoxicol Environ Saf 169:40–49

Badarau A, Dennison C (2011) Thermodynamics of copper and zinc distribution in the cyanobacterium *Synechocystis* PCC 6803. Proc Natl Acad Sci 108:13007–13012

Balaji S, Kalaivani T, Shalini M, Gopalakrishnan M, Rashith Muhammad MA, Rajasekaran C (2016) Sorption sites of microalgae possess metal binding ability towards Cr(VI) from tannery effluents—a kinetic and characterization study. Desalin Water Treat 57:14518–14529

Behler J, Vijay D, Hess WR, Akhtar MK (2018) CRISPR-based technologies for metabolic engineering in cyanobacteria. Trends Biotechnol 36:996–1010

Bhalara PD, Punetha D, Balasubramanian K (2014) A review of potential remediation techniques for uranium(VI) ion retrieval from contaminated aqueous environment. J Environ Chem Eng 2:1621–1634

Bhuyan PP, Nayak R, Jena M, Pradhan B (2023) Convoluted role of cyanobacteria as biofertilizer: an insight of sustainable agriculture. Vegetos 36:309–321

Bradley RW, Buck M, Wang B (2016) Tools and principles for microbial gene circuit engineering. J Mol Biol 428:862–888

Burut-Archanai S, Eaton-Rye JJ, Incharoensakdi A, Powtongsook S (2013) Phosphorus removal in a closed recirculating aquaculture system using the cyanobacterium *Synechocystis* sp. PCC 6803 strain lacking the SphU regulator of the pho regulon. Biochem Eng J 74:69–75

Can HK, Kavlak S, Gurbuz F, Odabaşı M (2022) Insights into the viscoelastic peculiarities of cyanobacterial extracellular polymeric substance (EPS). J Polym Environ 30:3055–3062

Cassier-Chauvat C, Blanc-Garin V, Chauvat F (2021) Genetic, genomics, and responses to stresses in cyanobacteria: biotechnological implications. Genes 12:500

Chakdar H, Thapa S, Srivastava A, Shukla P (2022) Genomic and proteomic insights into the heavy metal bioremediation by cyanobacteria. J Hazard Mater 424:127609

Chaurasia N, Mishra Y, Chatterjee A, Rai R, Yadav S, Rai LC (2017) Overexpression of phytochelatin synthase (*pcs*) enhances abiotic stress tolerance by altering the proteome of transformed *Anabaena* sp. PCC 7120. Protoplasma 254:1715–1724

Chen L, Zhu Y, Song Z, Wang J, Zhang W (2014) An orphan response regulator Sll0649 involved in cadmium tolerance and metal homeostasis in photosynthetic *Synechocystis* sp. PCC 6803. J Proteome 103:87–102

Chen S, Jin Y, Lavoie M, Lu H, Zhu K, Fu Z, Qian H (2016) A new extracellular von Willebrand A domain-containing protein is involved in silver uptake in *Microcystis aeruginosa* exposed to silver nanoparticles. Appl Microbiol Biotechnol 100:8955–8963

Chu LL (2022) CRISPR-Cas system in microbial hosts for terpenoid production. Crit Rev Biotechnol 42:1116–1133

Cui J, Sun T, Chen L, Zhang W (2020a) Engineering salt tolerance of photosynthetic cyanobacteria for seawater utilization. Biotechnol Adv 43:107578

Cui J, Sun T, Li S, Xie Y, Song X, Wang F, Chen L, Zhang W (2020b) Improved Salt tolerance and metabolomics analysis of *Synechococcus elongatus* UTEX 2973 by overexpressing *Mrp* antiporters. Front Bioeng Biotechnol 8

Cui J, Sun T, Chen L, Zhang W (2021) Salt-tolerant *Synechococcus elongatus* UTEX 2973 obtained via engineering of heterologous synthesis of compatible solute glucosylglycerol. Front Microbiol 12

Cuthbertson L, Mainprize Iain L, Naismith James H, Whitfield C (2009) Pivotal roles of the outer membrane polysaccharide export and polysaccharide copolymerase protein families in export of extracellular polysaccharides in gram-negative bacteria. Microbiol Mol Biol Rev 73:155–177

Dasgupta J, Sikder J, Chakraborty S, Curcio S, Drioli E (2015) Remediation of textile effluents by membrane based treatment techniques: a state of the art review. J Environ Manag 147:55–72

Debnath S, Muthuraj M, Bandyopadhyay TK, Bobby MN, Vanitha K, Tiwari ON, Bhunia B (2024) Engineering strategies and applications of cyanobacterial exopolysaccharides: a review on past achievements and recent perspectives. Carbohydr Polym 328:121686

Dong Z, Sun T, Zhang W, Chen L (2023) Improved salt tolerance of *Synechococcus elongatus* PCC 7942 by heterologous synthesis of compatible solute ectoine. Front Microbiol 14

Du H, Yang L, Wu J, Xiao L, Wang X, Jiang L (2012) Simultaneous removal of phosphorus and nitrogen in a sequencing batch biofilm reactor with transgenic bacteria expressing polyphosphate kinase. Appl Microbiol Biotechnol 96:265–272

Edwards CD, Beatty JC, Loiselle JBR, Vlassov KA, Lefebvre DD (2013) Aerobic transformation of cadmium through metal sulfide biosynthesis in photosynthetic microorganisms. BMC Microbiol 13:161

Eltanahy E, Torky A (2022) Integrated omics and mutation in algae. In: El-Sheekh MM, Abdullah N, Ahmad I (eds) Handbook of research on algae as a sustainable solution for food, energy, and the environment. IGI Global, Hershey

Francová A, Chrastný V, Šillerová H, Vítková M, Kocourková J, Komárek M (2017) Evaluating the suitability of different environmental samples for tracing atmospheric pollution in industrial areas. Environ Pollut 220:286–297

Gao J, Zhu X, Sun T, Chen L, Zhang W (2023) Advances in using adaptive laboratory evolution technology for engineering of photosynthetic cyanobacteria. Sheng wu gong cheng xue bao = Chinese journal of biotechnology 39:3075–3094

Gongi W, Cordeiro N, Pinchetti JLG, Ouada HB (2022) Functional, rheological, and antioxidant properties of extracellular polymeric substances produced by a thermophilic cyanobacterium *Leptolyngbya* sp. J Appl Phycol 34:1423–1434

Gu C, Kim GB, Kim WJ, Kim HU, Lee SY (2019) Current status and applications of genome-scale metabolic models. Genome Biol 20:121

Harada E, Von Roepenack-Lahaye E, Clemens S (2004) A cyanobacterial protein with similarity to phytochelatin synthases catalyzes the conversion of glutathione to γ-glutamylcysteine and lacks phytochelatin synthase activity. Phytochemistry 65:3179–3185

Hejazi M, Piotukh K, Mattow J, Deutzmann R, Volkmer-Engert R, Lockau W (2002) Isoaspartyl dipeptidase activity of plant-type asparaginases. Biochem J 364:129–136

Hendry JI, Bandyopadhyay A, Srinivasan S, Pakrasi HB, Maranas CD (2020) Metabolic model guided strain design of cyanobacteria. Curr Opin Biotechnol 64:17–23

Hervás M, López-Maury L, León P, Sánchez-Riego AM, Florencio FJ, Navarro JA (2012) ArsH from the cyanobacterium *Synechocystis* sp. PCC 6803 is an efficient NADPH-dependent quinone reductase. Biochemistry 51:1178–1187

Higo A, Isu A, Fukaya Y, Hisabori T (2017) Designing synthetic flexible gene regulation networks using RNA devices in cyanobacteria. ACS Synth Biol 6:55–61

Houot L, Floutier M, Marteyn B, Michaut M, Picciocchi A, Legrain P, Aude J-C, Cassier-Chauvat C, Chauvat F (2007) Cadmium triggers an integrated reprogramming of the metabolism of *Synechocystis* PCC6803, under the control of the Slr1738 regulator. BMC Genomics 8:350

Islam ST, Lam JS (2014) Synthesis of bacterial polysaccharides via the Wzx/Wzy-dependent pathway. Can J Microbiol 60:697–716

Karlyshev AV, Linton D, Gregson NA, Lastovica AJ, Wren BW (2000) Genetic and biochemical evidence of a *Campylobacter jejuni* capsular polysaccharide that accounts for Penner serotype specificity. Mol Microbiol 35:529–541. https://doi.org/10.1046/j.1365-2958.2000.01717.x

Kim J-E, Kim S-J, Lee B-H, Park R-W, Kim K-S, Kim I-S (2000) Identification of motifs for cell adhesion within the repeated domains of transforming growth factor-β-induced gene, βig-h3. J Biol Chem 275:30907–30915

Kim E-H, Nies Dietrich H, Mcevoy Megan M, Rensing C (2011) Switch or funnel: how RND-type transport systems control periplasmic metal homeostasis. J Bacteriol 193:2381–2387

Kumari S, Das S (2019) Expression of metallothionein encoding gene *bmtA* in biofilm-forming marine bacterium *Pseudomonas aeruginosa* N6P6 and understanding its involvement in Pb(II) resistance and bioremediation. Environ Sci Pollut Res 26:28763–28774

Lemaire S, Keryer E, Stein M, Schepens I, Issakidis-Bourguet E, Gérard-Hirne C, Miginiac-Maslow M, Jacquot J-P (1999) Heavy-metal regulation of thioredoxin gene expression in *Chlamydomonas reinhardtii*. Plant Physiol 120:773–778

Leong YK, Chang J-S (2020) Bioremediation of heavy metals using microalgae: recent advances and mechanisms. Bioresour Technol 303:122886

Li Y, Lin Y, Loughlin PC, Chen M (2014) Optimization and effects of different culture conditions on growth of *Halomicronema hongdechloris* – a filamentous cyanobacterium containing chlorophyll f. Front Plant Sci 5

Li J, Fang D, Ye R, Zhou C, Li P (2021) The released polysaccharide inhibits cell aggregation and biofilm formation in the cyanobacterium *Synechocystis* sp. PCC 6803. Eur J Phycol 56:119–128

Lin P-C, Zhang F, Pakrasi HB (2020) Enhanced production of sucrose in the fast-growing cyanobacterium *Synechococcus elongatus* UTEX 2973. Sci Rep 10:390

Lipok J, Wieczorek D, Jewgiński M, Kafarski P (2009) Prospects of in vivo ^{31}P NMR method in glyphosate degradation studies in whole cell system. Enzym Microb Technol 44:11–16

Liu K, Chen R, Yang R, Chen Y, Zhu C, Tang Y, Kong F (2022) Genome editing approaches applied to microalgae-based fuels. In: Jacob-Lopes E, Zepka LQ, Severo IA, Maroneze MM (eds) 3rd generation biofuels. Woodhead Publishing

López-Maury L, Florencio Francisco J, Reyes José C (2003) Arsenic sensing and resistance system in the cyanobacterium *Synechocystis* sp. strain PCC 6803. J Bacteriol 185:5363–5371

Lu L, Shen X, Sun X, Yan Y, Wang J, Yuan Q (2022) CRISPR-based metabolic engineering in non-model microorganisms. Curr Opin Biotechnol 75:102698

Maeda K, Okuda Y, Enomoto G, Watanabe S, Ikeuchi M (2021) Biosynthesis of a sulfated exopolysaccharide, synechan, and bloom formation in the model cyanobacterium *Synechocystis* sp. strain PCC 6803. eLife 10:e66538

Mahana A, Guliy OI, Mehta SK (2021) Accumulation and cellular toxicity of engineered metallic nanoparticle in freshwater microalgae: current status and future challenges. Ecotoxicol Environ Saf 208:111662

Mehta A, López-Maury L, Florencio FJ (2014) Proteomic pattern alterations of the cyanobacterium *Synechocystis* sp. PCC 6803 in response to cadmium, nickel and cobalt. J Proteome 102:98–112

Mendaña A, Santos-Merino M, Gutiérrez-Lanza R, Domínguez-Quintero M, Medina JM, González-Guerra A, Campa V, Ducos-Galand M, López-Igual R, Volke DC, Gugger M, Nikel PI, Mazel D, Cruz FDL, Fernández-López R (2024) Long-term evolution reveals the role of the circadian cycle in the environmental adaptation of cyanobacteria. bioRxiv, 2024.03.12.584591

Morohoshi T, Maruo T, Shirai Y, Kato J, Ikeda T, Takiguchi N, Ohtake H, Kuroda A (2002) Accumulation of inorganic polyphosphate in phoU mutants of *Escherichia coli* and *Synechocystis* sp. strain PCC6803. Appl Environ Microbiol 68:4107–4110

Morrison MS, Podracky CJ, Liu DR (2020) The developing toolkit of continuous directed evolution. Nat Chem Biol 16:610–619

Mota R, Pereira SB, Meazzini M, Fernandes R, Santos A, Evans CA, De Philippis R, Wright PC, Tamagnini P (2015) Effects of heavy metals on *Cyanothece* sp. CCY 0110 growth, extracellular polymeric substances (EPS) production, ultrastructure and protein profiles. J Proteome 120:75–94

Nesler A, Dalcorso G, Fasani E, Manara A, Di Sansebastiano GP, Argese E, Furini A (2017) Functional components of the bacterial CzcCBA efflux system reduce cadmium uptake and accumulation in transgenic tobacco plants. New Biotechnol 35:54–61

Newby R, Lee LH, Perez JL, Tao X, Chu T (2017) Characterization of zinc stress response in cyanobacterium *Synechococcus* sp. IU 625. Aquat Toxicol 186:159–170

Ng I-S, Keskin BB, Tan S-I (2020) A critical review of genome editing and synthetic biology applications in metabolic engineering of microalgae and cyanobacteria. Biotechnol J 15:1900228

Nishanth S, Bharti A, Gupta H, Gupta K, Gulia U, Prasanna R (2021) Cyanobacterial extracellular polymeric substances (EPS): biosynthesis and their potential applications. In: Das S, Dash HR (eds) Microbial and natural macromolecules. Academic Press

Niu T-C, Lin G-M, Xie L-R, Wang Z-Q, Xing W-Y, Zhang J-Y, Zhang C-C (2019) Expanding the potential of CRISPR-Cpf1-based genome editing Technology in the Cyanobacterium *Anabaena* PCC 7120. ACS Synth Biol 8:170–180

Nübel U, Garcia-Pichel F, Muyzer G (2000) The halotolerance and phylogeny of cyanobacteria with tightly coiled trichomes (*Spirulina Turpin*) and the description of *Halospirulina tapeticola* gen. nov., sp. nov. Int J Syst Evol Microbiol 50:1265–1277

Ohki K, Le NQT, Yoshikawa S, Kanesaki Y, Okajima M, Kaneko T, Thi TH (2014) Exopolysaccharide production by a unicellular freshwater cyanobacterium *Cyanothece* sp. isolated from a rice field in Vietnam. J Appl Phycol 26:265–272

Pandey S, Rai R, Rai LC (2012) Proteomics combines morphological, physiological and biochemical attributes to unravel the survival strategy of *Anabaena* sp. PCC7120 under arsenic stress. J Proteome 75:921–937

Pereira SB, Santos M, Leite JP, Flores C, Eisfeld C, Büttel Z, Mota R, Rossi F, De Philippis R, Gales L, Morais-Cabral JH, Tamagnini P (2019) The role of the tyrosine kinase Wzc (Sll0923) and the phosphatase Wzb (Slr0328) in the production of extracellular polymeric substances (EPS) by *Synechocystis* PCC 6803. Microbiol Open 8:e00753

Pickering IJ, Prince RC, George MJ, Smith RD, George GN, Salt DE (2000) Reduction and coordination of arsenic in Indian mustard. Plant Physiol 122:1171–1178

Pisciotta JM, Zou Y, Baskakov IV (2010) Light-dependent electrogenic activity of cyanobacteria. PLoS One 5:e10821

Potnis AA, Raghavan PS, Rajaram H (2021) Overview on cyanobacterial exopolysaccharides and biofilms: role in bioremediation. Rev Environ Sci Biotechnol 20:781–794

Prajapati R, Yadav S, Srivastava R, Atri N (2020) Nickel and arsenite responsive proteomic alterations in cyanobacterium *Anabaena* PCC7120. J Sci Res 64:8

Rana S, Upadhyay LSB (2020) Microbial exopolysaccharides: synthesis pathways, types and their commercial applications. Int J Biol Macromol 157:577–583

Ray K, Mukherjee C, Ghosh AN (2013) A way to curb phosphorus toxicity in the environment: use of polyphosphate reservoir of cyanobacteria and microalga as a safe alternative phosphorus biofertilizer for Indian agriculture. Environ Sci Technol 47:11378–11379

Rossi F, Li H, Liu Y, De Philippis R (2017) Cyanobacterial inoculation (cyanobacterisation): perspectives for the development of a standardized multifunctional technology for soil fertilization and desertification reversal. Earth Sci Rev 171:28–43

Rossi F, Mugnai G, De Philippis R (2022) Cyanobacterial biocrust induction: a comprehensive review on a soil rehabilitation-effective biotechnology. Geoderma 415:115766

Saffar B, Mehri Ghahfarrokhi A, Mahnam K, Mobini-Dehkordi M (2015) Improvement of Cd^{2+} uptake ability of SmtA protein by Lys/Cys mutation; experimental and theoretical studies. J Biomol Struct Dyn 33:2347–2359

Sakthivel K, Watanabe T, Nakamoto H (2009) A small heat-shock protein confers stress tolerance and stabilizes thylakoid membrane proteins in cyanobacteria under oxidative stress. Arch Microbiol 191:319–328

Sánchez-Riego AM, López-Maury L, Florencio FJ (2014) Genomic responses to arsenic in the cyanobacterium *Synechocystis* sp. PCC 6803. PLoS One 9:e96826

Santos M, Pacheco CC, Yao L, Hudson EP, Tamagnini P (2021a) CRISPRi as a tool to repress multiple copies of extracellular polymeric substances (EPS)-related genes in the cyanobacterium Synechocystis sp. PCC 6803. Life 11:1198

Santos M, Pereira Sara B, Flores C, Príncipe C, Couto N, Karunakaran E, Cravo Sara M, Oliveira P, Tamagnini P (2021b) Absence of KpsM (Slr0977) impairs the secretion of extracellular polymeric substances (EPS) and impacts carbon fluxes in *Synechocystis* sp. PCC 6803. mSphere 6. https://doi.org/10.1128/msphere.00003-21

Sanz-Luque E, Bhaya D, Grossman AR (2020) Polyphosphate: a multifunctional metabolite in cyanobacteria and algae. Front Plant Sci 11

Schmid J (2018) Recent insights in microbial exopolysaccharide biosynthesis and engineering strategies. Curr Opin Biotechnol 53:130–136

Sebesta J, Cantrell M, Schaedig E, Hou HJM, Pastore C, Chou KJ, Xiong W, Guarnieri MT, Yu J (2024) Polyphosphate kinase deletion increases laboratory productivity in cyanobacteria. Front Plant Sci 15

Shahid A, Khan AZ, Malik S, Liu C-G, Mehmood MA, Syafiuddin A, Wang N, Zhu H, Boopathy R (2021) Advances in Green Technologies for the Removal of effluent organic matter from the urban wastewater. Curr Pollut Rep 7:463–475

Show PL, Chai WS, Ling TC (2022) Microalgae for environmental biotechnology: smart manufacturing and industry 4.0 applications. CRC Press

Siebers N, Hofmann D, Schiedung H, Landsrath A, Ackermann B, Gao L, Mojzeš P, Jablonowski ND, Nedbal L, Amelung W (2019) Towards phosphorus recycling for agriculture by algae: soil incubation and rhizotron studies using ^{33}P-labeled microalgal biomass. Algal Res 43:101634

Singh PK, Shrivastava AK, Chatterjee A, Pandey S, Rai S, Singh S, Rai LC (2015) Cadmium toxicity in diazotrophic *Anabaena* spp. adjudged by hasty up-accumulation of transporter and signaling and severe down-accumulation of nitrogen metabolism proteins. J Proteome 127:134–146

Singh JS, Kumar A, Rai AN, Singh DP (2016) Cyanobacteria: a precious bio-resource in agriculture, ecosystem, and environmental sustainability. Front Microbiol 7

Snoeck N, De Mol ML, Van Herpe D, Goormans A, Maryns I, Coussement P, Peters G, Beauprez J, De Maeseneire SL, Soetaert W (2019) Serine integrase recombinational engineering (SIRE): a versatile toolbox for genome editing. Biotechnol Bioeng 116:364–374

Sode K, Yamamoto Y, Hatano N (1998) Construction of a marine cyanobacterial strain with increased heavy metal ion tolerance by introducing exogenic metallothionein gene. J Mar Biotechnol 6:174–177

Solovchenko A, Verschoor AM, Jablonowski ND, Nedbal L (2016) Phosphorus from wastewater to crops: an alternative path involving microalgae. Biotechnol Adv 34:550–564

Song K, Tan X, Liang Y, Lu X (2016) The potential of *Synechococcus elongatus* UTEX 2973 for sugar feedstock production. Appl Microbiol Biotechnol 100:7865–7875
Sun T, Xu L, Wu L, Song Z, Chen L, Zhang W (2017) Identification of a new target slr0946 of the response regulator Sll0649 involving cadmium tolerance in *Synechocystis* sp. PCC 6803. Front Microbiol 8
Sun T, Li S, Song X, Diao J, Chen L, Zhang W (2018) Toolboxes for cyanobacteria: recent advances and future direction. Biotechnol Adv 36:1293–1307
T V D, Chandwadkar P, Acharya C (2018) NmtA, a novel metallothionein of *Anabaena* sp. strain PCC 7120 imparts protection against cadmium stress but not oxidative stress. Aquat Toxicol 199:152–161
Tan X, Hou S, Song K, Georg J, Klähn S, Lu X, Hess WR (2018) The primary transcriptome of the fast-growing cyanobacterium *Synechococcus elongatus* UTEX 2973. Biotechnol Biofuels 11:218
Trovão M, Schüler LM, Machado A, Bombo G, Navalho S, Barros A, Pereira H, Silva J, Freitas F, Varela J (2022) Random mutagenesis as a promising tool for microalgal strain improvement towards industrial production. Mar Drugs 20:440
Victoria Angelo J, Cao E, Salmaso N, Segata N, Donati C (2018) Draft genome sequence of the cadmium-resistant strain JJU2, belonging to the family hapalosiphonaceae of the cyanobacteria. Microbiol Resourc Announce 7. https://doi.org/10.1128/mra.00876-18
Vörösmarty CJ, Mcintyre PB, Gessner MO, Dudgeon D, Prusevich A, Green P, Glidden S, Bunn SE, Sullivan CA, Liermann CR, Davies PM (2010) Global threats to human water security and river biodiversity. Nature 467:555–561
Voth W, Jakob U (2017) Stress-activated chaperones: a first line of defense. Trends Biochem Sci 42:899–913
Waditee-Sirisattha R, Kageyama H (2023) Halotolerance, stress mechanisms, and circadian clock of salt-tolerant cyanobacteria. Appl Microbiol Biotechnol 107:1129–1141
Wang PG, Guo H, Yi W, Song JK (2008) Current understanding on biosynthesis of microbial polysaccharides. Curr Top Med Chem 8:141–151
Wei W, Liu X, Sun P, Wang X, Zhu H, Hong M, Mao Z-W, Zhao J (2014) Simple whole-cell biodetection and bioremediation of heavy metals based on an engineered Lead-specific operon. Environ Sci Technol 48:3363–3371
Xia P-F, Ling H, Foo JL, Chang MW (2019) Synthetic biology toolkits for metabolic engineering of cyanobacteria. Biotechnol J 14:1800496
Xu C, Sun T, Li S, Chen L, Zhang W (2018) Adaptive laboratory evolution of cadmium tolerance in *Synechocystis* sp. PCC 6803. Biotechnol Biofuels 11:205
Yin X-X, Chen J, Qin J, Sun G-X, Rosen BP, Zhu Y-G (2011) Biotransformation and volatilization of arsenic by three photosynthetic cyanobacteria. Plant Physiol 156:1631–1638
Young R, Haines M, Storch M, Freemont PS (2021) Combinatorial metabolic pathway assembly approaches and toolkits for modular assembly. Metab Eng 63:81–101
Zhang X, Zhang X-F, Li H-P, Wang L-Y, Zhang C, Xing X-H, Bao C-Y (2014) Atmospheric and room temperature plasma (ARTP) as a new powerful mutagenesis tool. Appl Microbiol Biotechnol 98:5387–5396
Zhao C, Zhang Y, Chan Z, Chen S, Yang S (2015) Insights into arsenic multi-operons expression and resistance mechanisms in *Rhodopseudomonas palustris* CGA009. Front Microbiol 6
Zhou R, Wolk CP (2002) Identification of an Akinete marker gene in *Anabaena variabilis*. J Bacteriol 184:2529–2532
Zuniga EG, Boateng KKA, Bui NU, Kurnfuli S, Muthana SM, Risser DD (2020) Identification of a hormogonium polysaccharide-specific gene set conserved in filamentous cyanobacteria. Mol Microbiol 114:597–608

Chapter 6
Algae-Derived Food and Feed Products

Prakash Kumar Senapati, Meenakshi Singh, and Mohita Karmi

Abstract Algae include a diverse group of photosynthetic organisms with extraordinary metabolic profiles that could facilitate their wide-ranging applications in various sectors. They are richly augmented in vitamins, carbohydrates, lipids, minerals, and other high-value nutrients. Because of their nutritional properties, they are often used as dietary supplements for animals and humans. They are emerging as a substitute for traditional food sources because algae are used to upgrade the nutritional value of functional foods, bakery products, beverages, nutraceuticals, snacks, and smoothies. Besides developing functional foods, algae provide an alternative direction toward aquaculture feeds. They play a vital role in improving fish nutrition and sustainable aquaculture practices. This book chapter mainly highlights several algae species with the potential for human and animal food and feed products. Along with several success stories, the futuristic path toward sustainable food and feed industries is also discussed.

Keywords Algae · Functional foods · Nutraceuticals · Sustainable food and feed · Innovative algae food

6.1 Introduction

Algae, the thalloid, simplest autotrophic organisms, represent a fascinating polyphyletic group with a built-in potential for photosynthesis due to their chloroplasts. Depending on their remarkable size, they are categorized into macroscopic and microscopic (Suganya et al. 2016). Microalgae, the single celled, microscopic eukaryotic phytoplankton, are ubiquitous in diverse aquatic ecosystems, playing crucial roles in the food chain (Singh and Saxena 2015). Among them are diatoms

P. K. Senapati · M. Karmi
Department of Life Science, National Institute of Technology, Rourkela, Odisha, India

M. Singh (✉)
Department of Ecology and Biodiversity, Sustain Greens LLP, Vadodara, Gujarat, India

M. A. Mehmood et al. (eds.), *Cyanobacteria Biotechnology*,
https://doi.org/10.1007/978-3-031-70698-1_6

like Bacillariophyceae, green algae such as Chlorophyceae, and golden algae including Chrysophyceae, each with unique characteristics and ecological significance for animals and humans. On the other hand, Cyanobacteria, ancient microorganisms, possess an intermediate structure, sharing traits with both bacteria (such as the lack of a distinct membrane-bound nucleus, mitochondria, and chloroplasts) and plants due to their ability to perform oxygenic photosynthesis (Murphy and Cardona 2022). This unique blend of characteristics often leads to taxonomic debates, as Cyanobacteria are considered a polyphyletic group of prokaryotic microalgae. Taxonomists debate whether to categorize them as microalgae or bacteria, highlighting these fascinating organisms' complexity and evolutionary significance.

Microalgae and cyanobacteria represent promising resources for sustainable food and feed production due to their high nutritional value, rapid growth rates, and ability to thrive in diverse environmental conditions. Both have the potential to fulfil the current demand for global food and feed as a nutrition treasure. Microalgae and cyanobacteria are mainly used for different human foods such as bread, beverages, smoothies, pasta and cakes, and snacks. That is why they are considered a significant option for current and future food sources. This chapter highlights various microalgae and cyanobacteria species with promising human and animal nutrition applications. We delve into their potential uses and explore success stories where these microorganisms have been effectively utilized. Moreover, we take a forward-looking approach, envisioning the future of sustainable food and feed industries with microalgae and cyanobacteria at the forefront. Additionally, we cover innovative food products derived from these organisms and their wide-ranging applications. Last but not least, we present several case studies showcasing the role of microalgae and cyanobacteria in aquaculture, providing real-world examples of their impact and potential in this vital field. Through these discussions, we aim to highlight the multifaceted and exciting opportunities presented by microalgae and cyanobacteria in derived food and feed products.

6.2 Applications of Algae in Food and Feed Industries

Functioning as a natural treasure trove, microalgae and cyanobacteria enrich a myriad of functional food products, ranging from bakery items and beverages to algal-based nutraceuticals, snacks, and smoothies (Fig. 6.1) (Aly et al. 2023). Both microalgae and cyanobacteria are widely acknowledged as abundant sources of various valuable components, encompassing proteins, minerals, carbohydrates, lipids, vitamins (with a notable emphasis on water-soluble vitamins), enzymes, and pigments like chlorophyll, xanthophylls, carotenoids, and antioxidants (Hachicha et al. 2022).

Chlorella, Phaeodactylum, Arthrospira, Scenedesmus, Skeletonema, Tetraselmis, and Dunaliella are among the most extensively utilized microalgae in animal nutrition, particularly for the formulation of protein-rich feed supplements (Saadaoui et al. 2021; Udayan et al. 2021; Mavrommatis et al. 2023). Beyond their culinary

Fig. 6.1 Applications of microalgae and cyanobacteria in food and feeds

contributions, these microorganisms carve out an alternative path for aquaculture feeds, pivotal in bolstering fish nutrition and fostering sustainable aquaculture practices. Microalgae-based feeds are extensively utilized in both aquaculture and poultry farming, serving as sources of proteins and DHA (Saadaoui et al. 2021).

The primary purpose of food is to satisfy hunger while supplying the essential nutrients crucial for human survival (Burchi et al. 2011). As per the statistics given by the United Nations, the world's population is projected to surge by nearly 50% from 2000, reaching 9.5 billion by 2050. So, there is a need to explore and make advancements in the food and feed industries to obtain higher yields of microalgae and cyanobacterial food and feed products. Several successful attempts are being made in the microalgae and cyanobacterial food and feed industries to fulfil the requirements. For example, in the 1960s, Japan pioneered the large-scale cultivation of microalgae, notably employing *Chlorella* sp. as a food supplement (Benemann et al. 2018). Throughout the 1970s and 1980s, the industrial production of microalgae proliferated across countries like the USA, China, Taiwan, Australia, India, Israel, and Germany (Khatoon and Pal 2015). The focus primarily cantered on *Spirulina* sp., with *Chlorella* sp. gaining increased attention in recent years.

6.2.1 *Algae as A Nutritional Powerhouse*

Microalgae and cyanobacteria are nutritional powerhouses, boasting impressive compositions and a diverse array of vital nutrients (Siddhnath et al. 2024). These microscopic organisms rival conventional protein sources such as meat and dairy, providing complete proteins containing all essential amino acids crucial for human health. Additionally, they serve as abundant sources of beneficial fats, including omega-3 and omega-6 fatty acids like eicosapentaenoic acid (EPA) and docosahexaenoic acid (DHA), which are essential for cognitive function and heart health (Naik et al. 2024; Hosseinkhani et al. 2022). Packed with essential minerals such as iron, calcium, and B vitamins, they support various bodily functions, including immune response and energy metabolism. Furthermore, their rich antioxidant content, encompassing carotenoids and chlorophyll, aids in fighting oxidative stress (Pereira et al. 2023; Ravishankar and Ambati 2019). Moreover, their nutrients boast high bioavailability, ensuring efficient absorption and utilization within the body. Beyond their nutritional prowess, microalgae and cyanobacteria offer environmentally friendly solutions for food production, requiring minimal land, water, and resources compared to traditional farming methods, thus addressing global food security challenges sustainably (Nagappan et al. 2021; Diaz et al. 2023; Ahmad and Ashraf 2023).

In the current era of heightened nutrition and well-being awareness, the algae market is experiencing noteworthy growth. The surge is propelled by a collective emphasis on "superfoods," lauded for their exceptional nutritional benefits and disease-preventive qualities, acting as a significant catalyst in this burgeoning industry. *Spirulina* takes center stage among these superfoods, earning prestigious recognition from the World Health Organization as a premier superfood (Ravishankar and Ambati 2019). This acknowledgment has sparked a surge of exploration by both industries and researchers, delving into the as-yet untapped potential of algae as a promising frontier for cutting-edge functional foods. The growing acknowledgment of algae's potential as valuable sources for human and animal nutrition accentuates the need to prioritize enhancing the sustainability and economic feasibility of algae-based products (Kusmayadi et al. 2021). As industries and researchers dig deeper into the multifaceted applications of algae, the trajectory of their evolution promises to reshape nutritional landscapes and redefine conventional paradigms in the pursuit of well-being and sustainable nutrition.

In the realm of algal-derived products, specific varieties showcase impressive annual production volumes. *Spirulina*, leading the pack, reaches a remarkable 12,000 tons per year, closely followed by 3000 tons from *Dunaliella salina* per year and 1500 tons from *Aphanizomenon flosaquae*, contributing annually (Kusmayadi et al. 2021; Dhandayuthapani et al. 2021; Mu et al. 2019). In response to the surging demand for algae, notably *Spirulina*, in cosmetics and natural colorants, the anticipated Compound Annual Growth Rate (CAGR) for the global *Spirulina* market stands at an impressive 10%. This growth trajectory is expected to culminate in an estimated market value of US$2000 million by the year 2026. Zooming to the broader spectrum, 2023–2033 anticipates significant algae product market growth. According to Algae Products Global Market Report 2024, with an estimated value of US$5.41 billion in 2023, projections suggest it will surpass US$9.15 billion by 2033. However, despite the notable increase in algal production in recent years, biomass productivity still needs to meet escalating commercial demands. Algae feed costs remain relatively high compared to conventional alternatives, underscoring the need for price reduction to enhance competitiveness. Despite a notable increase in algal production in recent years, more than biomass productivity is needed to meet the growing commercial demands. As these technological strides continue to shape the landscape, the potential exists to meet escalating demands and drive the algal products market toward greater economic feasibility and sustainability.

6.2.2 Algae-Based Products for Human Consumption

Recently, algae-based products have garnered increasing attention for their potential to enhance the nutritional content of human food, thanks to their unique chemical composition (Barkia et al. 2019). There is a growing emphasis on exploring innovative applications of microalgae and cyanobacteria in various prepared foods, from bread and soup to noodles (Table 6.1). The main aim is to fulfil public interest and provide food and fodder options for human diets (Su et al. 2023). By incorporating microalgae and cyanobacteria into diverse food products, manufacturers, and researchers seek to harness their nutritional benefits while catering to evolving consumer preferences and dietary needs (Fig. 6.2). *Spirulina*, *Chlorella*, and *Haematococcus* species are already being used for human consumption. *Spirulina platensis* is used for biscuits to enhance texture, aroma, flavor, and color (Hemantkumar and Rahimbhai 2019; Egea et al. 2014). Similarly, for bread and snacks, *Spirulina* is also used to enhance protein composition. In other products like cheese, *Chlorella* is used as a quality enhancer. Moreover, *Dunaliella salina* is the commonly used algae for pasta protein, fats, and minerals composition enhancement. The cyanobacterial member *Nostoc commune* consumed raw, stir-fried, and in soups.

Table 6.1 List of algae-based available products used as human foods

Product	Species	Functions	References
Biscuits	*Spirulina platensis*	Used to enhance the biscuits' texture, aroma, flavor, and color	Abd El Baky et al. (2015)
Bread	*Spirulina*	Enhancement of protein composition of bread	Hernández-López et al. (2023)
Cheese	*Chlorella* sp.	Cheese quality enhancer	Lousada Falcão et al. (2023)
Cookies	*Haematococcus pluvialis*	Improve bioactive composition and decrease the rate of reducing sugar released	Hossain et al. (2017)
Snacks	*Arthrospira* sp.	Increase the nutritional contents (water solubility index, hardness, color, and expansion index)	Lucas et al. (2018)
Pasta	*Dunaliella salina*	Enhance the pasta nutrients composition	Sui and Vlaeminck (2020)

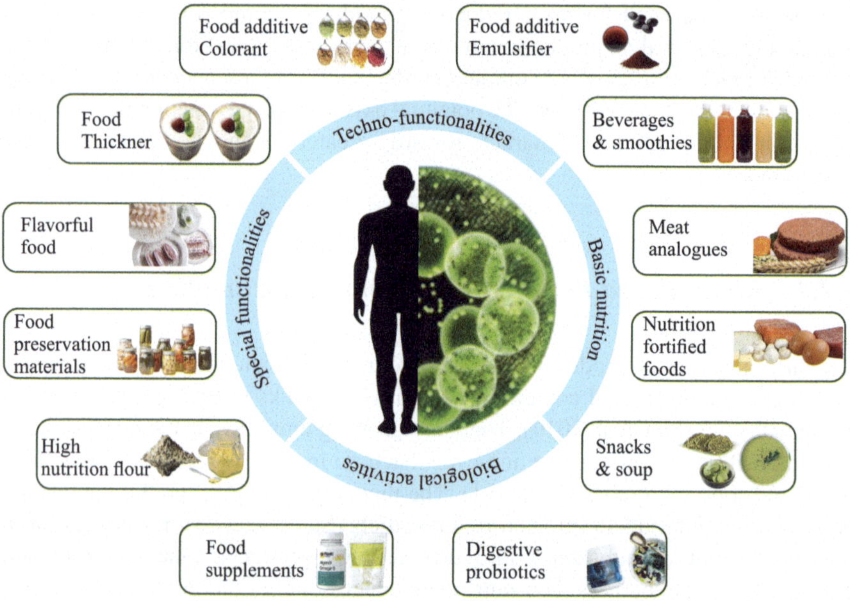

Fig. 6.2 Microalgae and cyanobacteria used for human feed

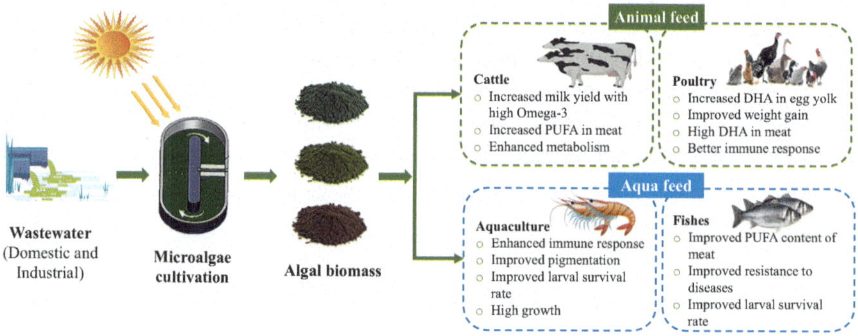

Fig. 6.3 Applications of algae biomass as animal and aquaculture feed

6.2.3 Algae as Animal Feed

Amid a growing awareness surrounding their nutritional richness, microalgae, and cyanobacteria are emerging as promising alternatives to traditional protein sources for both human and animal diets (Park et al. 2024). The versatility of various algae species has extended to supplements in animal feed, showcasing remarkable benefits (Fig. 6.3). An exemplary case involves the incorporation of *Desmosdesmus* sp. as a supplement for chickens, revealing positive outcomes such as improved body weight, enhanced feed efficiency, reduced feed intake, and an increase in antioxidant components found in both eggs and liver. Meanwhile, introducing the marine microalgae *Aurantiochytrium limacinum* at concentrations of 0.25% and 0.50% has demonstrated the ability to elevate the DHA (docosahexaenoic acid) content in pork loin and back fat in pigs. Moreover, integrating 10% algal biomass from *Desmodesmus* sp. into pig feed over 28 days has shown promising potential for enhancing growth. In another innovative approach, combining microalgae *Schizochytrium* sp. with a 3.89% addition of linseed and 5% algae have proven effective in enhancing meat quality and reducing feed ingestion in lambs. The algae *Porphyridium* sp. stands out for its significant impact on lowering blood cholesterol levels by up to 28%. Notably, its inclusion in chicken feed remarkably decreased egg yolk cholesterol, arachidonic acid, and linoleic acid levels by 10%, 29%, and 24%, respectively.

Furthering the spectrum of possibilities, incorporating *Spirulina* at a 10% level demonstrated a substantial increase, leading to the highest mean live weight in lambs, reaching approximately 41.9 ± 0.7 kg in sheep (Kusmayadi et al. 2021). This evolving landscape of nutritional exploration underscores the potential of microalgae and cyanobacteria as transformative elements in redefining dietary paradigms for animals and humans.

6.3 Use of Algae in Developing Functional Food Products

In the mid-1980s, in Japan, the term functional food was coined, referring to some processed foods that contain ingredients that provide specific physiological functions along with their nutritional properties (Corbo et al. 2014). The European Union (EU) divides food supplements into two broad categories based on categorization and coverage: first, the old and traditional category supplements, which contain mainly vitamins and minerals; second, the supplements with 'other substances', i.e., botanicals and other natural substances which are physiologically active (Vettorazzi et al. 2020). Besides these, additional sub-categories are also proposed, including amino acids, enzymes, probiotics, and many other substances. The latter category includes algae, fungi, plants, and their extracts.

Nowadays, there is a growing interest in new functional food products and beverages with microalgae and cyanobacteria in the market, and foods that can be fortified with them like bakery products, pasta, bread, noodles, biscuits, snack foods, candy, gums, functional bars, smoothies, drink mixes, soft drinks, sauces, and soups has dramatically motivated research in food industry field. In this section, the main aim is to explore the multifaceted roles of algae, especially microalgae including cyanobacteria, in the development of functional food and feed products, shedding light on their applications in bakery goods, beverages and smoothies, snacks, nutraceuticals, and their vital contributions to aquaculture feeds.

6.3.1 Microalgae and Cyanobacteria in Bakery Products

Microalgae and cyanobacteria have been used for human consumption since ancient times. Much global effort and attention have been paid to producing food products using microalgae and cyanobacteria in recent decades. Nowadays, efforts are placed to improve or enrich prepared foods and bakery products like breads, noodles, pasta, and more, with microalgae to increase public perception toward algal food products. Algae are becoming more significant as functional additives and improvers in baking. Their application aims to boost biological value, lengthen shelf life, enhance finished product quality, and improve dough quality. The enrichment of bread with various beneficial compounds that offer medicinal and preventative characteristics has received much attention globally in recent years.

Spirulina, C. vulgaris, and *Dunaliella salina* are majorly used in bakery products (Şahin 2020). According to a few studies, *Spirulina* and *Chlorella* have been evaluated with the potential to be used to make different food products like milkshakes, snacks, pasta, bread, and cookies (Şahin 2020). Another study shows that *Spirulina* has been used in products like rice pasta and Italian foods like sfogliatine, and grissini. *Spirulina* is used to make wheat flour cookies, and the sensory evaluation of 0, 5, 10, and 15% w/w was comparatively studied, and an increasing protein and energy ratio was recorded in all the experimental samples (Şahin 2020; Koli et al. 2022). The pigment phycocyanin from *S. platensis* is used as a colorant in desserts and sweet cake decoration (Athiyappan et al. 2024). Microalgae like *C. vulgaris, Schizochytrium* sp. and *Isochrysis galbana* are also being used in different bakery products (de Oliveira and Bragotto 2022; Kratzer and Murkovic 2021). DHA-rich microalgae oil from *Schizochytrium* sp. is used to make breads and rolls (Hakim 2012). *Chlorella vulgaris, Haematococcus pluvialis,* and *I. galbana* are used to make novel biscuits with sensory improvement and nutritional benefits (Mosibo et al. 2024). Long chain ω3 PUFA (LC-PUFAs-ω3), especially EPA and DHA of *I. galbana*, is used in biscuits (Matos et al. 2021). *C. vulgaris* is used in croissants stuffed with cheese. *Chloroidium ellipsoideum* is used in bread, rolls, and biscuits. Other microalgae called *Nannochloropsis oculata* are used in pasta and cookies (Sidari and Tofalo 2019). *D. salina* is a marine microalga used in pasta to improve quality and in powder form to make bread, rolls, cookies, and crackers (El-Baz et al. 2022). Kim et al., 2010 made a yellow layer cake using different amounts of *Chlorella* powder (0–4%), checked the quality characteristics, and found improved quality from textural, sensory, and functional points of view (Sidari and Tofalo 2019).

6.3.2 Beverages and Smoothies

Microalgae and cyanobacteria have shown their mettle in beverages and smoothies. These products represent a pivotal intersection of taste, nutrition, and visual appeal, with consumers seeking a delightful sensory experience and enhanced well-being.

Spirulina, a protein-rich cyanobacterium, has been strategically integrated into smoothies and beverages, infusing them with essential nutrients, including protein and essential fatty acids (Gogna et al. 2023). In the world of natural coloring, algae have carved out a niche, and *Spirulina* is renowned for its vibrant blue-green hue. The phycocyanin pigment present in *Spirulina* has played a significant role in crafting visually appealing, healthful beverages. This caters to consumers' sensory preferences and underscores the role of microalgae and cyanobacteria in enhancing the aesthetics of functional beverages (Bortolini et al. 2022). *S. platensis* is used as an extract or the pigment phycocyanin in producing sour milk and is incorporated with green tea (Marzieh Hosseini et al. 2013). *Spirulina* biomass is incorporated into a beverage called *Spirulina* beer or Anti-aging beer. Phycocyanin, derived from another microalga, *Porphyridium aerugineum,* is used in Pepsi and Bacardi Breezer (Sidari and Tofalo 2019).

6.3.3 Snacks and Functional Bars

Snacks are food items that are consumed between meals. They are often considered a source of energy and nutrients. Microalgae and cyanobacteria such as *Spirulina* are incorporated into snacks to provide additional nutritional benefits and make them functional foods (Habib et al. 2008; Marzieh Hosseini et al. 2013). *Spirulina* powder is used in snacks to make orange-flavored chewable wafers (Habib et al. 2008; Marzieh Hosseini et al. 2013). A filamentous cyanobacteria *Nostoc commune* can be consumed raw, stir-fried, in soups, or as a thickener for making other foods (Gantar and Svirčev 2008). Other cyanobacteria and microalgae that are eaten as dried filaments include *Nostoc flagelliforme*, which is used as food in China and is also known as "fa-cai" *Aphanotheca sacrum* or *Phylloderma sacrum* consumed in Japan, called "suizenji-nori (Gantar and Svirčev 2008). Dried biomass of four microalgae strains, *Spirulina platensis, Chlorella vulgaris, Tetraselmis suecica, and Phaeodactylum tricornutum* used in wheat crackers (Lafarga 2019; Batista et al. 2019). In 2013, F. Mohamadi-Alasti et al., examine the effect of *Spirulina platensis* powder on yoghurt containing spinach during cold storage on the starter culture viability and found that incorporation of *Spirulina* powder in the yoghurt is a good medium for probiotic lactic acid bacteria (Dinçoğlu et al. 2024). Microalgae and cyanobacteria are rich in carbohydrates, proteins, and lipids such as polyunsaturated fatty acids (PUFA) (Udayan et al. 2017). PUFA has high bioactivity. Recent experimental studies have shown that incorporating microalgae, such as *Chlorella vulgaris* (2–18% w/w), into 3D-printed snacks can enhance their nutritional profile, making them a protein and mineral source with enhanced antioxidant activity during storage and with a lower glycemic index (Letras et al. 2022). A thick pear snack contained juice by-products and *Euglena gracilis* microalgae called "Rocha" (Lomba-Viana et al. 2022). It showed interesting phenol content and was well accepted by the sensory panel, also rich in dietary fibers and various nutritional benefits (Lomba-Viana et al. 2022).

Functional bar in the context of food refers to a type of snack bar mainly formulated to provide additional health benefits beyond basic nutrition. For instance, the *Journal of Food Science and Technology* discussed formulating and evaluating a novel functional snack bar with ingredients like rolled oats, amaranth, and unripen banana peel powder. A study published in the *Journal of Functional Foods* investigated the incorporation of these microalgae into an energy bar at three levels of addition (0%, 2.5%, and 5%) and studied the nutritional, physicochemical, and sensory properties (Fanari et al. 2023). They found improved protein and vitamin B_{12} content and influenced the energy bars' color, flavor, and texture. Another study investigated 2% and 6% of *Spirulina* in snack bars and studied school children's acceptance of sensorial attributes (Franco Lucas et al. 2019). They conclude that *Spirulina* can be used as an ingredient in snack bars designed for infant feeding without changing sensorial characteristics (Franco Lucas et al. 2019).

6.4 Microalgae and Cyanobacteria in Aquaculture Feed

The role of microalgae and cyanobacteria in aquaculture feeds encompasses their potential as a source of essential nutrients for fish, their use in feeding strategies and formulation, and their contribution to sustainable aquaculture practices.

6.4.1 Role in Fish Feed

Due to the rapid global expansion in the population and food consumption and current increasing trends in seafood consumption, it has deliberately indicated that a large proportion of seafood must be procured from the aquaculture sector (Tidwell and Allan 2001). The declining natural harvest can be seen as a burden on wild seafood or freshwater aquatic food sources. It has been projected that the production in aquaculture needs to double by 2050 to fulfil the rate of consumption without being a burden on the wild sources (Hua et al. 2019; Tidwell and Allan 2001).

Health authorities recognize fish as a good diet source of essential fatty acids and protein. But, although the aquaculture industry is growing rapidly, the demand for quality fish still poses a challenge (Hua et al. 2019). Good nutrition is critical in fish culture or aquaculture for high-quality fish with good growth, survival, health, and composition (Hua et al. 2019; Pal et al. 2018). Nowadays, balanced, species-specific feed formulations and commercial diets have developed to support the growth of a more significant fish density, more than water's natural fertility. This rapid expansion resulted in the limited excess of the traditionally used key feedstuffs for aquaculture, e.g., fishmeal and fish oil. Additionally, feed cost and quality also affect the entire process. This, in turn, opens new ways to unconventional sources being investigated for fish nutrition (Nagappan et al. 2021). No wonder algae can be considered a potential alternative for conventional fish feeds like fishmeal because microalgae are the natural food source of fish and other aquatic animals like shrimp and bivalves (Nagappan et al. 2021).

Therefore, microalgae and cyanobacteria have the potential for the complete or partial replacement of fishmeal and commercial fish feeds or else can be incorporated into innovative and sustainable fish feeds (Ahmad et al. 2022; Nagappan et al. 2021). That drives the traditional culture of herbivorous and omnivorous fish rather than carnivorous fishlike catfishes, which rely on high protein sources. Proteins are by far the most essential nutrient for the growth of fish (Ahmad et al. 2022). Fishmeal is the traditionally preferred primary dietary source of protein and other nutrients for many farmed fish species. Besides, there are benefits to choosing microalgae for the aqua feed formulation, such as the low input cost, a low-carbon footprint, wastewater treatment options, and significant economic benefits (Ahmad et al. 2022; Nagappan et al. 2021). All these advantages make viable algae in aquaculture an alternative to traditional fishmeals besides their nutritional profile, high digestibility, mineral content, palatability, and growth factors (Nagappan et al. 2021). Different microalgae species are cultured for aquaculture are *Pavlova lutheri, Nannochloropsis* sp., *Tetraselmis suecica, Cyclotella cryptica, Chlorella* sp., *Crypthecodinium cohnii, Schizochytrium* sp., etc. using different culture systems like tanks, photobioreactors, and fermenters (Muller-Feuga 2000). *Spirulina* is the most common cyanobacteria in aquaculture (Mosha 2019; Rosas et al. 2019). *Chaetoceros, Tetraselmis, Isochrysis, Thalassiosira,* and *Nannochloropsis* are the microalgae used for fish nutrition directly or indirectly (by means of other tiny organisms like artemia, rotifers, and daphnia), feeding them to the cultured larvae (Becker 2003; Hemaiswarya et al. 2011; Morales-Ventura et al. 2012; Mosha 2019; Muller-Feuga 2000). *Scendesmus* sp. and *Chlorella* sp. are fed to artemia and rotifers as food sources, respectively (Morales-Ventura et al. 2012; Muller-Feuga 2000). Overall, algae are a promising and sustainable feed source for fish culture. They can be genetically engineered to produce various protein products, making them a valuable alternative for fish nutrition. However, more research is needed to increase the biomass productivity of algae and overcome production challenges to investigate their full potential as a primary feed source for fish nutrition in aquaculture.

6.4.2 Feeding Strategies and Formulations

Feeding strategies and formulation are crucial parts of aquaculture. They are employed as a range of practices to optimize performance, reduce waste generation, and ensure sustainability, one of the significant goals (Boyd et al. 2020; Goddard 1996). Feeding strategies vary based on geographical location, production unit (e.g., pond, tank, and cage), and farmed fish or aquaculture species. These are the planned approaches and methods to provide suitable nutrition, ensuring the health and growth of the cultured aquatic organisms (Goddard 1996). Efficient feeding strategies emphasize using high-quality feed, feeding only what the fish will eat, preventing overfeeding, and minimizing waste production. These strategies include the use of natural pond food organisms, artificial supplementary feeds, and agricultural by-products as a diet (Pailan and Biswas 2022). Feeding strategies encompass various approaches; for example, complete diet feeding involves providing a full diet to fish, including natural live feeds, i.e., live pond food organisms and artificial supplementary feeds. In particular, larvae and juveniles require live prey for optimal growth and development (Jobling 2016; Tacon 2002). Widely used feeding strategies in the grow-out stage include feeding the fish at 80–90% satiation every other day, at a high rate for several days, and then at a lower rate for several days (Davis and Hardy 2022; Yakupitiyage 2013). Microbial floc technology or bio floc technology is used in this process that utilizes microbial communities to recycle the nutrients and provide supplementary feed to the cultured species within the culture water (Crab 2010). Other feeding strategies include feeding by hand in smaller-scale aquaculture operations involving manual feed distribution to the culture organisms. It helps closely monitor feeding behavior and feed intake in sensitive species. Automated feeding systems, recirculating aquaculture systems, and integrated multi-trophic aquaculture are used for intensive polyculture systems where multiple species are cultured together (Jobling 2016). Advanced filtration and water treatment strategies are employed with microalgae like *Spirulina* and many more (Zhang et al. 2020). Overall, feeding strategies are tailored to meet specific nutritional requirements and environmental conditions and directly impact aquaculture's success, sustainability, and economy.

6.4.3 Sustainable Aquaculture Practices

Sustainable aquaculture is the practice that focuses on three primary areas or perspectives; environmental, economic, and social sustainability, to incorporate their spatial and temporal dimensions (Bunting 2013; Frankic and Hershner 2003). It aims to improve capacity building and utilize land effectively to produce aquatic organisms. The utilization of microalgae and cyanobacteria in aquaculture feeds aligns with the goal of sustainable aquaculture, as they offer potential benefits such as reduced reliance on fishmeal, improved feed efficiency, and the promotion of growth and health in fish (Han et al. 2019; Rahman Shah et al. 2018). These benefits support the development of environmentally friendly and economically viable aquaculture practices. Some sustainable aquaculture practices include the use of natural and formulated feeds, forced tangential aeration, elaborate technical assessments in site selection for pond construction, rearing fish at a higher stocking density and feeding them with natural food for a prolonged period, integration between aquatic plants and aquatic animals, and the use of plant-enriched diets to improve fish growth, immunity, and disease resistance (Han et al. 2019). Algal biotechnology plays a significant role in sustainable aquaculture. Algae are used in aquaculture for multiple purposes, including as a feed for fish and other culture species, as a natural wastewater purifier, and as a source of bioactive compounds for overall growth and development (Nagappan et al. 2021). One of the examples of microalgae or cyanobacteria being used for sustainable aquaculture is the production of fatty acids from them. Because fish are the traditional source of omega-3 fatty acids and protein, their reliance on marine algae for these nutrients has led to challenges due to declining fish stocks (Nagappan et al. 2021). By utilizing marine microalgae directly, the need for fish as an intermediary source of these essential nutrients can be bypassed, contributing to sustainable aquaculture and addressing the global food crisis.

6.5 Algae-Based Nutraceuticals

The nutraceutical industry harnesses microalgae and cyanobacteria due to their rich reservoirs of nutritional elements such as amino acids, fatty acids, vitamins, and minerals (Hachicha et al. 2022). A variety of microalgae, including *Tetraselmis, Chlorella, Chaetoceros, Spirulina, Nannochloropsis, Cryptothecodinium, and Dunaliella,* among others, are tapped for nutraceutical production owing to their capacity to generate bioactive compounds (Table 6.2) (Ahirwar et al. 2024; Shourie et al. 2022). *Spirulina* holds a significant place as a traditional food in regions like Lake Chad in central Africa and Lake Texcoco in Mexico (Damessa 2021; Koru 2012; Ohmori and Ehira 2013). *Spirulina*, a potent superfood, offers a protein content of 60–70% by weight, along with essential nutrients like vitamin B, phycocyanin, omega-6 fatty acids, vitamin E, and minerals. The C-phycocyanin pigment present in *Spirulina* possesses both antioxidant and anti-inflammatory properties (Wang et al. 2024; Ahirwar et al. 2024; AlFadhly et al. 2022; Lafarga et al. 2020; Matufi and Choopani 2020). Meanwhile, *S. platensis* has evolved into a crucial

Table 6.2 Annual revenue of several algae-based companies in 2021

Companies	Algal species	Key products	Annual revenue	References
E.I.D.—Parry (India) Limited	*Chlorella, Spirulina*	Nutraceuticals from microalgae	$22.7 million	Ratledge (2013), Kratzer and Murkovic (2021)
Algatechnologies Ltd—Israel, USA	*Haematococcus pluvialis*	AstaPure, a natural astaxanthin	$15 million	Balasubramaniam et al. (2021)
BlueBioTech Int. GmbH—Germany	*Spirulina platensis, Haematococcus pluvialis*	Phycocyanin, astaxanthin, and microalgae powder	$5.9 million	Martins et al. (2018)
Earthrise Nutritionals, LLC—California, USA	*Spirulina*	Spirulina-based products	$22 million	Sharma and Sharma (2018), Araújo et al. (2021)
Cyanotech Corporation—Hawaii, USA	*Haematococcus pluvialis, Spirulina*	Bioastin astaxanthin and *Spirulina* products	$32.3 million	Sharma and Sharma (2018)

industrial organic resource, serving as a health supplement, rich in β-carotene, and a natural coloring agent (Gentscheva et al. 2023; Pehlivanov et al. 2024). *Chlorella*, another microalga, served as an ideal space food due to its detoxifying properties and immune system stimulation. With over 50% protein, approximately 5% chlorophyll, around 20% dietary fiber, and a rich array of nutrients including provitamin A, thiamin B_1, riboflavin, niacin, vitamin E, B_6, B_{12}, biotin, folic acid, and pantothenic acid, it plays a vital role in regulating high blood pressure and providing relief for conditions like fibromyalgia and ulcerative colitis seen in a clinical trial (Bito et al. 2020; Merchant and Andre 2001). Astaxanthin, derived from *Haematococcus,* demonstrates significant gastroprotective, neuroprotective, ocular protective, anticancer, and hepatoprotective effects, aiding kidney function (Kumari et al. 2023). A human trial spanning 8 weeks was conducted to assess the efficacy and potential toxicity of astaxanthin extracted from *Haematococcus* (Satoh et al. 2009). Last but not least, *Aphanizomenon*, a prokaryotic cyanobacterium, contributes to the prevention of immunosuppressive and cardiovascular diseases (Hassan et al. 2022; Kumari et al. 2023).

6.6 Cyanobacteria-Derived Fish Feed

Toxic cyanobacterial blooms, known for generating cyclic heptapeptide toxins called microcystins, pose significant environmental challenges globally (Rastogi et al. 2015; Sivonen and Jones 1999). However, despite this drawback, cyanobacterial protein, constituting approximately 30–50% of its composition, has emerged as a potential substitute for aquaculture protein sources (Liang et al. 2015). In the study titled "Evaluation of Cyanobacterial Bloom from Lake Taihu as a Protein Substitute in Fish Diet-A Case Study on Tilapia," where case study shows that the toxin levels in cyanobacterial blooms from Lake Taihu are lower in June and increase during late summer and autumn (Huo et al. 2021). Diets incorporating low-toxicity cyanobacterial blooms did not adversely affect the growth of tilapia. Furthermore, tilapia muscle fed with 18.5% low-toxicity cyanobacterial bloom was deemed safe for human consumption (Huo et al. 2021). The case study suggests that selecting low-toxicity cyanobacterial blooms from large lakes can help overcome limitations in utilizing cyanobacterial biomass. The significance of cyanobacteria-derived fish feeds of the different case studies is highlighted in Table 6.3.

6.7 Innovative Food Products

The abundance of compounds in microalgae provides a foundation for developing an algal-based food industry dedicated to creating and utilizing microalgae to produce innovative, functional food products (Caporgno and Mathys 2018). Microalgae

Table 6.3 List of case studies on cyanobacteria-derived fish feed

Case studies	Results	References
Evaluation of algae-based feed in Goldfish (*Carassius auratus*) nutrition	Algae-based feed showed higher protein, lipid, and carotenoid content Fish fed with algae-based feed showed increased protein, glycogen, and carotenoid levels	Khatoon et al. (2010)
Effect of drying processes on biochemical contents of *Spirulina* platensis as a protein source for fish diet	Drying *Spirulina* platensis at 120 °C resulted in higher concentrations of chemical contents Feeding fish with *Spirulina* diet increased weight gain and daily gain	Abuo El-Kheir et al. (2016)
Effect of adding *Chlorella* sp. on some biological parameters and proximate analysis of common Carp *Cyprinus Carpio* L.	Feeding algae as a feed additive to fish significantly changed biological parameters *Chlorella* can be used as a protein-rich source in common carp diets	Abdulrahman et al. (2018)
The use of *S. platensis* and *Daphnia magna* as complementary protein and lipid sources in transitional diet for common carp (*Cyprinus carpio* L.) nursery	Combination of *S. platensis* and *D. magna* biomass was used as complementary protein and lipid sources in diet supplementation for common carp larvae Resulted in comparable fish growth, survival, and feed utilization in common carp culture at the nursery phase	Situmorang (2016)
Feeding *S. platensis* induced changes in production performance and quality of salmonid species	Adding *S. platensis* to fish burgers improved their nutritional composition. It was used as a nutritious additive to produce new fish-based products with high alimentary qualities	Rosenau et al. (2022)
Increasing the fatty acid content in muscle of common Carp (*Cyprinus carpio* L.)	Feeding common carp with algae can increase the fatty acid content in their muscles. Algae from wastewater treatment are nutritionally favorable	Kopp et al. (2019)

can enhance dairy products by introducing bioactive compounds (Hernández et al. 2022). Successful examples include the incorporation of *Chlorella* into yoghurts and cheeses (Hernández et al. 2022). Cookies and biscuits represent suitable categories for incorporating microalgae-based ingredients. *C. vulgaris*, used in cookies, is a coloring agent, potential antioxidant, and nutritional supplement (Batista et al. 2019). *Isochrysis galbana* is added to provide omega-3 polyunsaturated fatty acids (PUFAs) with advantageous effects on human health (Bonfanti et al. 2018; Lemahieu et al. 2015). Cookies and bread have been enriched with phycocyanin extracts and whole *S. platensis* to boost protein and fiber content, offering additional health benefits (Vieira et al. 2020). Similarly, incorporating *Haematococcus pluvialis* into cookies heightened antioxidant capacities and lowered the glycemic response (Mofasser Hossain et al. 2017). Like cookies, introducing microalgae in gluten-free bread, particularly adding *Spirulina*, substantially elevated protein content and enhanced bread quality by providing essential amino acids compared to non-supplemented bread (Caporgno and Mathys 2018). The green hue of microalgae poses a limitation in its application for everyday products, impacting consumers' perceptions of taste and quality. In items like pasta, where various colors are already accepted, the change in color does not significantly affect consumer perception. However, in certain products, adding microalgae led to a slightly unfavorable perception due to the introduction of a fish flavor (Caporgno and Mathys 2018).

6.8 Conclusion and Prospects

Both microalgae and cyanobacteria boast significant nutritional value, their integration into human diets may face challenges due to their unpleasant flavor and aroma (Yang et al. 2024). To overcome this hurdle, researchers need to focus on innovative techniques to mask or modify the taste and smell of microalgae and cyanobacteria-based products, thus enhancing their palatability and acceptance among consumers. Additionally, efforts are underway to develop novel processing methods that preserve the nutritional integrity of microalgae while minimizing undesirable sensory attributes. By addressing these challenges, researchers can unlock the full potential of microalgae and cyanobacteria as valuable sources of nutrition and sustainable food alternatives.

It is important to minimize the limitations and maximize biomass production to fulfil the current food and feed the needs of society. The main limitation of microalgae and cyanobacteria are their unpleasant smell and their toxicity (Khan et al. 2018; Sharma et al. 2013). So, it is important to alter or conceal the taste and smell of products derived from microalgae and cyanobacteria. Along with the evaluation of the toxicity level and exploring the new alternative microalgae and cyanobacteria need to be done to overcome the limitation.

The future direction toward utilizing microalgae and cyanobacteria for sustainable food and feed industries should prioritize biotechnological innovations, product diversification, and value chain integration (Fabris et al. 2020). By focusing on biotechnological advancements, we can enhance the productivity and efficiency of microalgae and cyanobacteria cultivation, ultimately improving the yield of these valuable resources (Fabris et al. 2020; Novoveská et al. 2019). Concurrently, by improving the quality of microalgae and cyanobacteria-derived food and feed products, we can cater to the nutritional needs of both human and animal populations, thus benefiting society. Moreover, product diversification of microalgae and cyanobacteria can open new market opportunities and enhance economic viability, ensuring the long-term sustainability of these industries. By expanding the range of products derived from microalgae and cyanobacteria, we can tap into various consumer preferences and demands, thereby bolstering market resilience and competitiveness. Furthermore, value chain integration plays a crucial role in maximizing resource efficiency and productivity. For instance, incorporating microalgae and cyanobacteria as feed supplements for livestock and aquaculture species can improve their health and growth outcomes while simultaneously reducing reliance on traditional feed sources. This holistic approach to value chain integration not only optimizes resource utilization but also fosters sustainability across the entire food and feed production process. In essence, by embracing biotechnological innovations, fostering product diversification, and integrating value chains effectively, we can unlock the full potential of microalgae and cyanobacteria as sustainable solutions for the food and feed industries, paving the way for a more resilient and environmentally friendly future.

References

Abd El Baky HH, El Baroty GS, Ibrahem EA (2015) Functional characters evaluation of biscuits sublimated with pure phycocyanin isolated from spirulina and spirulina biomass. Nutricion Hospitalaria 32(1):231–241. https://doi.org/10.3305/nh.2015.32.1.8804

Abdulrahman N, Abid SH, Khidir AA, Omer BB, Hama Rasheed DB, Baha Alddin LH (2018) Effect of adding microalgae *Chlorella* sp. on some biological parameters and proximate analysis of common carp Cyprinus Carpio L. Iranian. J Vet Med 12(3):199–206. https://doi.org/10.22059/ijvm.2018.244747.1004856

Abuo El-Kheir WS, Ibrahim EA, Abd El-Razek AB, Helal AM (2016) Effect of drying processes on biochemical contents of *Spirulina platensis* as a protein source for fish diet. J Environ Sci 32(1):1–23. https://doi.org/10.21608/jes.2016.24700

Ahirwar A, Rai A, Sirotiya V, Khandelwal P, Singh G, Jhadav D, Harish M, Schoefs B, Varjani S, Vinayak V (2024) Fermentation of algal biomass for its nutritional value: perspectives and revolutions in the food industry. Environ Technol Rev 13, Issue 1:1. https://doi.org/10.1080/21622515.2023.2283097

Ahmad A, Ashraf SS (2023) Sustainable food and feed sources from microalgae: food security and the circular bioeconomy. Algal Res 103185:103185

Ahmad A, Hassan W, Banat F (2022) An overview of microalgae biomass as a sustainable aquaculture feed ingredient: food security and circular economy. In: Bioengineered, vol 13, issue 4. https://doi.org/10.1080/21655979.2022.2061148

AlFadhly NKZ, Alhelfi N, Altemimi AB, Verma DK, Cacciola F, Narayanankutty A (2022) Trends and technological advancements in the possible food applications of spirulina and their health benefits: a review. In: Molecules, vol 27, issue 17. https://doi.org/10.3390/molecules27175584

Aly HF, Rizk MZ, Shams SGE (2023) Microalgae in food and feed: safety and toxicological aspects. In: Jacob-Lopes E, Queiroz MI, Maroneze MM, L Q B T-H of F and F. from M. Zepka (eds). Academic, pp 549–565. https://doi.org/10.1016/B978-0-323-99196-4.00004-8

Araújo R, Vázquez Calderón F, Sánchez López J, Azevedo IC, Bruhn A, Fluch S, Garcia Tasende M, Ghaderiardakani F, Ilmjärv T, Laurans M, Mac Monagail M, Mangini S, Peteiro C, Rebours C, Stefansson T, Ullmann J (2021) Current status of the algae production industry in Europe: an emerging sector of the blue bioeconomy. Front Mar Sci 7. https://doi.org/10.3389/fmars.2020.626389

Athiyappan KD, Routray W, Paramasivan B (2024) Phycocyanin from Spirulina: a comprehensive review on cultivation, extraction, purification, and its application in food and allied industries. Food Hum 2:100235. https://doi.org/10.1016/j.foohum.2024.100235

Balasubramaniam V, Gunasegavan RD-N, Mustar S, Lee JC, Mohd Noh MF (2021) Isolation of industrial important bioactive compounds from microalgae. Molecules (Basel, Switzerland) 26(4). https://doi.org/10.3390/molecules26040943

Barkia I, Saari N, Manning SR (2019) Microalgae for high-value products towards human health and nutrition. Mar Drugs 17(5). https://doi.org/10.3390/md17050304

Batista AP, Niccolai A, Bursic I, Sousa I, Raymundo A, Rodolfi L, Biondi N, Tredici MR (2019) Microalgae as functional ingredients in savory food products: application to wheat crackers. Food Secur 8(12). https://doi.org/10.3390/foods8120611

Becker W (2003) Microalgae for aquaculture: the nutritional value of microalgae for aquaculture. In: Handbook of microalgal culture. https://doi.org/10.1002/9780470995280.ch21

Benemann JR, Woertz I, Lundquist T (2018) Autotrophic microalgae biomass production: from niche markets to commodities. Ind Biotechnol 14(1):3–10. https://doi.org/10.1089/ind.2018.29118.jrb

Bito T, Okumura E, Fujishima M, Watanabe F (2020) Potential of chlorella as a dietary supplement to promote human health. In: Nutrients, vol 12, issue 9. https://doi.org/10.3390/nu12092524

Bonfanti C, Cardoso C, Afonso C, Matos J, Garcia T, Tanni S, Bandarra NM (2018) Potential of microalga *Isochrysis galbana*: bioactivity and bioaccessibility. Algal Res 29:242. https://doi.org/10.1016/j.algal.2017.11.035

Bortolini DG, Maciel GM, de Fernandes IAA, Pedro AC, Rubio FTV, Branco IG, Haminiuk CWI (2022) Functional properties of bioactive compounds from spirulina spp.: current status and future trends. Food Chem Molecular Sci 5:100134. https://doi.org/10.1016/j.fochms.2022.100134

Boyd CE, D'Abramo LR, Glencross BD, Huyben DC, Juarez LM, Lockwood GS, McNevin AA, Tacon AGJ, Teletchea F, Tomasso JR, Tucker CS, Valenti WC (2020) Achieving sustainable aquaculture: historical and current perspectives and future needs and challenges. In: Journal of the world aquaculture society. Blackwell publishing Inc., vol 51, issue 3, pp 578–633. https://doi.org/10.1111/jwas.12714

Bunting SW (2013) Principles of sustainable aquaculture: promoting social, economic and environmental resilience. In: Principles of sustainable aquaculture: promoting social, economic and environmental resilience, vol 9780203127438. https://doi.org/10.4324/9780203127438

Burchi F, Fanzo J, Frison E (2011) The role of food and nutrition system approaches in tackling hidden hunger. Int J Environ Res Public Health 8(2):358–373. https://doi.org/10.3390/ijerph8020358

Caporgno MP, Mathys A (2018) Trends in microalgae incorporation into innovative food products with potential health benefits. In: Frontiers in nutrition. Frontiers Media S.A., vol 5. https://doi.org/10.3389/fnut.2018.00058

Corbo MR, Bevilacqua A, Petruzzi L, Casanova FP, Sinigaglia M (2014) Functional beverages: the emerging side of functional foods: commercial trends, research, and health implications. Compr Rev Food Sci Food Saf 13(6):1192–1206

Crab R (2010) Bioflocs technology: an integrated system for the removal of nutrients and simultaneous production of feed in aquaculture

Damessa F (2021) Nutritional and functional values of microalgae (spirulina) naturally found in East Africa. The Nelson Mandela African Institution of Science and Technology

Davis DA, Hardy RW (2022) Feeding and fish husbandry. In: Fish nutrition. https://doi.org/10.1016/B978-0-12-819587-1.00015-X

de Oliveira APF, Bragotto APA (2022) Microalgae-based products: food and public health. Future Foods 6:100157

Dhandayuthapani K, Malathy S, Mulla SI, Gupta SK (2021) An insight into the potential application of microalgae in pharmaceutical and nutraceutical production. Algae: multifarious applications for a sustainable world, 135–179

Diaz CJ, Douglas KJ, Kang K, Kolarik AL, Malinovski R, Torres-Tiji Y, Molino JV, Badary A, Mayfield SP (2023) Developing algae as a sustainable food source. Front Nutr 9:1029841

Dinçoğlu AH, Akça SS, Çalışkan Z (2024) Effect of *Spirulina platensis* on probiotic, nutritional, and quality properties of yogurt. Int Food Res J 31(1):157

Egea BCB, Campos ALM, De Carvalho-Eliane JCM, Danesi DG (2014) Antioxidant and nutritional potential of cookies enriched with Spirulina platensis and sources of fibre. J Food Nutr Res 53(2):171–179

El-Baz F, Abdo S, Hussien A (2022) Microalgae Dunaliella salina for use as food supplement to improve pasta. Quality 45–51

Fabris M, Abbriano RM, Pernice M, Sutherland DL, Commault AS, Hall CC, Labeeuw L, McCauley JI, Kuzhiuparambil U, Ray P, Kahlke T, Ralph PJ (2020) Emerging technologies in algal biotechnology: toward the establishment of a sustainable, algae-based bioeconomy. Front Plant Sci 11. https://doi.org/10.3389/fpls.2020.00279

Fanari F, Comaposada J, Boukid F, Climent E, Coma AC, Guerrero L, Castellari M (2023) Enhancing energy bars with microalgae: A study on nutritional, physicochemical and sensory properties. J Funct Foods 109:105768

Franco Lucas B, Rosa A, Carvalho L, Morais M, Duarte Santos T, Costa JA (2019) Snack bars enriched with spirulina for schoolchildren nutrition. Food Sci Technol 40:146. https://doi.org/10.1590/fst.06719

Frankic A, Hershner C (2003) Sustainable aquaculture: developing the promise of aquaculture. Aquac Int 11(6):517–530. https://doi.org/10.1023/B:AQUI.0000013264.38692.91/METRICS

Gantar M, Svirčev Z (2008) Microalgae and cyanobacteria: food for thought (1). J Phycol 44(2):260–268. https://doi.org/10.1111/j.1529-8817.2008.00469

Gentscheva G, Nikolova K, Panayotova V, Peycheva K, Makedonski L, Slavov P, Radusheva P, Petrova P, Yotkovska I (2023) Application of Arthrospira platensis for medicinal purposes and the food industry: a review of the literature. Life 13(3). https://doi.org/10.3390/life13030845

Goddard S (1996) Feed management in intensive aquaculture. Feed management in intensive aquaculture. https://doi.org/10.1007/978-1-4613-1173-7

Gogna S, Kaur J, Sharma K, Prasad R, Singh J, Bhadariya V, Kumar P, Jarial S (2023) Spirulina – an edible cyanobacterium with potential therapeutic health benefits and toxicological consequences. J Am Nutr Assoc 42(6):559–572. https://doi.org/10.1080/27697061.2022.2103852

Habib MA, Parvin M, Huntington T, Hasan M (2008) A review on culture, production and use of spirulina as food for humans and feeds for domestic animals and fish

Hachicha R, Elleuch F, Ben-Hlima H, Dubessay P, De Baynast H, Delattre C, Pierre G, Hachicha R, Abdelkafi S, Michaud P, Fendri I (2022) Biomolecules from microalgae and cyanobacteria: applications and market survey. Appl Sci 12:1924. https://doi.org/10.3390/app12041924

Hakim AR (2012) The potential of heterotrophic microalgae (*Schizochytrium* sp.) as a source of DHA. Squalen Bull Mar Fish Postharvest Biotechnol 7(1):29–38

Han P, Lu Q, Fan L, Zhou W (2019) A review on the use of microalgae for sustainable aquaculture. In: Applied sciences (Switzerland), vol 9, issue 11. MDPI AG. https://doi.org/10.3390/app9112377

Hassan S, Meenatchi R, Pachillu K, Bansal S, Brindangnanam P, Arockiaraj J, Kiran GS, Selvin J (2022) Identification and characterization of the novel bioactive compounds from microalgae and cyanobacteria for pharmaceutical and nutraceutical applications. J Basic Microbiol 62(9):999. https://doi.org/10.1002/jobm.202100477

Hemaiswarya S, Raja R, Kumar RR, Ganesan V, Anbazhagan C (2011) Microalgae: a sustainable feed source for aquaculture. World J Microbiol Biotechnol 27(8):1737–1746. https://doi.org/10.1007/s11274-010-0632-z

Hemantkumar JN, Rahimbhai MI (2019) Microalgae and its use in nutraceuticals and food supplements (M. Vítová (ed.), p Ch. 5). IntechOpen. https://doi.org/10.5772/intechopen.90143

Hernández H, Nunes MC, Prista C, Raymundo A (2022) Innovative and healthier dairy products through the addition of microalgae: a review. In: Foods, vol 11, issue 5. https://doi.org/10.3390/foods11050755

Hernández-López I, Alamprese C, Cappa C, Prieto-Santiago V, Abadias M, Aguiló-Aguayo I (2023) Effect of spirulina in bread formulated with wheat flours of different alveograph strength. Foods (Basel, Switzerland) 12(20). https://doi.org/10.3390/foods12203724

Hossain AKMM, Brennan MA, Mason SL, Guo X, Zeng XA, Brennan CS (2017) The effect of Astaxanthin-rich microalgae "*Haematococcus pluvialis*" and wholemeal flours incorporation in improving the physical and functional properties of cookies. Foods (Basel, Switzerland) 6(8). https://doi.org/10.3390/foods6080057

Hosseinkhani N, McCauley JI, Ralph PJ (2022) Key challenges for the commercial expansion of ingredients from algae into human food products. Algal Res 64:102696

Hua K, Cobcroft JM, Cole A, Condon K, Jerry DR, Mangott A, Praeger C, Vucko MJ, Zeng C, Zenger K, Strugnell JM (2019) The future of aquatic protein: implications for protein sources in aquaculture diets. In: One Earth, vol 1, issue 3. https://doi.org/10.1016/j.oneear.2019.10.018

Huo Y, Li Y, Guo W, Liu J, Yang C, Li L, Liu H, Song L (2021) Evaluation of cyanobacterial bloom from lake taihu as a protein substitute in fish diet—a case study on Tilapia. Toxins 13(10). https://doi.org/10.3390/toxins13100735

Jobling M (2016) Fish nutrition research: past, present and future. Aquac Int 24(3):767. https://doi.org/10.1007/s10499-014-9875-2

Khan MI, Shin JH, Kim JD (2018) The promising future of microalgae: current status, challenges, and optimization of a sustainable and renewable industry for biofuels, feed, and other products. In: Microbial cell factories, vol 17, issue 1. BioMed Central Ltd. https://doi.org/10.1186/s12934-018-0879-x

Khatoon N, Pal R (2015) Microalgae in biotechnological application: a commercial approach BT – plant biology and biotechnology: volume II: plant genomics and biotechnology. In: Bahadur B, Venkat Rajam M, Sahijram L, Krishnamurthy KV (eds). Springer India, pp 27–47. https://doi.org/10.1007/978-81-322-2283-5_2

Khatoon N, Sengupta P, Homechaudhuri S, Pal R (2010) Evaluation of algae based feed in goldfish (Carassius auratus) nutrition. Proc Zool Soc 63(2):109–114. https://doi.org/10.1007/s12595-010-0015-3

Koli DK, Rudra SG, Bhowmik A, Pabbi S (2022) Nutritional, functional, textural and sensory evaluation of spirulina enriched green pasta: a potential dietary and health supplement. Foods (Basel, Switzerland) 11(7). https://doi.org/10.3390/foods11070979

Kopp R, Brabec T, Mareš J, Marsalkova E, Marsálek B (2019) The utilization of algae with the aim to increase the fatty acid content in muscle of common carp (Cyprinus carpio L.). Acta Universitatis Agriculturae et Silviculturae Mendelianae Brunensis 67:91–99. https://doi.org/10.11118/actaun201967010091

Koru E (2012) Earth food spirulina (*Arthrospira*): production and quality Standarts. Food Additive. https://doi.org/10.5772/31848

Kratzer R, Murkovic M (2021) Food ingredients and nutraceuticals from microalgae: main product classes and biotechnological production. Foods (Basel, Switzerland) 10(7). https://doi.org/10.3390/foods10071626

Kumari A, Garima, Bharadvaja N (2023) A comprehensive review on algal nutraceuticals as prospective therapeutic agent for different diseases. 3 Biotech 13(2). https://doi.org/10.1007/s13205-022-03454-2

Kusmayadi A, Leong YK, Yen H-W, Huang C-Y, Chang J-S (2021) Microalgae as sustainable food and feed sources for animals and humans–biotechnological and environmental aspects. Chemosphere 271:129800

Lafarga T (2019) Effect of microalgal biomass incorporation into foods: nutritional and sensorial attributes of the end products. Algal Res 41:101566

Lafarga T, Fernández-Sevilla JM, González-López C, Acién-Fernández FG (2020) Spirulina for the food and functional food industries. In: Food research international, vol 137. Elsevier. https://doi.org/10.1016/j.foodres.2020.109356

Lemahieu C, Bruneel C, Ryckebosch E, Muylaert K, Buyse J, Foubert I (2015) Impact of different omega-3 polyunsaturated fatty acid (n-3 PUFA) sources (flaxseed, *Isochrysis galbana*, fish oil and DHA gold) on n-3 LC-PUFA enrichment (efficiency) in the egg yolk. J Funct Foods 19:821. https://doi.org/10.1016/j.jff.2015.04.021

Letras P, Oliveira S, Varela J, Nunes MC, Raymundo A (2022) 3D printed gluten-free cereal snack with incorporation of Spirulina (Arthrospira platensis) and/or Chlorella vulgaris. Algal Res 68:102863. https://doi.org/10.1016/j.algal.2022.102863

Liang H, Zhou W, Zhang Y, Qiao Q, Zhang X (2015) Are fish fed with cyanobacteria safe, nutritious and delicious? A laboratory study. Sci Rep 5. https://doi.org/10.1038/srep15166

Lomba-Viana X, Raymundo A, Prista C, Alegria MJ, Sousa I (2022) Clean label "Rocha" pear (*Pyrus communis* L.) snack containing juice by-products and *Euglena gracilis* microalgae. Front Nutr 9:825999

Lousada Falcão R, Pinheiro V, Ribeiro C, Sousa I, Raymundo A, Nunes MC (2023) Nutritional improvement of fresh cheese with microalga Chlorella vulgaris: impact on composition, structure and sensory acceptance. Food Technol Biotechnol 61(2):259–270. https://doi.org/10.17113/ftb.61.02.23.7851

Lucas BF, de Morais MG, Santos TD, Costa JAV (2018) Spirulina for snack enrichment: nutritional, physical and sensory evaluations. LWT 90:270–276. https://doi.org/10.1016/j.lwt.2017.12.032

Martins AP, Zambotti-Villela L, Yokoya NS, Colepicolo P (2018) Biotechnological potential of benthic marine algae collected along the Brazilian coast. Algal Res 33:316–327

Marzieh Hosseini S, Shahbazizadeh S, Khosravi-Darani K, Reza Mozafari M (2013) *Spirulina platensis*: food and function. Curr Nutr Food Sci 9(3):189–193

Matos J, Afonso C, Cardoso C, Serralheiro ML, Bandarra NM (2021) Yogurt enriched with Isochrysis galbana: an innovative functional food. Foods (Basel, Switzerland) 10(7). https://doi.org/10.3390/foods10071458

Matufi F, Choopani A (2020) Spirulina, food of past, present and future. Health Biotechnol Biopharma 3(4)

Mavrommatis A, Tsiplakou E, Zerva A, Pantiora PD, Georgakis ND, Tsintzou GP, Madesis P, Labrou NE (2023) Microalgae as a sustainable source of antioxidants in animal nutrition, health and livestock development. In: Antioxidants, vol 12, issue 10. https://doi.org/10.3390/antiox12101882

Merchant RE, Andre CA (2001) A review of recent clinical trials of the nutritional supplement Chlorella pyrenoidosa in the treatment of fibromyalgia, hypertension, and ulcerative colitis. Altern Ther Health Med 7(3):79

Morales-Ventura J, Nandini S, Sarma SSS, Castellanos-Páez ME (2012) Demography of zooplankton (*Anuraeopsis fissa*, *Brachionus rubens* and *Moina macrocopa*) fed *Chlorella vulgaris* and *Scenedesmus acutus* cultured on different media. Revista de Biologia Tropical 60(3). https://doi.org/10.15517/rbt.v60i3.1750

Mosha SS (2019) The significance of spirulina meal on fishmeal replacement in aquaculture. A review. Sumerianz J Agric Vet 2(Issue 1)

Mosibo OK, Ferrentino G, Udenigwe CC (2024) Microalgae proteins as sustainable ingredients in novel foods: recent developments and challenges. Foods Basel 13(5). https://doi.org/10.3390/foods13050733

Mu N, Mehar JG, Mudliar SN, Shekh AY (2019) Recent advances in microalgal bioactives for food, feed, and healthcare products: commercial potential, market space, and sustainability. Compr Rev Food Sci Food Saf 18(6):1882–1897

Muller-Feuga A (2000) The role of microalgae in aquaculture: situation and trends. In. J Appl Phycol 12:527

Murphy DJ, Cardona T (2022) Photosynthetic life: origin, evolution, and future. Oxford University Press

Nagappan S, Das P, AbdulQuadir M, Thaher M, Khan S, Mahata C, Al-Jabri H, Vatland AK, Kumar G (2021) Potential of microalgae as a sustainable feed ingredient for aquaculture. J Biotechnol 341:1–20. https://doi.org/10.1016/j.jbiotec.2021.09.003

Naik B, Mishra R, Kumar V, Mishra S, Gupta U, Rustagi S, Gupta AK, Preet MS, Bhatt SC, Rizwanuddin S (2024) Micro-algae: revolutionizing food production for a healthy and sustainable future. J Agric Food Res 15:100939

Novoveská L, Ross ME, Stanley MS, Pradelles R, Wasiolek V, Sassi JF (2019) Microalgal carotenoids: a review of production, current markets, regulations, and future direction. In: Marine drugs, vol 17, issue 11. https://doi.org/10.3390/md17110640

Ohmori M, Ehira S (2013) Spirulina: an example of cyanobacteria as nutraceuticals. In: Cyanobacteria: an economic perspective. https://doi.org/10.1002/9781118402238.ch7

Pailan GH, Biswas G (2022) Feed and feeding strategies in freshwater aquaculture. Transforming Coastal Zone for Sustainable Food and Income Security, pp 455–475. https://doi.org/10.1007/978-3-030-95618-9_35

Pal J, Shukla B, Maurya AK, Verma HO, Pandey G, Amitha. (2018) A review on role of fish in human nutrition with special emphasis to essential fatty acid. Int J Fish Aquatic Stud 6(2)

Park S, Lee S-J, Noh W, Kim YJ, Kim J-H, Back S-M, Ryu B-G, Nam SW, Park S-H, Kim J (2024) Production of safe cyanobacterial biomass for animal feed using wastewater and drinking water treatment residuals. Heliyon 10(3):e25136. https://doi.org/10.1016/j.heliyon.2024.e25136

Pehlivanov I, Gentscheva G, Nikolova K, Andonova V (2024) Some applications of *Arthrospira platensis* and algae in pharmaceutical and food technologies. In: Biointerface research in applied chemistry, vol 14, issue 2. https://doi.org/10.33263/BRIAC142.032

Pereira L, Cotas J, Valado A (2023) A green revolution in antioxidant research: harnessing microalgae for wellness and longevity

Rahman Shah M, Antonio Lutzu G, Alam A, Sarker P, Kabir Chowdhury MA, Parsaeimehr A, Liang Y, Daroch M, Sarker PallabKSarker P, Kabir Chowdhury A, dartmouthedu M (2018) Microalgae in aquafeeds for a sustainable aquaculture industry. J Appl Phycol 30:197–213. https://doi.org/10.1007/s10811-017-1234-z/Published

Rastogi RP, Madamwar D, Incharoensakdi A (2015) Bloom dynamics of cyanobacteria and their toxins: environmental health impacts and mitigation strategies. Front Microbiol 6(NOV):156415. https://doi.org/10.3389/FMICB.2015.01254/BIBTEX

Ratledge C (2013) Microbial oils: an introductory overview of current status and future prospects. OCL 20(6). https://doi.org/10.1051/ocl/2013029

Ravishankar GA, Ambati RR (2019) Handbook of algal technologies and phytochemicals: volume I food, health and nutraceutical applications. CRC Press

Rosas VT, Poersch LH, Romano LA, Tesser MB (2019) Feasibility of the use of Spirulina in aquaculture diets. In: Reviews in aquaculture, vol 11, issue 4. https://doi.org/10.1111/raq.12297

Rosenau S, Ciulu M, Reimer C, Mott AC, Tetens J, Mörlein D (2022) Feeding green: spirulina (Arthrospira platensis) induced changes in production performance and quality of salmonid species. Aquac Res 53(12):4276–4287. https://doi.org/10.1111/are.15925

Saadaoui I, Rasheed R, Aguilar A, Cherif M, Al Jabri H, Sayadi S, Manning SR (2021) Microalgal-based feed: promising alternative feedstocks for livestock and poultry production. J Anim Sci Biotechnol 12(1):76. https://doi.org/10.1186/s40104-021-00593-z

Şahin OI (2020) Functional and sensorial properties of cookies enriched with Spirulina and Dunaliella biomass. J Food Sci Technol 57(10):3639–3646. https://doi.org/10.1007/s13197-020-04396-4

Satoh A, Tsuji S, Okada Y, Murakami N, Urami M, Nakagawa K, Ishikura M, Katagiri M, Koga Y, Shirasawa T (2009) Preliminary clinical evaluation of toxicity and efficacy of a new astaxanthin-rich *Haematococcus pluvialis* extract. J Clin Biochem Nutr 44(3):280. https://doi.org/10.3164/jcbn.08-238

Sharma N, Sharma P (2018) Industrial and biotechnological applications of algae: a review. J Adv Plant Biol 1:1. https://doi.org/10.14302/issn.2638-4469.japb-17-1534

Sharma NK, Rai AK, Stal LJ (2013) Cyanobacteria: an economic perspective. In: Cyanobacteria: an economic perspective. https://doi.org/10.1002/9781118402238

Shourie A, Vijayalakshmi U, Singh A (2022) Bioactive compounds from microalgae. An integration of phycoremediation processes in wastewater treatment, pp 503–528. https://doi.org/10.1016/B978-0-12-823499-0.00021-3

Sidari R, Tofalo R (2019) A comprehensive overview on microalgal-fortified/based food and beverages. Food Rev Intl 35(8):778–805

Siddhnath, Surasani VKR, Singh A, Singh SM, Hauzoukim, Murthy LN, Baraiya KG (2024) Bioactive compounds from micro-algae and its application in foods: a review. Discov Food 4(1):27

Singh J, Saxena R (2015) An introduction to microalgae: diversity and significance. Diversity and significance. Handbook of marine microalgae. Biotechnol Adv:11–24. https://doi.org/10.1016/B978-0-12-800776-1.00002-9

Situmorang M (2016) The use of cyanobacteria Arthrospira platensis and Cladoceran Daphnia magna as complementary protein and lipid sources in transitional diet for common carp (Cyprinus carpio L.) nursery. Nat Resourc 7:423–433. https://doi.org/10.4236/nr.2016.77037

Sivonen K, Jones G (1999) Cyanobacterial toxins. The New York Times:7–10. https://researchportal.helsinki.fi/en/publications/cyanobacterial-toxins-2

Su M, Bastiaens L, Verspreet J, Hayes M (2023) Applications of microalgae in foods, pharma and feeds and their use as fertilizers and biostimulants: legislation and regulatory aspects for consideration. Food Secur 12(20). https://doi.org/10.3390/foods12203878

Suganya T, Varman M, Masjuki HH, Renganathan S (2016) Macroalgae and microalgae as a potential source for commercial applications along with biofuels production: A biorefinery approach. Renew Sust Energ Rev 55:909–941. https://doi.org/10.1016/j.rser.2015.11.026

Sui Y, Vlaeminck SE (2020) Dunaliella microalgae for nutritional protein: an undervalued asset. Trends Biotechnol 38(1):10–12. https://doi.org/10.1016/j.tibtech.2019.07.011

Tacon AGJ (2002) Thematic review of feeds and feed management practices in shrimp aquaculture. Intellect Prop

Tidwell JH, Allan GL (2001) Fish as food: Aquaculture's contribution. Ecological and economic impacts and contributions of fish farming and capture fisheries. EMBO Rep 2(11):958. https://doi.org/10.1093/embo-reports/kve236

Udayan A, Arumugam M, Pandey A (2017) Nutraceuticals from algae and cyanobacteria. In: Algal green chemistry, pp 65–89. Elsevier

Udayan A, Pandey AK, Sharma P, Sreekumar N, Kumar S (2021) Emerging industrial applications of microalgae: challenges and future perspectives. Syst Microbiol Biomanufact 1(4):411–431. https://doi.org/10.1007/s43393-021-00038-8

Vettorazzi A, López de Cerain A, Sanz-Serrano J, Gil AG, Azqueta A (2020) European regulatory framework and safety assessment of food-related bioactive compounds. Nutrients 12(3). https://doi.org/10.3390/nu12030613

Vieira MV, Oliveira SM, Amado IR, Fasolin LH, Vicente AA, Pastrana LM, Fuciños P (2020) 3D printed functional cookies fortified with *Arthrospira platensis*: evaluation of its antioxidant potential and physical-chemical characterization. Food Hydrocoll 107:105893. https://doi.org/10.1016/j.foodhyd.2020.105893

Wang N, Xiang Y, Ma Y, Zhang P, Zhou X, Zhu H, Zhang Z, Li Z, Xiao X, He M, Mehmood MA, Zhu H (2024) Fermented *Spirulina platensis* alleviated DSS-induced ulcerative colitis by regulating gut microbiota and MyD88/TLR4 signaling pathway. Asian J Agric Biol 1(1):2023232. https://doi.org/10.35495/ajab.2023.232

Yakupitiyage A (2013) On-farm feeding and feed management strategies in tropical aquaculture. In: Hasan MR, New MB (eds) On-farm feeding and feed management in aquaculture. FAO fisheries and aquaculture technical paper No. 583. FAO, Rome

Yang S, Wang Y, Wang J, Cheng K, Liu J, He Y, Zhang Y, Mou H, Sun H (2024) Microalgal protein for sustainable and nutritious foods: a joint analysis of environmental impacts, health benefits and consumer's acceptance. Trends Food Sci Technol 143:104278. https://doi.org/10.1016/j.tifs.2023.104278

Zhang F, Man YB, Mo WY, Wong MH (2020) Application of Spirulina in aquaculture: a review on wastewater treatment and fish growth. In: Reviews in aquaculture, vol. 12, issue 2. https://doi.org/10.1111/raq.12341

Chapter 7
Microalgae and Cyanobacteria for Resource Conservation, Recycling, and Preservation

Iqra Akbar, Neha Razzaque, Muhammed Abdullah, Payam Ghorbannezhad, and Achmad Syafiuddin

Abstract Despite the escalating human needs that drive the consumption of natural resources, there needs to be more emphasis on developing sustainable approaches for resource recycling and reuse. Resource conservation and recycling are the fundamental pillars for the sustainable development of the ecosystem, keeping the ecological balance of resources between present and future generations. Microalgae and cyanobacteria are versatile microorganisms that have the potential to address emerging environmental problems through resource recovery, recycling, and conservation. This chapter highlights the multifaceted roles of these microorganisms across different sectors of industry and ecosystem. Their ability to assimilate nutrients from wastewater streams and agricultural and industrial waste facilitates nutrient recycling, turning the waste into valuable biomass with different industrial applications. This approach fosters a circular bioeconomy paradigm by utilizing the valuable biomass for biofuel, biogas, biofertilizers, and high-value compound production. Moreover, microalgae showed remarkable potential in carbon sequestration; therefore, it can be integrated into carbon capture and utilization systems, harnessing dual benefits. Furthermore, integrating microalgae cultivation in saline or brackish water bodies utilizing non-arable land to conserve natural resources minimizes competition with freshwater resources and food crops. So far, these microorganisms present a sustainable pathway to address global environmental challenges and ecosystem restoration, contributing to biodiversity conservation and habitat preservation.

I. Akbar · N. Razzaque · M. Abdullah
Bioenergy Research Center, Department of Bioinformatics and Biotechnology, Government College University Faisalabad, Faisalabad, Pakistan

P. Ghorbannezhad
Department of Biorefinery Engineering, Faculty of New Technologies Engineering, Shahid Beheshti University, Tehran, Iran

A. Syafiuddin (✉)
Center for Environmental Health of Pesantren, Universitas Nahdlatul Ulama Surabaya, Surabaya, Indonesia
e-mail: achmadsyafiuddin@unusa.ac.id

M. A. Mehmood et al. (eds.), *Cyanobacteria Biotechnology*, https://doi.org/10.1007/978-3-031-70698-1_7

Keywords Resource depletion · Environmental mitigation · Resource conservation and restoration · Net zero emissions

7.1 Introduction

Global population increase and food habitat changes drive the demand for natural resources such as water, land, and energy. Due to this, the demand for mineral and fossil fuels is increasing; however, the extraction and depletion of resources seriously affect the environment (Freedman 2018). Industrial activities are using these resources as a constant and endless supply just following the approach of take-use-dispose, which is affecting the ecosystem (Smol et al. 2020). Due to the direct or indirect influence of these industrial and anthropogenic activities, small and large ecosystems are at risk, leading to insufficient water availability. It is estimated that over 40% of the global population is going to face extreme water scarcity by 2035 (Strong and Kuzma 2020). Climate change-driven water depletion will intensify in the countries around mid-latitude North Africa, Texas, China, North Chile, and North Africa (Maddocks et al. 2019). Lack of sufficient water resources to fulfill portable water is certainly a limitation for the economic growth, human well-being, ecosystem, and conservation of the environment (Cosgrove and Loucks 2015). Meanwhile, in September 2015, the United Nations outlined 17 sustainable development goals, with SDG6 focusing on the sustainable availability of clean water and sanitation. Therefore, it is essential to develop eco-friendly wastewater treatment methods facilitating higher resource recovery, thus improving water recycling (Cf 2015).

Cultivation of non-food crops using arable lands to meet fuel and energy demands is a potential threat to food security and nutrition (Renzaho et al. 2017). According to some reports, the European Union is estimated to produce 16.5 billion liters of biodiesel on average, whereas the United States is expected to generate 14 billion liters annually on average between 2023 and 2025 (Chia et al. 2022). Sustainable and affordable energy is one of the important factors associated with the economic development of a country. To overcome this problem, energy production has been shifted toward other resources such as waste material, waste oils, greases, and microorganisms, but these failed to fulfill the demand for petroleum-based fuels (Da Silva et al. 2014). Conservation and recycling of these resources are crucial to meet the food, energy, and water demands of the increasing population, achieving sustainable development goals.

Microalgae and cyanobacteria are potential candidates for the conservation and recycling of resources due to their various properties. Cyanobacteria are involved in various biogeochemical cycles like oxygen, nitrogen, and carbon, and they are responsible for bringing oxygen into the world for the first time through oxygenic photosynthesis (Sharma et al. 2011). It can efficiently assimilate CO_2 10–50 times more than vascular plants without competing with food crops (Prasad et al. 2021).

It interacts with other microbes, and these interactions play a crucial role in the stability and development of various ecosystems.

As an effective biological tool, algae showed exceptional wastewater bioremediation efficacy and nutrient recovery due to high adaptability (Mohsenpour et al. 2021). It primarily relies upon nitrogen, phosphorus, and carbon from wastewater and is given light and dark conditions to synthesize essential biochemical constituents in its biomass. This, in turn, makes the water relatively safe to be disposed of in water bodies (lakes, rivers, oceans, etc.) (Mohsenpour et al. 2021). Moreover, it showed exceptional ability to grow in saline, brackish, and seawater, producing valuable biomass. Extracted biomass can be transformed into useful products through fermentation, pyrolysis, and anaerobic digestion (Oliveira et al. 2021). In addition, it plays an essential role in food industries by producing many critical amino acids, fatty acids, and secondary metabolites.

7.2 Algae in Sustainable Resource Conservation and Management

Algae represent a promising alternative among many sustainable resources to fossil fuels and play a role in achieving important Sustainable Development Goals (SDGs). Utilizing algae biomass is crucial to promoting global sustainable development since it presents the potential for the sustainable growth of several industries. Algae are found in almost any water body that holds inorganic nutrients like carbon, nitrogen, and phosphorus and light (Oliveira et al. 2022). Algal biomass can be transformed into valuable products through different processes like fermentation, pyrolysis, and anaerobic digestion (Oliveira et al. 2021). Algae play an important role in the conservation of resources that are present on land, air, or water.

7.2.1 Algae for the Restoration of Degraded Lands/Soil

The primary goal of restoring any degraded land is to use various effective and sustainable techniques to reduce undesirable qualities like structural damage, chemical imbalance, and biological soil sickness and improve desirable traits like fertility and related properties. (1) The chemistry of the amender utilized and (2) the physical, chemical, and biological states of the land being treated are the two main determinants of the restoration efficacy of applied soil amenders (Singh et al. 2023). Algae represent the most convenient resource for converting carbon dioxide into prospective food ingredients suitable for animal as well as human use. Compared to most terrestrial plants, algae have a 10–20% higher photosynthetic efficiency (Singh et al. 2023). Because algae can absorb pollutants as nutrients and destroy them enzymatically, they are ideal for water cleanup (Mustafa et al. 2021). Algae are fast-growing aquatic organisms found naturally in a wide range of environments across the ocean, including freshwater and saltwater ecosystems (Goh et al. 2022; Santos et al. 2020). Microalgae that are grown in wastewater can be used directly as whole biomass or as lipid-extracted biomass and indirectly as biofertilizers for restoring degraded lands such as salt-affected lands, mine spoils, arid lands, and heavy metal-polluted lands (Table 7.1).

Restoration of salt-affected lands for basic properties may possibly be effective if degradation in its physical, chemical, and biological properties can be reversed to the level needed for highly productive agriculture, forestry, and agroforestry. Using algae as a biofertilizer increases nitrogen and phosphorus concentration by solubilizing inorganic phosphorus. It also helps to reduce pH while improving soil physical stability and cation exchange balance. For example, *C. vulgaris* survived at a high soil salinity level (1% NaCl) with 0.8 million cells per mL of soil suspension (Singh et al. 2023).

Severe environmental conditions such as drought, sandstorms, heatwaves, and floods result in the loss of ecological functions mainly due to their harmful effect on vegetation and soil fertility. Algal biomass rich in water, organic matter, and such organic compounds, principally exopolysaccharides, which bind soil particles and enhance soil's water-holding capacity, can play a crucial part in effectively restoring such lands (Alvarez et al. 2021). For instance, the applied mixture of algae species, *Microcoleus vaginatus* and *Scytonema javanicum*, improved water availability and its conservation through exopolysaccharides in sandy soils (Colica et al. 2014).

Fly ash (FA) bioremediation with microalgae has recently become a novel method for cleaning up toxins and pollutants in the air, soil, and water (Viegas et al. 2021). In addition, some proteins, polysaccharides, and cell surfaces of algae can be employed as chelating agents to effectively extract heavy metals (HM) from FA and other HM-polluted areas. According to laboratory research, applying algae (blue-green algae) can improve fertility and fly ash particle binding in rice fields by facilitating biomass decomposition (Pandey 2020).

Mine spoil exhibits unfavorable environmental conditions that must be corrected for proper ecosystem functioning and a lack of fertility, soil structure, and functions. Fertilizing soil to change its chemical properties and adding soil additives and plants to increase soil biological activity are some strategies to recover mine wastes. While microbial inoculations, selective planting, and soil amendments have been the primary methods used to restore the fertility of mine spoils, more information is still needed on the application of microalgae in the restoration of mined land. Furthermore, a restricted amount of research was conducted to treat acid mine drainage and remediate metalliferous mine waste using microalgae species (Manikandan et al. 2022). But algal species can play a competent role in nitrogen-fixing, nutrient cycling, overall accretion of soil fertility (de Siqueira Castro et al. 2020) and to reinstate the biogeochemical cycle. Fertilization of bioenergy plants and their growth on mined lands can potentially increase carbon stock above- and belowground carbon stock and help mining industries to offset carbon emissions (Singh et al. 2023).

Table 7.1 Comprehensive table on the use of algae and their effects on degraded agricultural soil

Species	Mode of application	Type of degraded soil	Effect on degraded soil	References
C. sorokiniana	Live cells	Salt-affected soils	15% increase in total soil carbon	Marks et al. (2019)
Coelastrella sp., *Aspergillus* sp., and *Penicillium* sp.	Consortium	Mining lands	Improved soil carbon, cation exchange capacity, soil moisture, sucrase activity, and microbial biomass carbon by 55.41%, 27.18%, 6.29%, 471%, and 216%, respectively; decreased pH from 9.09 to 8.95; and decreased bioavailable Cu content and bulk density by 15.43% and 6.30%, respectively	Wu et al. (2022)
Nostoc commune	Live cells	Arid and semi-arid lands	36% increase in biomass; increases soil organic carbon and exopolysaccharide content of 8.9 g kg^{-1} and 2.2 mg g^{-1}, respectively	Roncero-Ramos et al. (2019)
Anabaena doliolum	Live cells	Fly ash (FA) and heavy metal (HM) polluted lands	Improve soil nutrients and reduce heavy metal concentration	Singh and Das (2020)
Desmodesmus asymmetricus	NA	Mining lands	Decrease in pH, heavy metals, and increase in oxidation-reduction potential, alkali earth elements and sulfur concentration	Levett et al. (2023)
Azotobacter beijerinckii and *C. pyrenoidosa*	Consortium, either dried cells or live cells	Salt-affected soils	Increase in wheat plant biomass, improving salt tolerance, and soil fertility	Zhou et al. (2023)
Chlorococcum spp.	Live cells	Fly ash (FA) and heavy metal (HM) polluted lands	Biomass production and nutrient recovery from fly ash	Levett et al. (2023)
Nannochloropsis oculata	Live cells	Mining land	99.92% removal of copper from coal mine	Martínez-Macias et al. (2019)
C. pyrenoidosa	Live cells	Salt-affected soils	Increase in soil available N, P, and K contents and 30% increase in germination of *Chenopodium quinova* seeds	Ma et al. (2022)

7.2.2 *Conservation of Food Crops*

The main concerns brought about by the energy extraction from energy crops are deforestation, heightened food costs, and the use of land resources. Additionally, the ongoing subsidence of drinking water and the overuse of fertilizers and pesticides on agricultural land are causing environmental imbalances, such as higher concentrations of heavy metals in water bodies and food webs. Obtaining energy from crops like maize and corn causes an increase in the prices of these crops worldwide. Another concern related to the biofuel production from energy crops is the constant reduction of arable land (Katiyar et al. 2017).

To solve the problem of land shrinkage for non-agricultural purposes, algae-based biofuel production projects have been suggested as a partial solution because algae can be grown on wastelands even in harsh environmental conditions. Compared to other fuel-producing crops, algae require less than 2–13% of all agricultural areas to produce enough biofuel to meet all fuel demand derived from petroleum. This does not compete with land usage for food crop cultivation. Increased fertility of less or non-fertile land after algae cultivation for other energy crops is an additional advantage. Algae feedstock is considered an environmental regulator in terms of carbon-neutral energy production, it also regulates ecological balance, water balance and food web (Katiyar et al. 2017).

According to carbon balance, the energy sector is the primary source of CO_2 emission. As a result, cost-effective measures, for instance, emissions from the oil and gas industry and generation of electricity from alternative sources such as solar panels and wind farms, implemented in the energy sector can lower greenhouse gas emissions up to 100–200 million t year^{-1} (Hickem and Siew 2019). Algae, fulfilling all the criteria above, can be a promising feedstock for renewable third and fourth-generation biofuels. To cope with the world's increasing energy demand, phototrophic microorganisms can be used to make sustainable liquid and gaseous biofuels due to their vast metabolic capacity (Peng et al. 2020). These fuels offer multiple advantages over carbon neutrality, such as high biodegradability and non-toxicity and the capacity to co-produce additional food, cosmetic, agricultural, and medicinal bioproducts from the leftover biomass (Fig. 7.1) (Sadvakasova et al. 2023).

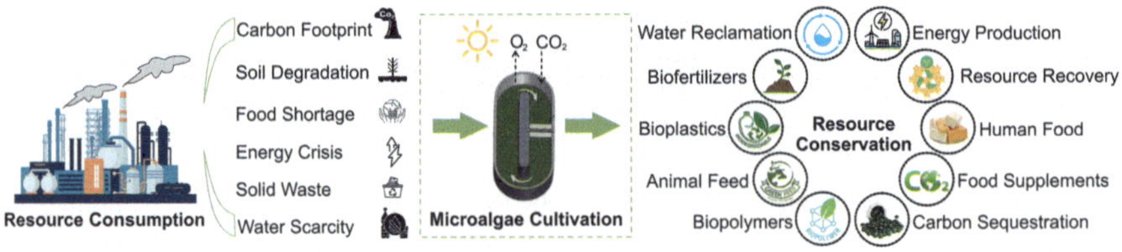

Fig. 7.1 Microalgae and cyanobacteria for conserving natural resources to produce valuable bioproducts

7.2.3 Algae as an Alternate Resource for Sustainable Energy Production

There are significant ecological and economic drawbacks to the current use of fossil fuels. The energy consumed worldwide is projected to rise by 57% between 2002 (435 trillion MJ) and 2025 (681 trillion MJ). Currently, the primary agricultural products used to manufacture biofuels are sugarcane, corn, sugar beetroot, rapeseed, soybean, and jatropha, which demand high-energy usage. The energy conversion efficiency for corn and switchgrass is roughly 0.25% and 1%, respectively. However, the algal biomass can absorb up to 10% of solar radiation, making the system more efficient (Sharma et al. 2011). Developing bioenergy and biofuels that "provide a net energy gain, having environmental benefits, being economically competitive, and being producible in large quantities without risking food security" has been a major challenge, even though these solutions have been proposed to address these societal concerns.

To address these issues, algae and cyanobacteria have recently been promoted as "third-generation feedstocks" for bioenergy and biofuels. Most prevalent and common algae among the algal species, including *Dunaliella* sp., *Neochloris* sp., *Chlorella* sp., *Chlamydomonas* sp., and *Botryococcus* sp., are of great interest because of their high oil content, which ranges from 20 to 70% dry weight (DW) (Taher et al. 2015). Algal biomass can also attain high productivity of organic compositions, which can be converted into biofuel or high-value chemicals. Based on different biomass transformation methods like pyrolysis, gasification, fermentation, and solvent extraction, microalgae can be converted into various sustainable biofuels such as biohydrogen, biogas, biodiesel, and bio-crude (He et al. 2023).

The following fatty acids are good options for producing biodiesel: palmitic acid, stearic acid, oleic acid, linoleic acid, and linolenic acid. Growing at 25 and 30 °C, *Chaetoceros* sp. FIKU035 demonstrated more than 80% of SFAs (short-chain fatty acids) and MUFAs (monounsaturated fatty acids), supporting the biodiesel's higher cetane number, oxidative stability, and thermal stability (Chaisutyakorn et al. 2018). Biodiesel is mainly produced from transesterification of microalgae. Glycerol and biodiesel are produced when lipids, particularly triacylglycerides (TAG), are transesterified in the presence of an alkali or an acid. An experimental study on *Chlorella* species revealed that they contain a very high amount of fatty acid content and can produce 34.53–230.38 mg^{-1} L^{-1} day^{-1} of biodiesel in Malaysia.

The synthetic algae bloom produced an environment that was comparable to that of a natural algae bloom, producing 1.3 kg of lipid and 8 kg of biomass with 10,000 L of tap water and a 2:1 urea to single super phosphate ratio (Chia et al. 2022). According to one study, there was an increase in the productivity of lipids by 113.7% because of the co-limitation of nitrogen and silicone on the algae-blooming species (Gao et al. 2019).

Various methods such as acidic, alkaline, enzymatic milling, microwave, thermal, and ultrasonic are used in combination to extract algal biomass. Cyanobacteria have a particular type of cell called a heterocyst, which provides the microaerobic environment needed to fix nitrogen. The main enzyme in filamentous heterocystous cyanobacteria that produces hydrogen is nitrogenase. The protein-rich cell of *Arthospira platensis* under microwave-assisted pre-treatment produced more carbohydrates to enzyme glucoamylase, leading to the catalyzation of α-1,4 and α-1,6 glycosidic bonds of glycogen to produce reduced sugars (Wang and Yin 2018).

Production of biohydrogen from microalgae has become a potential approach for the production of green energy (Singh and Das 2020). When *C. vulgaris* FSPE biomass was fermented in the dark with *C. butyricum* CGS5, 2.87 mmol H_2/g of biohydrogen from *Chlorella* sp. (~11.6 mL/gVS) was produced (Lunprom et al. 2019). Recent studies have shown that a *Tertaspora* sp. CU2551 produces 182 nmol/mg/h oh H_2 (Maswanna et al. 2020), *Spirulina platensis* produces 1.92 mmol/g of H_2 (Pandey et al. 2021), and *C. reinhardtii* produces 2.8–2.9 mL/h of H_2 (Mona et al. 2020).

7.3 Nutrient Recycling

The rapid increase in population has resulted in an intensification of urbanization and industrialization, leading to the deterioration of agricultural lands and overexploitation of freshwater resources. Along with the high demand for freshwater for domestic, agricultural, and industrial purposes, excess wastewater is being produced and discharged directly into the environment (Abdelfattah et al. 2023). High nutrient concentration in urban wastewater leads to eutrophication of local freshwater and marine environments, making it unfit for life. Different wastewater, depending upon their source, maybe industrial, domestic or agricultural, and has been treated by chemical, biological, physical, and advanced approaches for the efficient removal of primary and secondary pollutants, but every technique has its advantages and limitations to some extent (Liu and Hong 2021).

Algae-based resource recovery from wastewater is one of the most effective techniques to recycle the nutrients from wastewater and produce valuable biomass comprising ecologically important metabolites that can be valorized into significant bio-based products (Table 7.2) (Usman et al. 2023). Microalgae and cyanobacteria used for wastewater bioremediation demonstrate exceptional efficacy as they have high adaptability to their environment and primarily rely upon nitrogen, phosphorus, and carbon from wastewater for the synthesis of essential biochemical constituents in their biomass (Shahid et al. 2021).

The composition of wastewater varies from source to location to season, thus conferring variation in biomass productivity and affecting the biochemical pathways referring to a respective biochemical constituent in the biomass. So, a modified biological approach, microalgae-cyanobacteria co-culture, would not only increase the effectiveness of resource recycling from wastewater and reduce the demand for freshwater resources but aid in diminishing greenhouse gas emissions by carbon sequestration as the innate ability of microalgae/cyanobacteria to

Table 7.2 Product recovery from algae biomass by using different valorization methods

Algae species	Conversion process	Valorization method	Targeted product	Biorefinery approach	References
Chlamydomonas Mexicana	Chemical conversion	Transesterification	Biodiesel	High-throughput fermentation of biomass followed by transesterification	El-Dalatony et al. (2019)
Acaryochloris marina, *Oscillatoria* sp., and *Pleurocapsa* sp.	Chemical conversion	Transesterification	Biodiesel	Wastewater treatment Pigment-extracted biomass applied for biodiesel production	Shahid et al. (2021)
C. vulgaris	Thermal conversion	Gasification	Syngas	Gasification of residual biomass integrated to biodiesel production	Raheem et al. (2021)
C. vulgaris, *Spirulina* sp., and *Nannochloropsis* sp.	Thermal conversion	Pyrolysis	Biochar and bio-oil	Pyrolysis of biomass	Binda et al. (2020)
Plectonema terebrans	Biological conversion	Fermentation	α-Amylase	Wastewater-nutrient recycling Fungal fermentation of lipid-extracted biomass	Haider et al. (2022)
C. vulgaris and *S. obliquus*	Biological conversion	Photohydrogen production	Biohydrogen	Wastewater treatment Hydrogen production	Ruiz-Marin et al. (2020)
Chroococcidiopsis sp.	Thermal conversion	Hydrothermal liquefaction	Bio-crude oil	Pigment and carbohydrate-extracted biomass used for bio-crude and biochar production	Das et al. (2023)
S. obliquus	Thermal conversion	Pyrolysis	Bio-oil	Pyrolysis of biomass	Mustapha et al. (2023)
Pseudoscillatoria coralii	Chemical conversion	Transesterification	Biodiesel	Wastewater nutrients recycling	Zafar Khan et al. (2022)

assimilate carbon dioxide from its surrounding during photosynthesis, and along with enhanced biomass productivity, the cost of harvesting could be lowered, one of the significant challenges in algae cultivation (Debeni Devi et al. 2023).

Major biochemical constituents of algal biomass are carbohydrates, proteins, lipids, and pigments, assimilated according to the nutrition and environmental conditions provided. Carbohydrate content in algae is mainly comprised of starch, cellulose, and polysaccharides, which can be valorized to bioethanol, bioplastics, and biopolymers. At the same time, the protein content can be utilized to produce value-added chemicals that can be used in animal feed, dietary supplements, and in health sector due to its antioxidative, anti-inflammatory, and antibacterial properties (Wang et al. 2024). In addition, lipid content mainly comprises phospholipids, triacylglycerols, and glycolipids, which can be used to obtain biodiesel, a potential substitute for fossil fuels, oleochemicals, and biosurfactants. Other than these, photosynthetic and secondary pigments including chlorophyll a, carotenoids and phycobilins (only specific to cyanobacteria), are essential for nutraceutical and pharmaceutical applications (Trivedi et al. 2015; Chandra et al. 2019; Sarwer et al. 2022).

Algae-based biorefinery systems are efficient and eco-friendly solutions to several environmental problems, including atmospheric, water and soil pollution, fossil fuel depletion, and escalated industrialization, by optimization of upstream and downstream processes along with comprehensive utilization of important metabolites present in the biomass to be valorized for nutraceutical, animal feed, biofuel, health, and agricultural products. These biorefinery systems comprise methods for pre-treatment of biomass and valorization to valuable bioproducts; and are classified into biological (fermentation and anaerobic digestion), thermal (pyrolysis, direct combustion, gasification and liquefaction) and chemical (transesterification and enzyme-based catalysis) conversion mechanisms which can be integrated or sequentially processed alongside carbon sequestration, wastewater treatment, bioremediation in a zero-waste cascading biorefinery approach (Malik et al. 2022; Franco-Morgado et al. 2023; Kandasamy et al. 2022).

7.3.1 Recycling of Aquaculture Wastewater

Marine aquacultures require a high amount of freshwater and fractional renewal of the water added after a specific time, leading to high organic matter, nitrate, and phosphate content accumulation. This marine wastewater has high salinity and excessive nutrients and needs to be treated efficiently and in a sustainable manner (Nie et al. 2020). Algae-based technology is considered the most efficient and eco-friendly treatment method that helps recycle nutrient load from this wastewater and makes the withdrawn water reusable for marine aquaculture, thus reducing stress on freshwater reservoirs (Tom et al. 2021). The algal biomass obtained is suitable for agricultural purposes, such as improving soil quality or aiding in crop production by acting as a biofertilizer and biostimulant (Alvarez et al. 2021). The algae-bacteria consortium is also significant in the bioremediation of wastewater. The symbiotic relationship between the two leads to higher organic compounds metabolism and better nutrient recycling (Perales-Pérez et al. 2023).

7.3.2 Recycling of Dairy Wastewater

Despite its varying composition, dairy wastewater comprises high concentrations of nitrogen- and phosphorus-based compounds and organic compounds, resulting in high biological oxygen demand (BOD) and chemical oxygen demand (COD) values. The wastewater treatment plants exhibit utilization of expensive, non-sustainable, and environment-hazardous physical and chemical processes (Brar et al. 2019a). So, algae-based biological treatment systems could be used to efficiently recycle this wastewater by reducing the consumption of freshwater-based resources, treating the wastewater to be re-utilized for various industrial or agricultural purposes and producing value-added metabolites in algal biomass (Gramegna et al. 2020). Algae and bacteria, when co-cultivated in dairy wastewater, act profoundly in wastewater remediation and recycling the nutrients to produce lipids (potential candidates to produce biodiesel) and industrially significant enzymes (Biswas et al. 2021).

7.3.3 Recycling of Lipid-Rich Biomass into Bioproducts

Algae-based biofuel is the most common valorized product obtained by conversion of algal lipids via thermal or chemical processes. Biodiesel produced from lignin-free algal biomass is considered an auspicious substitute to primitive crops-based biodiesel due to the potential of algae to accumulate high lipid content in its biomass (Anto et al. 2020). On the other hand, lipid-rich algal biomass comprising of long-chain polyunsaturated fatty acids (PUFAs), monounsaturated fatty acids (MUFAs), phytosterols, omega-3, and omega-6 fatty acids can also be used in nutraceutical industries and cosmetic industries as a preferable alternative to plant oil- and animal fat-based additives (De Luca et al. 2021). The lipid-extracted biomass still consists of carbohydrates that can be fermented to produce industrially valuable enzymes or bioethanol, which can be used as a substitute for fossil fuel-based petroleum. Furthermore, pyrolysis of the residual biomass can yield biogas and biochar based on the biomass composition, which can be used as energy sources or in biofertilizers respectively. Besides carbohydrates, a significant portion of the biomass contains other valuable compounds such as pigments, proteins, and bioactive compounds. The residual biomass, thus, after extraction of lipids, can be used directly as biofertilizers or in the food feed industry as a food additive.

7.3.4 Recycling of Carbohydrate-Rich Biomass into Bioproducts

Carbohydrates are one of the significant primary metabolites produced in algae and consist of many ecologically essential polysaccharides and monosaccharides such as starch, cellulose, glucose, galactose, xylose, and arabinose (Ahmad et al. 2022b). Starch-like polysaccharides produced in algae make them significant for food\feed supplementary actions while the exopolysaccharides, profoundly β-glucan, produced by algal species show antibiotic properties making them essential in pharmaceutical industries (Ahmad et al. 2022a). Other than these applications, carbohydrate-rich algal biomass is considered a potential substrate to be valorized to syngas when treated with steam at high temperature and pressure under a controlled oxygen supply, which can be used as a fuel alternative directly or after liquefaction (Lam et al. 2019). Carbohydrate-abundant microalgal biomass can also be used as a potential feedstock to be anaerobically digested in the presence of a diverse group of microbes for biogas generation comprising of biomethane, biohydrogen, carbon dioxide and, in some instances, hydrogen sulfide (Almomani et al. 2023). Pigment-extracted biomass can also be bioprocessed to get bioethanol through an integrative approach. Residual biomass can be pyrolyzed to obtain bio-oil and biochar.

7.4 Algae-Based Resource Preservation

7.4.1 Preservation of Freshwater Bodies

Numerous industrial activities depend on natural resources as a constant and abundant source, creating a linear economy characterized by a mindset of extraction, utilization, and disposal (Fuentes-Grünewald et al. 2021). To address water scarcity, it is crucial to significantly increase water-use efficiency across all sectors and ensure the development of a sustainable approach to the use of sea and wastewater. Utilizing different water sources such as urban wastewater, brackish, and seawater for energy production and resource recovery can play a crucial role in circular sustainability. Soon, the wastewater treatment plants will be an "environmentally sustainable" center for water recycling and reuse (Flores 2018).

Algae are suitable alternatives that can be grown using non-freshwater resources (brackish, wastewater, and seawater) unsuitable for standard crops. This will help reduce the cost of added nutrients, water pumping, and freshwater footprints (Table 7.3). A study reported that the life cycle assessment of biodiesel production via algae could lessen the demand for freshwater and nutrient usage by 84% and 55%. It is reported that in the semi-arid regions of Brazil, 70% of wells contain brackish or saline water, and groundwater is the sole source available to communities. The cultivation of *C. vulgaris* and *Ankistrodesmus fusiformis* in BG11 with

Table 7.3 Resource recovery by various kinds of algae from different types of wastewaters

Wastewater source	Algae species	Resource recovery potential (%)								References
		NH_4^+	PO_4^{3-}	Nitrates	COD	BOD	TDS	TN	TP	
Urban wastewater	*Scenedesmus* spp.	95	–	–	69	–	–	91	81	Álvarez-González et al. (2022)
Urban wastewater	*Chlorella* spp.	93	–	–	70	–	86	–	77	Li et al. (2021)
Urban wastewater	*Chlorella* spp. and *Scenedesmus* spp.	98	95	96	83	–	–	94	–	Silambarasan et al. (2021)
Rice industry wastewater	*C. pyrenoidosa*	69.4	64.7	–	–	–	–	–	–	Umamaheswari and Shanthakumar (2021)
Agricultural wastewater	*Tetradesmus* spp., *Chlorella* spp., and *Nostoc* spp.	79–92	90–98	–	62–79	–	–	–	–	Ferreira et al. (2021)
Municipal wastewater	*C. sorokiniana*	–	–	53–96	17–47	60–80	–	12–100	59–92	Nagarajan et al. (2020)
Urban wastewater	*Scytonema ambikapurensis*	96	70	98	–	–	68	–	–	Jaiswal et al. (2023)
Dairy industry wastewater	*C. pyrenoidosa*	99	97	–	85	88	–	–	–	Yadavalli and Heggers (2013)
Agricultural wastewater	*Chlamydomonas reinhardtii*	70–90	–	–	20–42	–	–	32.3–52.4	78–93	Qi et al. (2017)
Poultry slaughterhouse wastewater	*Phormidium* sp.	–	–	–	97.6	–	–	85.5	–	Dos Santos et al. (2017)
Textile industry wastewater	*Scenedesmus abundans* NCIM 2897	–	70.79	68.8	86.8	13.9	–	–	–	Brar et al. (2019b)

NH_4^+: Ammonium ion, PO_4^{3-}: Phosphate ion, COD: Chemical Oxygen Demand, BOD: Biochemical Oxygen Demand, TDS: Total Dissolved Solids, TN: Total Nitrogen, TP: Total Phosphorus dilutions of concentrated saline water showed both strains performed well under the condition of 75% saline water with the production of 0.96 g L^{-1} and 0.98 g L^{-1}, respectively (Deamici et al. 2024). Another study reported the bioremediation and recycling of brackish water collected from fish water by algae cultivation. At first, ten strains were grown, but based on preliminary data, *Haematococcus* sp., *Monoraphidium* sp., *Neochloris* sp., and *Nostoc* sp. were selected for outdoor cultivation. Although *Nostoc* could not grow in wastewater, the other three performed well, assimilating 70.5% of TN and 93.5% of TP. Similarly, the biomass yield of these strains was between 0.15 and 0.52 g L^{-1}. This study provided the approach for sustainable bioremediation and recycling of brackish water, which produces valuable biomass that could be used as fish feed. The potential outcome of utilizing the brackish water for algae cultivation will prevent environmental-related problems of wastewater discharge, enhance food security, and provide a transition toward a circular economy (Kashem et al. 2023).

7.4.2 Cultivation on Non-arable Land to Preserve the Agricultural Land

An increase in the population demands a plentiful supply of food, energy production, water, and other resources, consequently driving an increase in industrialization. Currently, the food crop is the primary means of meeting the demand for food and energy. The extraction of energy from food crops raises a lot of concerns, including food prices, the depletion of forests and the exertion of land resources (Duncan 2011). Furthermore, first- and second-generation biofuel energy production competes with agricultural land. These approaches will likely impair food security and the degradation of ecosystems, particularly in developing countries. Besides, the ongoing depletion of drinking water and the excessive application of pesticides and fertilizers on agricultural land are contributing to the environmental imbalance (Searchinger et al. 2015). To avoid the direct and indirect use of agricultural land, there is a need to find sustainable alternatives for biofuel production to achieve future targets in energy demands. The center of the Agriculture Board in India, under its 2007 policy, proposed the utilization of wasteland (non-arable) to save cultivation land for agricultural purposes. Algae are the suitable alternatives because of their non-dependence on freshwater and fertile land, higher biofuel production per unit area, reduced land footprints, and can occupy areas with low biodiversity (Correa et al. 2019).

The development of an algal biorefinery system reduces production costs by producing high-value co-products such as food and animal feed along with biofuel (Ruiz et al. 2016). These technologies reduce the energy input and enhance nutrient recycling, making the system more cost-effective (Ruiz et al. 2016). A study was performed to identify the cost-effective area to meet the 30% energy demand of each country in 2016–2050 by estimating the lipid productivity and resource availability. Interestingly, less than 1.1% of the global land area will be enough to achieve this target, mainly drier low-latitude areas or lowlands in each country. The most promising countries for microalgal biofuel production are primarily located in the Middle East, the Caribbean, North and East Africa, Oceania, and western South America (Correa et al. 2021).

Another research reported the biofuel production in four Neotropical countries (Colombia, Ecuador, Panama, and Venezuela) to meet the 30% energy demand. The results were compared with those of the best area for growing oil palm and sugarcane crops to fulfill the same energy target. Results suggested that the microalgae biofuel production within the Neotropics meets the energy demand by reducing the land area requirement by 52% compared to oil palm and sugarcane (Correa et al. 2020). Similarly, the large-scale cultivation of *Chlorella sorokiniana* in flat, open, and shallow modes was performed to evaluate biomass production and critical parameters. Results suggested that in fed-batch mode, the microalgae density reached 6.40 g L^{-1}, and the highest volumetric and areal productivity were 0.37 g L^{-1} day^{-1} and 9.08 g m^{-2} day^{-1} in batch mode. Interestingly, the yield remains stable till the eleventh time harvested from the continuous batch. Based on data, this study upgraded the system to 4000 L outdoor cultivation in summer autumn and spring with productivities of 7.60, 6.80, and 7.20 g m^{-2} day^{-1}, respectively. Techno-economic analysis showed that this system is more sustainable and has great potential for deployment on non-arable land for massive algal production (Wang et al. 2022).

7.4.3 Preservation of Resources to Protect Ecosystem

Due to the increasing population, the world faces two closely interconnected environmental crises: (1) climate change and (2) biodiversity loss. Different human activities, such as pollution, habitat destruction, excessive use of resources, and climate change, pose significant threats to ecosystems. Various factors are involved in ecosystem degradation, such as water pollution, soil erosion, loss of biodiversity, and ecological imbalances (Folke et al. 2004). Algae interact with other microbes, and these interactions play a crucial role in the stability and development of various ecosystems microorganisms (Hom et al. 2015). Algal systems provide a habitat for aquatic microorganisms; they support their growth and survival, which can affect the aquatic ecosystem. Studies showed that algal systems or communities significantly impact the food web and ecosystem. When algae consortium works together, it forms colonies, biofilms, and mat-like structures, which provide sites for attachment or shelter for various organisms. These structures are helpful to enhance the diversity of the ecosystem and develop microhabitats. These microhabitats support the reproduction and survival of various organisms such as fish, microbes, invertebrates, and zooplankton (Nava and Leoni 2021).

Algae were found to reduce the concentration of inorganic compounds, heavy metals, and emerging contaminants from wastewater, making them a suitable candidate to protect the ecosystem. Algal systems use three significant pathways to bioremediate the pollutants: biosorption, biodegradation, and bioadsorption. The main advantage of using algal systems for bioremediation is the ability to recover the resources for reuse opting biorefinery approach (Fig. 7.2) to transform algal biomass into value-added products such as animal feed, fertilizer, biofuel, and bio-plastics and bioplastics (Sutherland and Ralph 2019). Algae can provide a food source for fish, while bacteria and fungi can help maintain the waste and water quality. Algal–fungal interactions can be used to produce novel food products with high nutritional and functional properties.

7.4.4 Carbon Sequestration to Mitigate Climate Change

Applying algal systems in carbon fixation and wastewater treatment has been considered a promising technique. The atmospheric carbon is increasing day by day due to human activities. So, there is a need to fix more and more carbon through biological systems. Algal systems fix the inorganic carbon from the atmosphere into organic compounds, which are used to store chemical energy. Algal culture is a viable technique for CO_2 capture from the environment due to its photosynthetic potential (Kong et al. 2021). Carbon fixation efficiency depends on various factors like suitable strain selection and cultivation conditions, including pH, temperature, and concentration of CO_2 supplied. Optimization of these factors helps to maximize the rate of carbon fixation and enhance the overall activity of algal systems (Zhou et al. 2017).

All cyanobacteria and microalgae utilize the Calvin cycle and ribulose-1,5-bisphosphate carboxylase/oxygenase (Rubisco) as the primary mechanisms for inorganic carbon assimilation. Mechanisms found in cyanobacteria and algae include the evolution of Rubiscos with kinetics more favorable to carboxylase activity and/or mechanisms that raise the CO_2 concentration at the Rubisco active site.

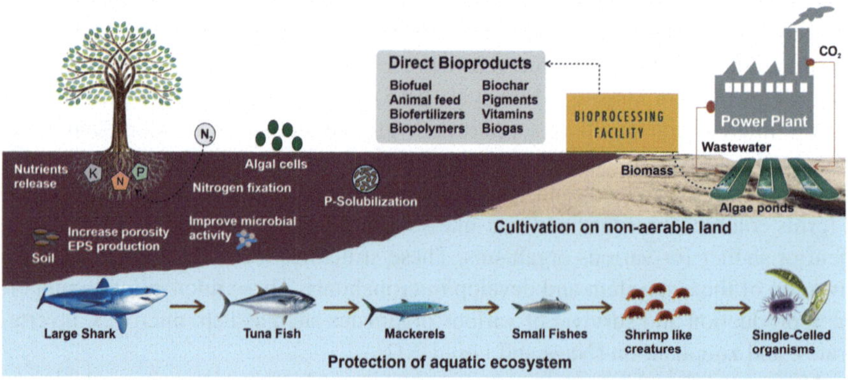

Fig. 7.2 Role of algae to protect the ecosystem, and preservation of natural resources

When CO_2 is accumulated within carboxysomes that contain all the cellular Rubisco and carbonic anhydrase, it occurs in cyanobacteria via active transport of HCO_3^- at the plasmalemma and/or downhill CO_2 entry with energized conversion of CO_2 to HCO_3^- at the thylakoid membrane. Active HCO_3^--transport in eukaryotic microalgae occurs at the chloroplast envelope, the plasmalemma, or both. In algae that have them, CO_2 is finally concentrated in pyrenoids that contain rubisco (Beardall and Raven 2020). The stroma serves as the primary carboxylation site in high-CO_2 cells, whereas Rubisco is primarily found in pyrrole compounds. In low carbon dioxide environments, this might be important for carbon dioxide fixation (Xu et al. 2019).

Nevertheless, developing novel strains of these algae and bacteria that can withstand high temperatures and CO_2 concentrations is crucial to advancing algae-based carbon capture. Upon reaching 40 °C, Leptolyngbya sp. and *Picochlorum* sp. exhibited robust growth, achieving respective productivities of 106.6 ± 10.0 and 87.5 ± 2.1 mg biomass L^{-1} day^{-1}. *Tetraselmis* sp. isolate had the highest carbon capture rate (278.8 ± 23.9 mg CO_2 L^{-1} day^{-1}) and biomass productivity (157.7 ± 10.3 mg biomass L^{-1} day^{-1}) at 30 °C. Both *Tetraselmis* and *Picochlorum* isolates showed a more than twofold increase in biomass productivity under 20% CO_2, reaching 333.8 ± 41.1 and 244.7 ± 29.5 mg biomass L^{-1} day^{-1} (Schipper et al. 2019).

7.5 Conclusion and Prospects

The concept of sustainable conservation and recycling of natural resources is a key factor in achieving environmental sustainability and combating the impact of climate change. In other words, natural resources should be consumed in a balanced way to meet the needs of the present with the needs of the future. With increasing population and global warming, progress should be made in conserving and recycling resources to protect the ecosystem, conserve energy, and reduce emissions. Microalgae and cyanobacteria are the organisms that can significantly contribute to the conservation, recycling, and preservation of resources. Extracted biomass can produce sustainable energy to protect food crops and agricultural land, improve soil fertility, and be used as food and feed. Microalgae showed the potential to grow in wastewater, generating various forms of bioenergy, justifying their economic and ecological benefits. Algae facilitates wastewater treatment plants for resource recovery and energy generation, achieving circular sustainability. Algae seem to be a viable, sustainable, and economical alternative approach that can contribute multifunctionally. This multipurpose application of microalgae to produce biofuel and other bioproducts supports the socioeconomic welfare of the population. Microalgae-based products have the potential to emerge as feedstock for several industries, such as the automotive, pharmaceuticals, chemicals, and food industries. So far, microalgae are expected to be the best candidate for conservation and recycling of energy, water, land and other resources.

References

Abdelfattah A, Ali SS, Ramadan H et al (2023) Microalgae-based wastewater treatment: mechanisms, challenges, recent advances, and future prospects. Environ Sci Ecotechnol 13:100205. https://doi.org/10.1016/j.ese.2022.100205

Ahmad A, Banat F, Alsafar H et al (2022a) Algae biotechnology for industrial wastewater treatment, bioenergy production, and high-value bioproducts. Sci Total Environ 806:150585. https://doi.org/10.1016/j.scitotenv.2021.150585

Ahmad A, Hassan S, Banat F (2022b) An overview of microalgae biomass as a sustainable aquaculture feed ingredient: food security and circular economy. Bioengineered 13:9521–9547. https://doi.org/10.1080/21655979.2022.2061148

Almomani F, Hosseinzadeh-Bandbafha H, Aghbashlo M et al (2023) Comprehensive insights into conversion of microalgae to feed, food, and biofuels: current status and key challenges towards implementation of sustainable biorefineries. Chem Eng J 455:140588. https://doi.org/10.1016/j.cej.2022.140588

Alvarez AL, Weyers SL, Goemann HM et al (2021) Microalgae, soil and plants: a critical review of microalgae as renewable resources for agriculture. Algal Res 54:102200. https://doi.org/10.1016/j.algal.2021.102200

Álvarez-González A, Uggetti E, Serrano L et al (2022) Can microalgae grown in wastewater reduce the use of inorganic fertilizers? J Environ Manag 323:116224. https://doi.org/10.1016/j.jenvman.2022.116224

Anto S, Mukherjee SS, Muthappa R et al (2020) Algae as green energy reserve: technological outlook on biofuel production. Chemosphere 242:125079. https://doi.org/10.1016/j.chemosphere.2019.125079

Beardall J, Raven JA (2020) Acquisition of inorganic carbon by microalgae and cyanobacteria. Microbial Photosynthesis 151–168. https://doi.org/10.1007/978-981-15-3110-1_8

Binda G, Spanu D, Bettinetti R et al (2020) Comprehensive comparison of microalgae-derived biochar from different feedstocks: a prospective study for future environmental applications. Algal Res 52:102103. https://doi.org/10.1016/j.algal.2020.102103

Biswas T, Bhushan S, Prajapati SK et al (2021) An eco-friendly strategy for dairy wastewater remediation with high lipid microalgae-bacterial biomass production. J Environ Manag 286:112196. https://doi.org/10.1016/j.jenvman.2021.112196

Brar A, Kumar M, Pareek N (2019a) Comparative appraisal of biomass production, remediation, and bioenergy generation potential of microalgae in dairy wastewater. Front Microbiol 10. https://doi.org/10.3389/fmicb.2019.00678

Brar A, Kumar M, Vivekanand V et al (2019b) Phycoremediation of textile effluent-contaminated water bodies employing microalgae: nutrient sequestration and biomass production studies. Int J Environ Sci Technol 16:7757–7768. https://doi.org/10.1007/s13762-018-2133-9

Cf O (2015) Transforming our world: the 2030 agenda for sustainable development. United Nations, New York. https://doi.org/10.1057/s41271-016-0002-7

Chaisutyakorn P, Praiboon J, Kaewsuralikhit C (2018) The effect of temperature on growth and lipid and fatty acid composition on marine microalgae used for biodiesel production. J Appl Phycol 30:37–45. https://doi.org/10.1007/s10811-017-1186-3

Chandra R, Iqbal HMN, Vishal G et al (2019) Algal biorefinery: a sustainable approach to valorize algal-based biomass towards multiple product recovery. Bioresour Technol 278:346–359. https://doi.org/10.1016/j.biortech.2019.01.104

Chia SR, Nomanbhay SBHM, Chew KW et al (2022) Algae as potential feedstock for various bioenergy production. Chemosphere 287:131944. https://doi.org/10.1016/j.chemosphere.2021.131944

Colica G, Li H, Rossi F et al (2014) Microbial secreted exopolysaccharides affect the hydrological behavior of induced biological soil crusts in desert sandy soils. Soil Biol Biochem 68:62–70. https://doi.org/10.1016/j.soilbio.2013.09.017

Correa DF, Beyer HL, Possingham HP ct al (2019) Global mapping of cost-effective microalgal biofuel production areas with minimal environmental impact. GCB Bioenergy 11:914–929. https://doi.org/10.1111/gcbb.12619

Correa DF, Beyer HL, Possingham HP et al (2020) Freeing land from biofuel production through microalgal cultivation in the Neotropical region. Environ Res Lett 15:094094. https://doi.org/10.1088/1748-9326/ab8d7f

Correa DF, Beyer HL, Possingham HP et al (2021) Microalgal biofuel production at national scales: reducing conflicts with agricultural lands and biodiversity within countries. Energy 215:119033. https://doi.org/10.1016/j.energy.2020.119033

Cosgrove WJ, Loucks DP (2015) Water management: current and future challenges and research directions. Water Resour Res 51:4823–4839. https://doi.org/10.1002/2014WR016869

Da Silva TL, Gouveia L, Reis A (2014) Integrated microbial processes for biofuels and high value-added products: the way to improve the cost effectiveness of biofuel production. Appl Microbiol Biotechnol 98:1043–1053. https://doi.org/10.1007/s00253-013-5389-5

Das P, Khan S, Abdulquadir M et al (2023) Biocrude oil production from a self-settling marine cyanobacterium, *Chroococcidiopsis* sp., using a biorefinery approach. Renew Energy 203:1–9. https://doi.org/10.1016/j.renene.2022.12.044

De Luca M, Pappalardo I, Limongi AR et al (2021) Lipids from microalgae for cosmetic applications. Cosmetics 8:52. https://doi.org/10.3390/cosmetics8020052

De Siqueira CJ, Calijuri ML, Ferreira J et al (2020) Microalgae based biofertilizer: a life cycle approach. Sci Total Environ 724:138138. https://doi.org/10.1016/j.scitotenv.2020.138138

Deamici KM, Pessôa LC, Mata SN et al (2024) Enhancing microalgal cultures in desalinized wastewater from semiarid regions: an assessment of growth dynamics and biomass accumulation. J Appl Phycol 1–8. https://doi.org/10.1007/s10811-023-03175-w

Debeni Devi N, Sun X, Hu B et al (2023) Bioremediation of domestic wastewater with microalgae-cyanobacteria co-culture by nutritional balance approach and its feasibility for biodiesel and animal feed production. Chem Eng J 454:140197.https://doi.org/10.1016/j.cej.2022.140197

Dos Santos AM, Roso GR, De Menezes CR et al (2017) The bioeconomy of microalgal heterotrophic bioreactors applied to agroindustrial wastewater treatment. Desalin Water Treat 64:12–20. https://doi.org/10.5004/dwt.2017.20279

Duncan G-R (2011) Beyond food versus fuel. Nature 474:6–8. https://doi.org/10.1038/474S06a

El-Dalatony MM, Salama E-S, Kurade MB et al (2019) Whole conversion of microalgal biomass into biofuels through successive high-throughput fermentation. Chem Eng J 360:797–805. https://doi.org/10.1016/j.cej.2018.12.042

Ferreira A, Melkonyan L, Carapinha S et al (2021) Biostimulant and biopesticide potential of microalgae growing in piggery wastewater. Environ Adv 4:100062. https://doi.org/10.1016/j.envadv.2021.100062

Flores CC (2018) Towards circular economy–a wastewater treatment perspective, the Presa Guadalupe case Cesar Casiano Flores, Hans Bressers, Carina Gutierrez, Cheryl de Boer. https://doi.org/10.1108/MRR-02-2018-0056

Folke C, Carpenter S, Walker B et al (2004) Regime shifts, resilience, and biodiversity in ecosystem management. Annu Rev Ecol Evol Syst 35:557–581. https://doi.org/10.1146/annurev.ecolsys.35.021103.105711

Franco-Morgado M, Amador-Espejo GG, Pérez-Cortés M et al (2023) Microalgae and cyanobacteria polysaccharides: important link for nutrient recycling and revalorization of agro-industrial wastewater. Appl Food Res 3:100296. https://doi.org/10.1016/j.afres.2023.100296

Freedman B (2018) Resources and sustainable development. Environ Sci Can Perspect:1–37

Fuentes-Grünewald C, Ignacio Gayo-Peláez J, Ndovela V et al (2021) Towards a circular economy: a novel microalgal two-step growth approach to treat excess nutrients from digestate and to produce biomass for animal feed. Bioresour Technol 320:124349. https://doi.org/10.1016/j.biortech.2020.124349

Gao G, Wu M, Fu Q et al (2019) A two-stage model with nitrogen and silicon limitation enhances lipid productivity and biodiesel features of the marine bloom-forming diatom Skeletonema costatum. Bioresour Technol 289:121717. https://doi.org/10.1016/j.biortech.2019.121717

Goh PS, Lau WJ, Ismail AF et al (2022) Microalgae-enabled wastewater treatment: a sustainable strategy for bioremediation of pesticides. Water 15:70. https://doi.org/10.3390/w15010070

Gramegna G, Scortica A, Scafati V et al (2020) Exploring the potential of microalgae in the recycling of dairy wastes. Bioresour Technol Rep 12:100604. https://doi.org/10.1016/j.biteb.2020.100604

Haider MN, Liu C-G, Tabish TA et al (2022) Resource recovery of the wastewater-derived nutrients into algal biomass followed by its cascading processing to multiple products in a circular bioeconomy paradigm. Fermentation 8:650. https://www.mdpi.com/2311-5637/8/11/650

He S, Barati B, Hu X et al (2023) Carbon migration of microalgae from cultivation towards biofuel production by hydrothermal technology: a review. Fuel Process Technol 240:107563. https://doi.org/10.1016/j.fuproc.2022.107563

Hickem T, Siew S (2019) Carbon capture utilisation and storage. Carbon Capture Utilisation and Storage for Singapore

Hom EFY, Aiyar P, Schaeme D et al (2015) A chemical perspective on microalgal–microbial interactions. Trends Plant Sci 20:689–693. https://doi.org/10.1016/j.tplants.2015.09.004

Jaiswal TP, Chakraborty S, Sharma S et al (2023) Prospects of a hot spring–originated novel cyanobacterium, Scytonema ambikapurensis, for wastewater treatment and exopolysaccharide-enriched biomass production. Environ Sci Pollut Res 30:53424–53444. https://doi.org/10.1007/s11356-023-26032-2

Kandasamy S, Zhang B, He Z et al (2022) Microalgae as a multipotential role in commercial applications: current scenario and future perspectives. Fuel 308:122053. https://doi.org/10.1016/j.fuel.2021.122053

Kashem AHM, Das P, Abdulquadir M et al (2023) Microalgal bioremediation of brackish aquaculture wastewater. Sci Total Environ 873:162384. https://doi.org/10.1016/j.scitotenv.2023.162384

Katiyar R, Gurjar BR, Biswas S et al (2017) Microalgae: an emerging source of energy based bio-products and a solution for environmental issues. Renew Sust Energ Rev 72:1083–1093. https://doi.org/10.1016/j.rser.2016.10.028

Kong W, Shen B, Lyu H et al (2021) Review on carbon dioxide fixation coupled with nutrients removal from wastewater by microalgae. J Clean Prod 292:125975. https://doi.org/10.1016/j.jclepro.2021.125975

Lam MK, Loy ACM, Yusup S et al (2019) Biohydrogen production from algae. Biohydrogen. Elsevier

Levett A, Gagen EJ, Levett I et al (2023) Integrating microalgae production into mine closure plans. J Environ Manag 337:117736. https://doi.org/10.1016/j.jenvman.2023.117736

Li J, Lens PNL, Ferrer I et al (2021) Evaluation of selenium-enriched microalgae produced on domestic wastewater as biostimulant and biofertilizer for growth of selenium-enriched crops. J Appl Phycol 33:3027–3039. https://doi.org/10.1007/s10811-021-02523-y

Liu X-Y, Hong Y (2021) Microalgae-based wastewater treatment and recovery with biomass and value-added products: a brief review. Curr Pollut Rep 7:227–245. https://doi.org/10.1007/s40726-021-00184-6

Lunprom S, Phanduang O, Salakkam A et al (2019) A sequential process of anaerobic solid-state fermentation followed by dark fermentation for bio-hydrogen production from Chlorella sp. Int J Hydrog Energy 44:3306–3316. https://doi.org/10.1016/j.ijhydene.2018.06.012

Ma C, Cui H, Ren C et al (2022) The seed primer and biofertilizer performances of living Chlorella pyrenoidosa on Chenopodium quinoa under saline-alkali condition. J Appl Phycol 34:1621–1634. https://doi.org/10.1007/s10811-022-02699-x

Maddocks A, Otto B, Luo T (2019) The future of fresh water

Malik S, Shahid A, Haider MN et al (2022) Prospects of multiproduct algal biorefineries involving cascading processing of the biomass employing a zero-waste approach. Curr Pollut Rep 8:147–158. https://doi.org/10.1007/s40726-022-00213-y

Manikandan A, Suresh Babu P, Shyamalagowri S et al (2022) Emerging role of microalgae in heavy metal bioremediation. J Basic Microbiol 62:330–347. https://doi.org/10.1002/jobm.202100363

Marks EA, Montero O, Rad C (2019) The biostimulating effects of viable microalgal cells applied to a calcareous soil: increases in bacterial biomass, phosphorus scavenging, and precipitation of carbonates. Sci Total Environ 692:784–790. https://doi.org/10.1016/j.scitotenv.2019.07.289

Martínez-Macias MDR, Correa-Murrieta MA, Villegas-Peralta Y et al (2019) Uptake of copper from acid mine drainage by the microalgae Nannochloropsis oculata. Environ Sci Pollut Res 26:6311–6318. https://doi.org/10.1007/s11356-018-3963-1

Maswanna T, Lindblad P, Maneeruttanarungroj C (2020) Improved biohydrogen production by immobilized cells of the green alga Tetraspora sp. CU2551 incubated under aerobic condition. J Appl Phycol 32:2937–2945. https://doi.org/10.1007/s10811-020-02184-3

Mohsenpour SF, Hennige S, Willoughby N et al (2021) Integrating micro-algae into wastewater treatment: a review. Sci Total Environ 752:142168. https://doi.org/10.1016/j.scitotenv.2020.142168

Mona S, Kumar SS, Kumar V et al (2020) Green technology for sustainable biohydrogen production (waste to energy): a review. Sci Total Environ 728:138481. https://doi.org/10.1016/j.scitotenv.2020.138481

Mustafa S, Bhatti HN, Maqbool M et al (2021) Microalgae biosorption, bioaccumulation and biodegradation efficiency for the remediation of wastewater and carbon dioxide mitigation: prospects, challenges and opportunities. J Water Process Eng 41:102009. https://doi.org/10.1016/j.jwpe.2021.102009

Mustapha SI, Mohammed UA, Rawat I et al (2023) Production of high-quality pyrolytic bio-oils from nutrient-stressed *Scenedesmus obliquus* microalgae. Fuel 332:126299. https://doi.org/10.1016/j.fuel.2022.126299

Nagarajan D, Lee D-J, Chen C-Y et al (2020) Resource recovery from wastewaters using microalgae-based approaches: a circular bioeconomy perspective. Bioresour Technol 302:122817. https://doi.org/10.1016/j.biortech.2020.122817

Nava V, Leoni B (2021) A critical review of interactions between microplastics, microalgae and aquatic ecosystem function. Water Res 188:116476. https://doi.org/10.1016/j.watres.2020.116476

Nie X, Mubashar M, Zhang S et al (2020) Current progress, challenges and perspectives in microalgae-based nutrient removal for aquaculture waste: a comprehensive review. J Clean Prod 277:124209. https://doi.org/10.1016/j.jclepro.2020.124209

Oliveira CYB, D'alessandro EB, Antoniosi Filho NR et al (2021) Synergistic effect of growth conditions and organic carbon sources for improving biomass production and biodiesel quality by the microalga Choricystis minor var. minor. Sci Total Environ 759:143476. https://doi.org/10.1016/j.scitotenv.2020.143476

Oliveira CYB, Jacob A, Nader C et al (2022) An overview on microalgae as renewable resources for meeting sustainable development goals. J Environ Manag 320:115897. https://doi.org/10.1016/j.jenvman.2022.115897

Pandey VC (2020) Phytomanagement of fly ash. Elsevier

Pandey A, Sinha P, Pandey A (2021) Hydrogen production by sequential dark and photofermentation using wet biomass hydrolysate of Spirulina platensis: response surface methodological approach. Int J Hydrog Energy 46:7137–7146. https://doi.org/10.1016/j.ijhydene.2020.11.205

Peng S, Chen Q, Liu E (2020) The role of computational fluid dynamics tools on investigation of pathogen transmission: prevention and control. Sci Total Environ 746:142090. https://doi.org/10.1016/j.scitotenv.2020.142090

Perales-Pérez Á, Macías-Sánchez MD, Ruiz J et al (2023) Process for nutrient recycling from intensive aquaculture through microalgae-bacteria consortium. Sci Total Environ 901:165848. https://doi.org/10.1016/j.scitotenv.2023.165848

Prasad R, Gupta SK, Shabnam N et al (2021) Role of microalgae in global CO2 sequestration: physiological mechanism, recent development, challenges, and future prospective. Sustain For 13:13061. https://doi.org/10.3390/su132313061

Qi F, Xu Y, Yu Y et al (2017) Enhancing growth of *Chlamydomonas reinhardtii* and nutrient removal in diluted primary piggery wastewater by elevated CO_2 supply. Water Sci Technol 75:2281–2290. https://doi.org/10.2166/wst.2017.111

Raheem A, Abbasi SA, Mangi FH et al (2021) Gasification of algal residue for synthesis gas production. Algal Res 58:102411. https://doi.org/10.1016/j.algal.2021.102411

Renzaho AM, Kamara JK, Toole M (2017) Biofuel production and its impact on food security in low and middle income countries: implications for the post-2015 sustainable development goals. Renew Sust Energ Rev 78:503–516. https://doi.org/10.1016/j.rser.2017.04.072

Roncero-Ramos B, Román J, Gómez-Serrano C et al (2019) Production of a biocrust-cyanobacteria strain (Nostoc commune) for large-scale restoration of dryland soils. J Appl Phycol 31:2217–2230. https://doi.org/10.1007/s10811-019-1749-6

Ruiz J, Olivieri G, De Vree J et al (2016) Towards industrial products from microalgae. Energy Environ Sci 9:3036–3043. https://doi.org/10.1039/C6EE01493C

Ruiz-Marin A, Canedo-López Y, Chávez-Fuentes P (2020) Biohydrogen production by *Chlorella vulgaris* and *Scenedesmus obliquus* immobilized cultivated in artificial wastewater under different light quality. AMB Express 10:191. https://doi.org/10.1186/s13568-020-01129-w

Sadvakasova AK, Kossalbayev BD, Bauenova MO et al (2023) Microalgae as a key tool in achieving carbon neutrality for bioproduct production. Algal Res 72:103096. https://doi.org/10.1016/j.algal.2023.103096

Santos KBD, Almeida VOD, Weiler J et al (2020) Removal of pollutants from an AMD from a coal mine by neutralization/precipitation followed by "in vivo" biosorption step with the microalgae Scenedesmus sp. Fortschr Mineral 10:711. https://doi.org/10.3390/min10080711

Sarwer A, Hamed SM, Osman AI et al (2022) Algal biomass valorization for biofuel production and carbon sequestration: a review. Environ Chem Lett 20:2797–2851. https://doi.org/10.1007/s10311-022-01458-1

Schipper K, Al Muraikhi M, Alghasal GSH et al (2019) Potential of novel desert microalgae and cyanobacteria for commercial applications and CO 2 sequestration. J Appl Phycol 31:2231–2243. https://doi.org/10.1007/s10811-019-01763-3

Searchinger TD, Estes L, Thornton PK et al (2015) High carbon and biodiversity costs from converting Africa's wet savannahs to cropland. Nat Clim Chang 5:481–486. https://doi.org/10.1038/nclimate2584

Shahid A, Usman M, Atta Z et al (2021) Impact of wastewater cultivation on pollutant removal, biomass production, metabolite biosynthesis, and carbon dioxide fixation of newly isolated cyanobacteria in a multiproduct biorefinery paradigm. Bioresour Technol 333:125194. https://doi.org/10.1016/j.biortech.2021.125194

Sharma NK, Tiwari SP, Tripathi K et al (2011) Sustainability and cyanobacteria (blue-green algae): facts and challenges. J Appl Phycol 23:1059–1081. https://doi.org/10.1007/s10811-010-9626-3

Silambarasan S, Logeswari P, Sivaramakrishnan R et al (2021) Removal of nutrients from domestic wastewater by microalgae coupled to lipid augmentation for biodiesel production and influence of deoiled algal biomass as biofertilizer for *Solanum lycopersicum* cultivation. Chemosphere 268:129323. https://doi.org/10.1016/j.chemosphere.2020.129323

Singh H, Das D (2020) Biohydrogen from microalgae. Handbook of microalgae-based processes and products. Elsevier

Singh K, Ansari FA, Ingle KN et al (2023) Microalgae from wastewaters to wastelands: leveraging microalgal research conducive to achieve the UN sustainable development goals. Renew Sust Energ Rev 188:113773. https://doi.org/10.1016/j.rser.2023.113773

Smol M, Adam C, Preisner M (2020) Circular economy model framework in the European water and wastewater sector. J Mater Cycles Waste Manage 22:682–697. https://doi.org/10.1007/s10163-019-00960-z

Strong C, Kuzma S (2020) It could only cost 1% of GDP to solve global water crises

Sutherland DL, Ralph PJ (2019) Microalgal bioremediation of emerging contaminants - opportunities and challenges. Water Res 164:114921. https://doi.org/10.1016/j.watres.2019.114921

Taher H, Al-Zuhair S, Al-Marzouqi A et al (2015) Growth of microalgae using CO2 enriched air for biodiesel production in supercritical CO2. Renew Energy 82:61–70. https://doi.org/10.1016/j.renene.2014.08.013

Tom AP, Jayakumar JS, Biju M et al (2021) Aquaculture wastewater treatment technologies and their sustainability: a review. Energy Nexus 4:100022. https://doi.org/10.1016/j.nexus.2021.100022

Trivedi J, Aila M, Bangwal DP et al (2015) Algae based biorefinery—how to make sense? Renew Sust Energ Rev 47:295–307. https://doi.org/10.1016/j.rser.2015.03.052

Umamaheswari J, Shanthakumar S (2021) Paddy-soaked rice mill wastewater treatment by phycoremediation and feasibility study on use of algal biomass as biofertilizer. J Chem Technol Biotechnol 96:394–403. https://doi.org/10.1002/jctb.6551

Usman M, Amin M, Kamal I et al (2023) Algae-mediated resource recovery from urban wastewater. Curr Pollut Rep 9:243–258. https://doi.org/10.1007/s40726-023-00254-x

Viegas C, Nobre C, Mota A et al (2021) A circular approach for landfill leachate treatment: chemical precipitation with biomass ash followed by bioremediation through microalgae. J Environ Chem Eng 9:105187. https://doi.org/10.1016/j.jece.2021.105187

Wang J, Yin Y (2018) Fermentative hydrogen production using pretreated microalgal biomass as feedstock. Microb Cell Factories 17:1–16. https://doi.org/10.1186/s12934-018-0871-5

Wang X, Zhou Y, Peng Q et al (2022) Development of plastic flatbed-based algal culture system deployable on non-arable land. Algal Res 66:102814. https://doi.org/10.1016/j.algal.2022.102814

Wang N, Xiang Y, Ma Y et al (2024) Fermented *Spirulina platensis* alleviated DSS-induced ulcerative colitis by regulating gut microbiota and MyD88/TLR4 signaling pathway. Asian J Agric Biol 2024:2023232. https://doi.org/10.35495/ajab.2023.232

Wu W, Ke T, Zhou X et al (2022) Synergistic remediation of copper mine tailing sand by microalgae and fungi. Appl Soil Ecol 175:104453. https://doi.org/10.1016/j.apsoil.2022.104453

Xu X, Gu X, Wang Z et al (2019) Progress, challenges and solutions of research on photosynthetic carbon sequestration efficiency of microalgae. Renew Sust Energ Rev 110:65–82. https://doi.org/10.1016/j.rser.2019.04.050

Yadavalli R, Heggers GRVN (2013) Two stage treatment of dairy effluent using immobilized *Chlorella pyrenoidosa*. J Environ Health Sci Eng 11:36. https://doi.org/10.1186/2052-336X-11-36

Zafar Khan A, Haider MN, Zhao X-Q et al (2022) Evaluation of resource recovery potential of the *Pseudoscillatoria coralii* BERC01 under variable compositions of wastewater to produce biomass for cyanobacterium biorefinery. Sustain Energy Technol Assess 54:102804. https://doi.org/10.1016/j.seta.2022.102804

Zhou W, Wang J, Chen P et al (2017) Bio-mitigation of carbon dioxide using microalgal systems: advances and perspectives. Renew Sust Energ Rev 76:1163–1175. https://doi.org/10.1016/j.rser.2017.03.065

Zhou L, Liu W, Duan H et al (2023) Improved effects of combined application of nitrogen-fixing bacteria Azotobacter beijerinckii and microalgae Chlorella pyrenoidosa on wheat growth and saline-alkali soil quality. Chemosphere 313:137409. https://doi.org/10.1016/j.chemosphere.2022.137409

Chapter 8
Cyanobacteria-Based Carbon Capture and Storage

Hira Ashfaq, Fatima Tahir, Hafiza Aroosa Malik, Nayab Nazir, Peter J. Ralph, and Nature Poddar

Abstract The rising levels of carbon dioxide (CO_2) emissions are a significant cause of climate change, contributing greatly to the greenhouse effect and global warming. This chapter discusses the severe problem of CO_2 emissions, mainly from industries and buildings. Reducing our carbon footprint is essential to protect the environment and public health. The chapter explores various Carbon Capture, Storage, or Utilization (CCS/U) technologies, such as pre-, post-, and Oxyfuel combustion methods, highlighting the need for energy-efficient and cost-effective ways to reduce CO_2 emissions. The chapter also covers different techniques for separating CO_2, including adsorption, absorption, membrane separation, and cryogenic distillation. Each technique has its own strengths and weaknesses in terms of economic and technical parameters. One promising technology is cyanobacteria-based CCU. This method uses cyanobacteria, extraordinary microorganisms that can capture and use CO_2 through photosynthesis, turning it into high-value chemicals and sustainable energy. This approach has advantages like needing less space and being suitable for large-scale use in photobioreactors. Despite its challenges, this approach is a widely used, eco-friendly, and nature-inspired CCU technology that offers significant advantages over other methods. We can develop more sustainable and efficient carbon capture solutions by optimizing and integrating these biological systems. Reviewing traditional and new CCS technologies, this chapter highlights the potential of innovative methods like cyanobacteria-based carbon capture to tackle the urgent problem of CO_2 emissions and support global efforts to fight climate change.

Keywords Carbon capture · Carbon storage and utilization · Cyanobacteria · Photobioreactors

H. Ashfaq · F. Tahir · H. A. Malik · N. Nazir
Bioenergy Research Center, Department of Bioinformatics and Biotechnology, Government College University Faisalabad, Faisalabad, Pakistan

P. J. Ralph · N. Poddar (✉)
Climate Change Cluster, University of Technology Sydney, Ultimo, NSW, Australia
e-mail: nature.poddar@uts.edu.au

M. A. Mehmood et al. (eds.), *Cyanobacteria Biotechnology*,
https://doi.org/10.1007/978-3-031-70698-1_8

8.1 Introduction

The increasing CO_2 emissions worldwide has resulted in greenhouse effects and other environmental issues. It has been shown that carbon dioxide (CO_2) is the primary greenhouse gas in the atmosphere and that it accounts for around 66% of global warming (Yao et al. 2023). According to the International Energy Agency (IEA), the CO_2 emissions from the building and industrial sectors have been restored to their 2019 levels (Mabhaudhi et al. 2022). This resurgence is a result of rising emissions in both emerging markets and advanced economies, accounting for 46% of the increase in emissions worldwide (Mabhaudhi et al. 2022). The leading cause of an increase in anthropogenic emissions is the rapid growth in the use of fossil fuels, such as oil, natural gas, lignite, and stone coal, particularly in the energy sector, which is expected to continue to be the primary source of greenhouse gases, particularly CO_2 (Allan et al. 2023).

About 22% of the world's CO_2 emissions come from heavy sectors like petrochemicals, iron and steel, cement, and oil refining. Among these industries, oil refineries account for 4–6% (Güleç et al. 2020). The combustion of energy and industrial processes caused CO_2 emissions to reach a record high in 2021. However, it should be emphasized that the COVID-19 epidemic significantly influenced energy usage in 2020, leading to a 5.2% reduction in worldwide CO_2 emissions (Fig. 8.1). It can

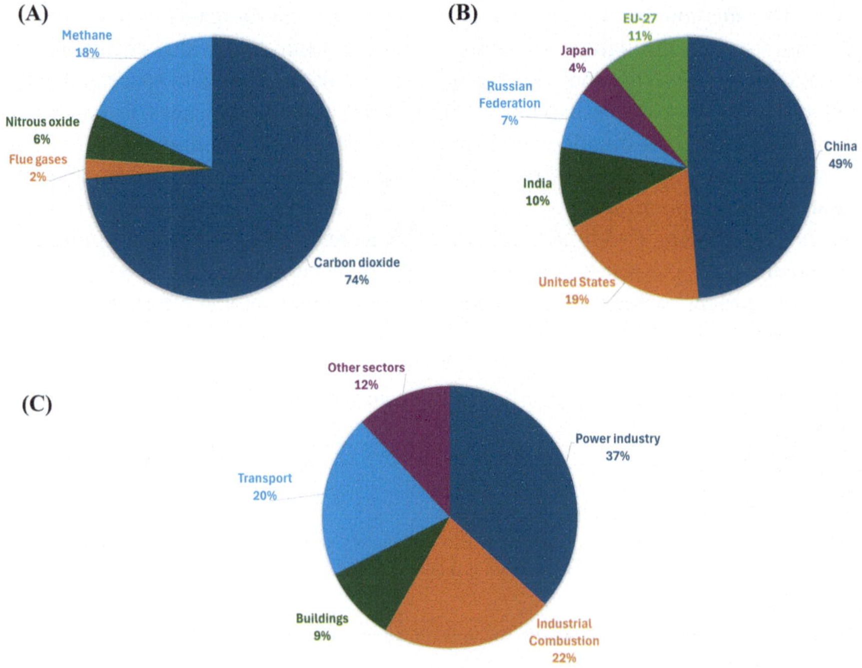

Fig. 8.1 (**a**) Greenhouse gas emission scenario in 2018, (**b**) global scenario of CO_2 emission from fossil fuel combustion in 2020, (**c**) global emission of CO_2 in 2020 by sector (Crippa et al. 2021)

be observed that the emissions were very low before the Industrial Revolution, and emission growth was prolonged until the middle of the twentieth century (Bello et al. 2023). The emissions of greenhouse gases are referred to as "carbon footprints," and their decrease is essential because it mitigates the effects of climate change, improving both plant and animal diversity (Castellanos Diaz et al. 2022). Additionally, it also promotes the world economy by generating innovative and ecologically friendly solutions (Bello et al. 2023). Hence, keeping an eye on the drastic effects of carbon emissions is essential. For instance, recently, in Texas, the climatic change considered to be the cause of extreme weather along with the destruction of water infrastructure and loss of life (Busby et al. 2021). However, there are methods for carbon separation, carbon storage, and carbon utilization to reduce the carbon footprint and address climate change. However, these methods have some challenges of high-energy demand and a separating unit specifically for separating CO_2 from other gaseous mixtures.

To address this problem, large absorption areas are needed, which are often achieved by employing sustainable green materials or functionalized porous materials. However, space limitations provide a practical constraint on CO_2 emissions. One effective strategy is to convert carbon dioxide into sustainable energy and valuable chemicals by growing microalgae and cyanobacteria. Cyanobacteria, with the promising potential of photosynthesis, can recover the carbon polluted environment with their natural Carbon Capture and utilization (CCU) technology. Cyanobacteria have versatile carbon metabolism. Various reports have explored the large-scale cultivation of cyanobacteria in photobioreactor systems fed with inorganic carbon and CO_2 from compressed gas cylinders, flue gas, bicarbonate salts, etc. This method is notable for requiring less space than carbon capture methods (Satya et al. 2024).

8.2 CCUS Technologies to Limit CO_2 Emission

The increasing level of atmospheric CO_2 due to the growing population and their demand for energy consumption are polluting the climate to a drastic level. To overcome this pollution, clean coal technologies (CCT) are developed to clean the carbon-based fuel before usage and discharging it into the environment. Among CCT, CCUS technologies are widely applied to recover anthropogenic environments. CCUS technology has three different carbon capture configurations (Fig. 8.2). The configurations, pre-, post-, and Oxyfuel combustion capture are followed by the carbon separation power plants (Dziejarski et al. 2023). The power plants mainly separate the concentrated carbon from the fuel combustion for further storage and utilization. In addition to these three main configurations, there are methods to directly capture carbon from the atmosphere and from the industrial process stream that are more efficient than the others (Lau et al. 2021).

Different CO_2 separation technologies, such as adsorption, absorption, cryogenic distillation, and membrane separation, follow these carbon capture configurations. Before carbon capture, transportation, storage, and utilization there is a need to

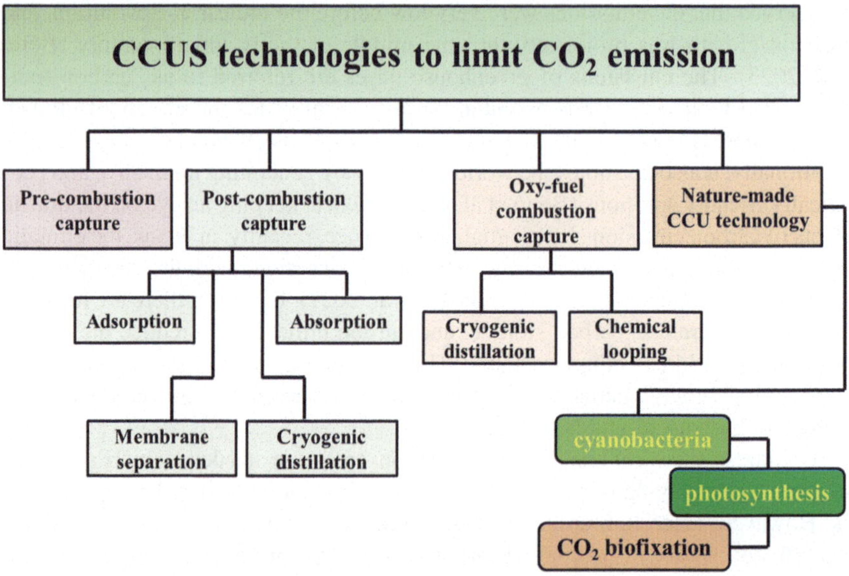

Fig. 8.2 Carbon capture, utilization, and storage (CCUS) technologies to reduce CO_2 emission

separate it from the biomass or fuel in its appropriate concentration (Dziejarski et al. 2023). Carbon separation or capture is the main and significant part of the carbon capture and storage (CCS) strategy as it costs 75–80% of the whole CCS process. Therefore, carbon capture demands an efficient capture configuration with a competitive method of carbon separation for providing the required level of CO_2 for further processing of CCS (Blomen et al. 2009).

8.2.1 Pre-combustion Capture

Pre-combustion CO_2 capture is mainly based on the gasification and conditioning process of the solid coal fuel (Irons et al. 2007). In this technology, the heavy fractions of petroleum are initially converted into a gaseous form, which is hydrogen-rich through gasification (Olajire 2010). Briefly, a reaction between fuel and air generates a synthesis gas (a mixture of carbon monoxide and hydrogen). The gaseous mixture is then moved toward the catalytic reactor where another reaction, water–gas shift reaction (WGSR), takes place in which the gaseous mixture reacts with the steam that eventually transforms CO into CO_2 (Gibbins and Chalmers 2008; Theo et al. 2016). This carbon capture method captures about 50–60% of the CO_2, much higher than the concentration obtained from the combustion process from the exhaust gasses with a minute number of impurities. Additionally, this is a cheap method of CO_2 capture because the higher concentration of CO_2 itself and the high pressure of the reactor make it able to separate from the gaseous mixture with very low energy expenditure.

After gasification, there is a conditioning process where CO_2 is finally separated from the hydrogen gas and stored in its pure form (Lau et al. 2021). Briefly, some physical solvents (such as rectisol and selexol) are used to pure the captured CO_2, where at the first step there is a condensation of gas mixture, then the moisture produced from condensation is removed, the solid fractions are separated, and finally providing high or low temperature to the gas for further cleaning of the other gaseous impurities and lastly the compression of pure CO_2 for its storage (Dziejarski et al. 2023).

However, there are still some challenges related to this method as the amount of energy generated throughout this process is difficult to recover so the system needs improvements for efficient energy recovery (Osman et al. 2021). The system also loses efficiency during the water-gas shift reaction (Olajire 2010). On the other hand, operating the catalytic reactor at high pressure is challenging with no control on excessive emission of NO_x (Elhenawy et al. 2020; Omodolor et al. 2020). The main pitfall of the process is its requirement for a separate air unit or an extensive supporting system (Elhenawy et al. 2020).

8.2.2 Post-combustion Capture

In localized combustion power plants, the production of CO_2 is much lower than its concentration obtained in pre- and post-combustion capture processes. Post-combustion capture is a conventional method that could be applied to old and new coal power plants already operating. It involves the direct separation of CO_2 from the flue gas in the presence of oxygen, which is supplied by the air (Rajabloo et al. 2023). The low concentration of CO_2 in flue gas makes the post-combustion capture a promising methodology to control the CO_2 emission from other processes and create the infrastructure of the CCS (Balraj et al. 2022). Additionally, flue gas contains oxides of nitrogen and sulfur along with a low concentration of CO_2, so it is subjected to the cleaning of these gases before carbon capture and separation. After this primary cleaning, flue gas containing s mixture of CO_2, N_2, and H_2O is moved toward the separation unit. The CO_2 separation units follow different separation methods that are decided based on the physiochemical properties of the exhaust gases, the concentration of CO_2, and the size of the stream of gas mixture.

Different typical carbon separation methods, such as chemical and physical absorption, as well as adsorption, membrane separation, and cryogenic distillation, are applied while separating the captured carbon via post-combustion capture technology (Dziejarski et al. 2023). Additionally, there is a biological method to capture and separate the CO_2 from biomass or fuel which is microalgae and cyanobacteria based because of their photosynthetic nature. Cyanobacteria-based carbon capture and utilization method is the recovery of anthropogenic climate and is also a kind of post-combustion capture technology with promising bio-fixation potential (Benson and Cole 2008; Osman et al. 2021). A detailed review of the cyanobacteria-based carbon capture and storage method will be described in relation to its typical carbon capture technologies.

8.2.3 Adsorption

The term adsorption refers to the creation of adhesion of atoms, ions, or molecules to a surface in a liquid, gaseous, or solid state. Adsorbates are the ions, atoms, or particles that adhere to a substance's surface and form a film; the substance on which they attach is referred to as an adsorbent. Adsorption occurs on the exterior surface, whereas absorption takes up the entire volume of the material. Sorption is associated with the couplet modes, whereas desorption is considered the reverse of the adsorption process or a counter-reaction (Benson and Cole 2008).

The adsorption and desorption of CO_2 are the two independent processes that make up the cycle adsorption process. During adsorption, an experimental apparatus is loaded with the adsorbent. Following that, the CO_2-containing flue gas is moved through it. The solid sorbent material is then brought to the surface of the gas mixture, where it adsorbs CO_2 molecules and allows other gases to pass through its structure. Following adsorption, CO_2 is extracted from the adsorbent layer through desorption, which involves regenerating the adsorbent for future use. The extracted CO_2 is compressed, cleaned, and transported (Dziejarski et al. 2023).

The choice of a specific adsorbent depends on the type of CO_2 capture, where temperature is a crucial factor. Flue gas temperatures are relatively low for post-combustion capture; for coal-fired flue gas, they are roughly 60–150 °C, and for natural gas-fired flue gas, they are around 100 °C, in that case, most adsorbent types can be employed. However, CO_2 capture occurs before combustion when the gas is produced by gasification or the post-water-gas shift reaction of the fuel has a considerably higher temperature, between 500 and 1800 °C and 250 and 550 °C, respectively (Dziejarski et al. 2023).

The two primary forms of adsorption can be alternately referred to as stages in the development of technology due to the adsorption progress in the market. Physical adsorption, such as that of activated carbon, silica, MOFs, alumina, or zeolites, is currently more frequently used in the chemical, cement, iron and steel sectors. Physical adsorption is possible; this requires inefficient Van der Waals forces. Similarly, it might occur chemically, requiring covalent bonding (chemisorption) or caused by electrostatic attraction (Osman et al. 2021). Chemical adsorption (amine-based adsorbents, metal oxides, metal salts, hydrotalcite) is in a much earlier technological stage, mainly in the research and development phase.

8.2.4 Physical and Chemical Absorption

The process of physical absorption involves dissolving the absorbed carbon dioxide in a chemically inert solvent that does not react with CO_2. The process is based on Henry's law, which states that the concentration of gas dissolved in the absorption liquid is proportional to the partial pressure of the gas above the liquid. Typical substances that act as a solvent in physical absorption are selexol (polyethylene glycol dimethyl ethers), and rectisol (cold methanol). Physical absorption has widespread use in specialized industrial applications such as the synthesis of methanol, ammonia manufacture, and the generation of high-value compounds.

The most advanced method of separating CO_2 was created in the 1930s and has been applied in industry for many years: chemical absorption. The reaction of a solvent with CO_2 to generate a new intermediate during a reversible or irreversible chemical transformation is the basis for the CO_2 separation process. Monoethanolamine (MEA), diethanolamine (DEA), N-methyl diethanolamine (MDEA), and di-2-propanolamine (DIPA) are the most commonly used solvents. Temperature increases facilitate regeneration; therefore, the intermediated compound breaks down into the CO_2 stream and the primary solvent. However, the recovery of CO_2 varies depending on the specific situation and the type of chemical reaction (Dziejarski et al. 2023).

8.2.5 Membrane Separation

The membrane technique is an innovative method for removing CO_2 from the flue gas mixture. Membrane separation is mainly based on a membrane, a thin layer of impermeable barrier material that separates a specific gaseous medium when a driving force is applied such as electric potential, temperature differential, or pressure difference on both sides of the membrane. The membrane splits the gas stream (feed) into a retained stream (retentate) and a permeate gas stream.

Three general properties can be utilized to classify membranes used for CO_2 separation: origin, morphology, and structure. Many membranes used in industry are organic membranes which show a wide range in terms of material from which they are made and physical structure. They are resilient to hostile process conditions such as high pressure and temperature, as well as the chemically reactive characteristics of the exhaust gas mixture. Polymeric membranes and their hybrid system have been recognized as an efficient CO_2 separation method because of their selectivity performance, simplicity in regulating membrane pore size during the formation process, and excellent permeability (Dziejarski et al. 2023).

Membranes can be categorized into four classes based on their structure: homogenous solids, porous solids, solids that carry electric charge, and liquids that contain selected carriers. One advantage of this separation technique is that it does not result in flooding, foaming, entrainment, weeping, or any other usual problems associated with employing a packed column. Furthermore, membranes have better control over gas flow rates and a greater surface area.

Membrane separation can be classified into two methods: gas separation membrane and gas absorption membrane, based on the CO_2 capture process and the emerging transport mechanism through the membrane structure. Between these two types of mechanisms based on considerably different methods of CO_2 separation by membrane structure, only gas separation membranes used in natural gas processing are at the commercial separation phase.

8.2.6 *Cryogenic Distillation*

To separate CO_2 depending on the dew point or sublimation for a particular component, cryogenic carbon capture (CCC) systems compress the exhaust gas mixture and cool it to the proper temperature at high pressure in numerous stages, causing a phase change only of carbon dioxide. Compared to previous separation technologies, this approach can achieve better CO_2 recovery (99.99%) and purity (99.99%). Moreover, its primary application is the separation of gaseous streams containing high carbon dioxide concentrations, usually greater than 50%. In the case of low CO_2 concentrations (equivalent to boiler exhaust gas), the use of this technology is impractical due to the large amount of energy needed for the cooling processes and compression (Dziejarski et al. 2023).

To generate phase changes in CO_2 in flue gases and, often, other components in the mixture, this separation process involves multiple phases of compression and cooling of gas mixtures. The separated carbon dioxide is extracted either directly in the liquid phase or as a solid (dry ice). The primary cause of the cryogenic method's selectivity is the variation in the dew/sublimation points for the distinct exhaust gas components that need to be separated, including CO, SO_2, NOx, H_2O, CH_4, and NH_3. They cause corrosion, fouling, and clogging and seriously reduce cooling (Dziejarski et al. 2023).

8.2.7 *Oxyfuel Combustion*

The oxygen-fueled combustion process of carbon-based fuel is when pure oxygen from an oxygen separation unit separates the CO_2 from the gas mixture. Oxyfuel combustion provides an easy route of CO_2 separation, but the high cost of the O_2 separation unit makes the whole process costly. On a large scale, the cost of obtaining pure O_2 increases to a higher extent because of the demand for large amounts of O_2, which is only obtained by cryogenic distillation, an energy-intensive approach (Chen et al. 2018). However, in terms of efficiency, it is considered the most efficient of all three: pre-, post-, and Oxyfuel combustion (Wienchol et al. 2020).

Its applicability in power systems could simultaneously decrease exhaust gas emissions and increase boiler efficiency. Additionally, some ionic-electronic mixed conducting membranes were designed to fulfill the energy demand of cryogenic air separation. These membranes provide 20–50% economic savings to the whole process of Oxyfuel combustion as compared to post-combustion capture (Kosheleva et al. 2019). Besides cryogenic distillation, chemical looping is also considered an efficient method to facilitate Oxyfuel combustion and enhance its performance to approximately 3% compared to the facilitation provided by the air separation unit. Interestingly, chemical looping reduces the capital cost of the operating power plant along with the cost of electrical energy utilized to run the overall process (Wei et al. 2020).

Oxyfuel combustion is considered an advanced technology among pre-, post-, and Oxyfuel combustion capture because of the replacement of conventional methods with an integrated system of bioenergy production along with carbon capture and storage (Cabral et al. 2019; Wei et al. 2020). It is a synergistic method with an integration of carbon dioxide cycle with air separation and purification units for heat recovery and CO_2 compression, respectively, along with carbon storage methodology [bioenergy with carbon capture and storage (BECCS)] (Cabral et al. 2019; Chen et al. 2019).

8.3 Cyanobacteria-Based Carbon Capture

All the above-mentioned strategies to cope up with the remediation of carbon polluted environment are no doubt efficient in their working. However, due to their high running cost and not being ecofriendly make their use less common. Therefore, an ecofriendly and cost-effective strategy for the recovery of an anthropogenic environment should be preferred by using the best microbial candidates, such as cyanobacteria. Cyanobacterial potential of carbon capture and resource recovery makes them highly applicable for the recovery of polluted environment. Their physiological adaptations and mechanisms make them survive in harsh and changing climatic conditions.

Cyanobacteria adapt to the carbon-flooded environment because of the flexibility of their carbon metabolic pathways (Damrow et al. 2016). Cyanobacteria primarily capture carbon from the environment through glycolysis and further metabolize it mainly through the pentose phosphate pathway, the Calvin Benson-Bassham (CBB) cycle, and the tricarboxylic acid (TCA) cycle (Xiong et al. 2015). All these metabolic pathways are highly interlinked, so utilizing the product of one pathway as the substrate for the other allows the conversion of captured carbon into valuable metabolic precursors and ends up with the higher production of cyanobacterial biomass (Kumar et al. 2022).

8.3.1 *Natural Carbon Capture and Utilization (CCU) Technology*

In nature, there exists a CCU technology that is directly related to the photosynthetic potential of cyanobacteria or the evolution of chloroplast, the carbon capture bio-factories, for the bio-fixation of CO_2 from the anthropogenic environment (Daneshvar et al. 2022). This natural CCU technology assimilates CO_2 by the synchronization of the light and dark reactions inside the cyanobacterial chloroplast (Pang et al. 2019). Chloroplast is a specialized organelle with disk-shaped thylakoids in its bilayer membrane, providing the central platform for photosynthesis reactions (light reactions). Thylakoids generate energy through light reactions for light-independent reactions where energy is required for the bio-fixation of CO_2 (Anwar et al. 2019).

8.3.2 *Bio-Machinery Used to Generate Energy for CO_2 Bio-fixation (Light Reaction)*

As mentioned in the above section, the photosynthetic organelle (chloroplast) in cyanobacteria contains the bio-machinery for the operation of light reactions to harvest energy for dark reactions (Daneshvar et al. 2022). Bio-machinery located in thylakoids (Fig. 8.3) which consists of photosystems (PSI and PSII), ATP synthase enzyme, light-harvesting antennae, cytochrome b6/f (cyt b6/f) (Middepogu et al. 2018). The light-harvesting antennae absorb light energy and fuel the Photosystem II (PSII) to activate its reaction center. PSII, having chlorophyll dimer with a photon absorbing capacity of the maximum wavelength of 680 nm (P680), finally absorbs the light energy captured by light-harvesting antennae. After light absorption, P680

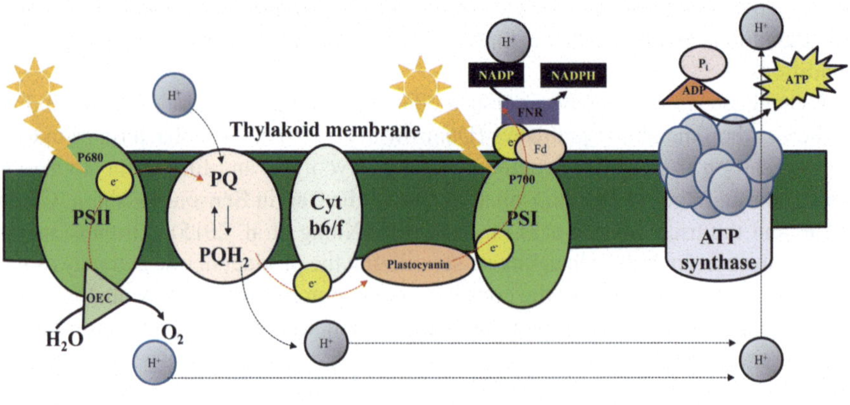

Fig. 8.3 Photosynthetic activity in the thylakoid membrane: Bio-machinery for generating the energy molecules, $NADPH_2$ and ATP, by using light for light-independent reactions

oxidized and donate a pair of electrons to the pheophytin molecule (Hüner and Grodzinski 2011). This molecule further activates the electron transport chain (ETC) where it reduces the primary electron acceptor (PQ-A) and secondary electron acceptor (PQ-B) (Su 2021). PSII further reduces PQ into plastoquinol (PQH_2), which is then oxidized by cyt b6/f while the slow and continuous flow of ETC (Laisk et al. 2015; Su 2021). PSII then transfers the electron through plastocyanin (PC) to PSI, with the reaction center having an absorbing capacity of 700 nm (P700). This electron transfer is an energy-consuming step that automatically generates a gradient to pump H^+ inside the thylakoids (lumen) from the stroma. The accepted electrons inside P700 are then excited by absorbing the light and transferred to the ferredoxin where ferredoxin-$NADP^+$ reductase (FNR) catalyzes the reduction of a final metabolite of ETC ($NADPH_2$) (Anwar et al. 2019; Su 2021). On the other hand, the accumulated H^+ ions inside the lumen provide energy to the ATPase enzyme for the phosphorylation of ADP to produce the high-energy molecule ATP (Su 2021). In this way, the light reaction inside the cyanobacterial chloroplast prepares the energy molecules ATP and $NADPH_2$ as primary and secondary energy sources for dark reactions, respectively, for CO_2 bio-fixation.

8.3.3 CO_2 Intake Via HCO_3^- Transport Systems

Cyanobacteria have three different organized transport systems (transporters) in their plasma membrane to intake inorganic carbon or HCO_3^-. One is ATP-dependent and named BCT1, while the other, BicA and SbtA, are activated by generating a sodium electrochemical gradient inside the cell (Fig. 8.4) (Klanchui et al. 2017). Under sodium electrochemical gradient BicA, monomeric transporter protein binds HCO_3^- and Na^+ on its N-terminal (transmembrane domain) and transports HCO_3^- via the Sulfate Transporter and Antisigma Factor Antagonist (STAS) domain on C-terminal into the cytoplasm (Wang et al. 2019). It is a low-affinity dimeric transporter that is significant for the overall transportation activity of BicA along with the dimerization of HCO_3^- (Wang et al. 2019). Unlike BicA, SbtA has a transmembrane domain and is regulated by its partner protein, SbtB, via the allosteric regulation process. It is a high-affinity HCO_3- transporter expressed in the cyanobacterial cell with a low inorganic carbon concentration. In addition to this, the partner protein of SbtA that is SbtB, acts as a signaling protein and upregulates the SbtA when it senses the high concentration of cyclic adenosine monophosphate (cAMP) due to a higher accumulation level of HCO_3^- (Du et al. 2014; Liu et al. 2021). Hence, this transporter is not only a way of transferring the surrounding carbon into the cell but is a proper signaling molecule, the activation of which is dependent on the intracellular concentration of inorganic carbon (Liu et al. 2021; Rottet et al. 2021). The captured inorganic carbon and CO_2 make a carbonic pool in the cytoplasm and are then further processed for its fixation, concentration and interconversions.

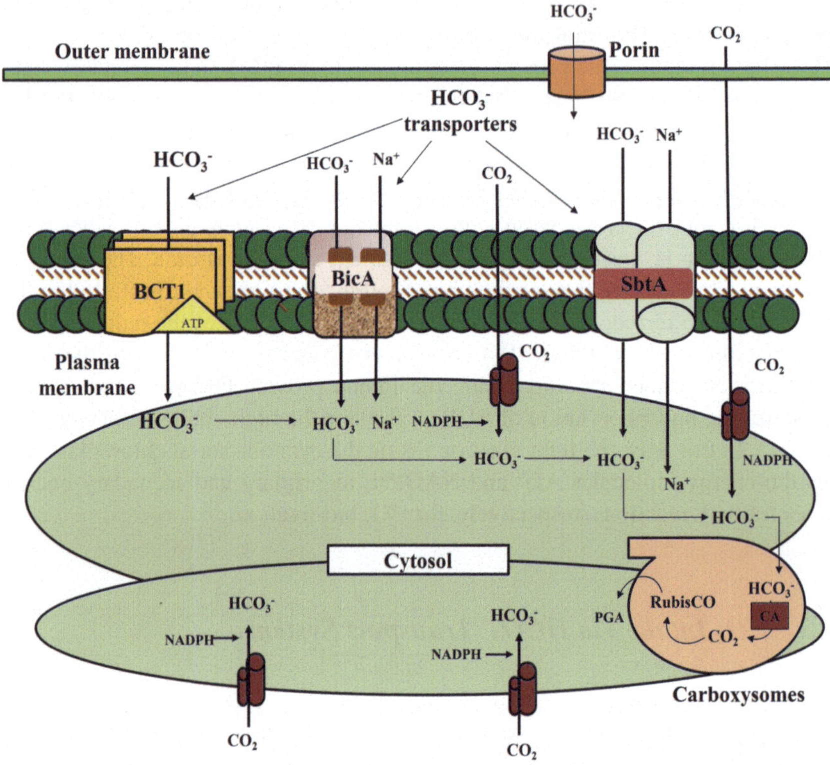

Fig. 8.4 HCO_3^- transport systems (transporters) for CO_2 intake in cyanobacteria from the carbon polluted environment, interconversions between HCO_3^- and CO_2 and the advent of carboxysomes for carbon-concentrating mechanism to avoid the leakage of CO_2 from the cell

8.3.4 CO_2 Fixation and Concentration

The microcompartments and carboxysomes contain two specific enzymes encapsulated in their membrane, such as RUBISCO and Carbonic Anhydrase (CA), for fixing and concentrating CO_2, respectively. The inorganic carbon or HCO_3 intake by cyanobacteria is then catalyzed to CO_2 by the catalytic activity of RUBISCO. This generates a high flux of CO_2, which then acts as a substrate for RUBISCO to overcome its oxygenation ability by enhancing its affinity toward CO_2 (Mangan and Brenner 2014; Iñiguez et al. 2020). Conversely, CA performs the interconversions where CO_2 hydration and HCO_3^- dehydration activity catalyzes side by side. However, the uncatalyzed interconversions in a typical living cell at physiological pH are not sufficient to fulfill the physiological demand of the cell. That is why CA with highly efficient interconversion ability is used to ensure the requirement of metabolic CO_2 and HCO_3^- (Talekar et al. 2022). This catalytic activity not only converts HCO_3^- into CO_2 but also ensures the pH balance of the cell. After the fixation of captured CO_2, the high flux of fixed CO_2 is then concentrated by CCM to avoid leakage from the CO_2 accumulation zone (Rae Benjamin et al. 2013). Also, CCM makes cyanobacteria able to concentrate 1000-fold higher inorganic carbon than the accumulation capacity of the surrounding medium (Kumar et al. 2022). The efficiency of CCMs avoids the leakage of CO_2 by accumulating it in a less permeable form, HCO_3^-, and the deactivation of CA to prevent the wasteful conversion of HCO_3^- into CO_2. This carbon-concentrating potential of cyanobacteria makes them a promising candidate for recovering carbon polluted environment via natural CCU technology. Ultimately, the CO_2 is pumped back to the thylakoids to recycle into a bicarbonate pool for carbon utilization.

8.3.5 *Carbon Metabolism in the Light-Independent Reactions of the Calvin Cycle*

Carbon metabolism in cyanobacteria mainly follows a three-step biochemical process: carboxylation, reduction, and regeneration inside the chloroplast (stroma) (Vecchi et al. 2020). Briefly, carboxylation is the addition of CO_2 into ribulose 1, 5-bisphosphate (RUBP) that produces a six-carbon unstable molecule, breaking into two molecules of 3-phosphoglyceric acid (3-PGA) for stability. Ribulose bisphosphate carboxylase/oxygenase (RUBISCO) is the mainly involved enzyme in carboxylation (Yang et al. 2017).

Carboxylation follows the reduction process where the 3-PGA molecule is converted into glyceraldehyde-3-phosphate (GA3P), a sugar molecule with three carbons. This conversion is a two-step process where initially ATP is utilized to phosphorylate 3-PGA molecule into 1,3-bisphosphoglyceric acid (1,3BPG), which is an unstable intermediate so convert into GA3P by the reducing power of $NADPH_2$ in Calvin cycle (Daneshvar et al. 2022).

The third biochemical process is regeneration or recycling, during which the utilized molecules such as ATP, $NADPH_2$, and RUBP are restored by catalyzing the interconversions again (Fig. 8.5). Briefly, in this phase, most of the GA3P molecules are catalyzed by complex enzymatic reactions and produce several intermediates with 3-,4-,5-,6-, and 7-carbon sugar phosphates (Gerotto et al. 2020). Some additional GA3P molecules, such as starch, are converted into storage molecules and stored in the chloroplast. In photosynthesis, carbohydrates are obtained as the final product, but in the case of carbon capture and utilization, CO_2 is inserted into other compounds like fatty acids, amino acids, and organic acids (Masojídek et al. 2013).

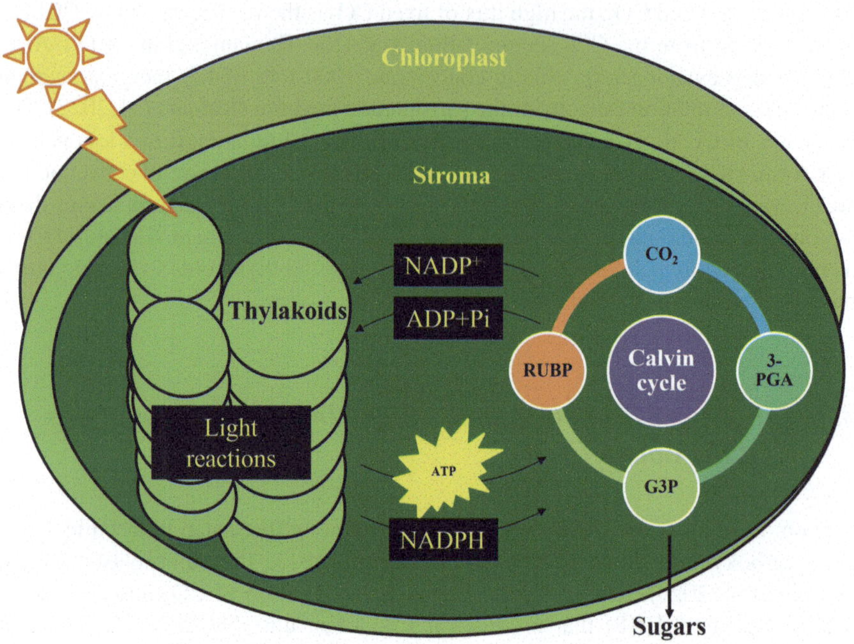

Fig. 8.5 CO_2 fixation and concentration under light-independent chloroplast reactions with the utilization of ATP and NADPH generated by light-dependent reaction of oxygenic photosynthesis

8.4 CO_2 Utilization: A Climate Recovery Approach

CCU technology is not only about the carbon capture and fixation; instead, it ends up with an additional potential of utilizing that captured CO_2 by developing a proper synergism between the capturing, fixation, and utilization pathways. In this phase, cyanobacterial biomass with higher carbon content drives different value-added products (VAPs) after its incorporation into the metabolites, macromolecules, and some biochemical compounds. Hence, it is a synergistic approach to climate recovery and valorizing the biomass for value creation (Mohan et al. 2016; Peter et al. 2021).

The utilization phase ends with the concept of biorefinery, a zero-waste approach where every single side product is valorized while generating the targeted product (Cherubini et al. 2009). Cyanobacterial biomass has a significant fraction of lipids compared to other metabolites. However, the targeted end products, such as lipids, proteins, and carbohydrates, are easily enhanced by optimizing the growth conditions to facilitate the higher production of the specific product. Most of all the variation in light density, CO_2 concentration, and nitrogen stress leads to the higher production of lipids (Mata et al. 2010). These lipids could be a good source of biodiesel production, but this depends further on the nature and characteristics of fatty acids (Bhattacharya and Goswami 2020).

Another value-added product of cyanobacterial biomass is amino acids and proteins that have nutritional value and could be a food source for humans and animals. Such as cyanobacterial strains like *S. maxima* and *S. platensis,* with a high protein content of 46–60%, are used in food, pharmaceuticals, and cosmetics (Montalvo et al. 2019). Additionally, protein-rich biomass could be used to extract protein-derived bioactive compounds with broad applications in the pharmaceutical and food industries. For example, a study reported the extraction of three peptides from the biomass of *A. maxima* OF15 with antimicrobial and anti-inflammatory characteristics (Montalvo et al. 2019).

Similarly, carbohydrates are also the main biochemical compounds of cyanobacterial biomass, which are not only extracted from the biomass as the end product but play a significant role in the biofilm formation of cyanobacteria during cultivation and make harvesting very easy; hence are extremely helpful in an industrial scale. During cultivation, cyanobacteria produce extracellular polymeric substances (EPS) that are polysaccharides or proteins in nature, so they are valuable for making the whole cultivation process of cyanobacteria cost-effective. However, the extracted carbohydrates from the biomass are used as substrates for the production of monosaccharides such as glucose for the final production of starches and cellulose (Daneshvar et al. 2022).

A study valorized algal biomass to produce nutritional fish-feed supplements by feeding fishes in rectangular tanks with algal biomass. The carotenoids of algal biomass benefit muscle pigmentation and the beautiful skin of fishes. Additionally, some dietary pigments, such as de novo, that fishes could not synthesize are provided through the algal biomass. Hence, microalgae-biomass could be an excellent alternative food source for fishes (Yadav et al. 2020a, b).

8.5 Factors Affecting the Photosynthetic Efficiency of Cyanobacteria

Cyanobacteria are no doubt a promising microbe for climate recovery with their photosynthetic potential, but different environmental factors influence the efficiency of this potential. Factors like temperature, seasonal variations, pH, salinity, turbidity, and light penetration into the culture medium and most importantly the nutrient composition of the growth medium. However, these factors vary from strain to strain and have a direct impact on the CO_2 bio-fixation efficiency of cyanobacteria as shown in Table 8.1.

Table 8.1 Impact of different growth factors on the CO_2 bio-fixation efficiency of various species of algae

Species	CO_2%	Temp (°C)	Culture medium	CO_2 bio-fixation rate (g L^{-1} day^{-1})	References
Scenedesmus sp.	10% mixed with N_2	20	BG11	0.347	Rodas-Zuluaga et al. (2021)
	20% mixed with N_2	20	BG11	2.177	Chen and Xu (2021)
Dunaliella salina	Ambient air			0.055	
	6% mixed with air			0.067	
	20% mixed with air			0.016	
	20% mixed with N_2			0.043	
Botryococcus braunii	10% mixed with N_2	20	BG11	0.356	Rodas-Zuluaga et al. (2021)
	20% mixed with N_2	20	BG11	0.532	
Chlorella vulgaris P12	2% mixed with air	30	Synthetic medium	1.5	Anjos et al. (2013)
	6% mixed with air	30	Synthetic medium	2.29	
	10% mixed with air	30	Synthetic medium	1.93	
Scenedesmus obtusiusculus	4–5% (flue gas)	28	BG11	0.1114	Estrada-Graf et al. (2020)
Chlorella sp.	10% (flue gas)	28	Domestic wastewater	0.261	Yadav et al. (2020a, b)
Heynigia riparia SX01	0.03% (ambient air)	28	BG11	0.27	Jin et al. (2021)
	5% mixed with air	28	BG11	0.37	
	10% mixed with air	28	BG11	0.46	
	15% mixed with air	28	BG11	0.71	
Chlorella vulgaris ESP-31 (wild type)	25% (simulated flue gas)	40	BG11	0.05335	Chou et al. (2019)
Chlorella sp. C-1 (control set)	Ambient air	30	3 N-BBM	0.357	Yadav et al. (2020a, b)
Chlorella sp. LAMB 122	40% mixed with air	26	BG11	0.017	Li et al. (2020)

8.6 Conclusion and Prospects

The rising CO_2 emissions significantly damage the climate, and the growing energy demands of an increasing population make it difficult to reduce these emission levels. However, this issue can be addressed through various carbon capture and utilization (CCU) technologies that can reduce CO_2 emissions directly into the environment. While these technologies are effective, their high-energy requirements limit their widespread application. On the other hand, biologically mediated carbon capture and utilization present a promising alternative. This approach leverages the natural photosynthetic abilities of microorganisms, such as cyanobacteria, to capture CO_2 and convert it into high-value products for the industrial, pharmaceutical, and health sectors. Although this approach presents certain challenges, it stands out as a widely used, eco-friendly, and nature-inspired CCU technology compared to other methods. Enhancing and integrating these biological systems can achieve more sustainable and efficient carbon capture solutions.

Enhancing the efficiency of cyanobacteria-based carbon capture and utilization technology is essential. This can be achieved by optimizing the entire process, from selecting the most effective cyanobacterial strains to refining cultivation methods and bioreactor designs. Additionally, integrating this technology into existing CO_2 emission control systems, particularly at industrial scales in cement factories, steel mills, and power plants, will be crucial. Moreover, advancing cyanobacterial engineering to improve efficiency will be vital for mitigating carbon pollution and producing valuable industrial and environmental products. This involves enhancing the cyanobacteria's carbon capture capabilities and leveraging genetic and metabolic engineering to maximize the yield of high-value byproducts. By addressing these aspects, cyanobacteria-based technologies could significantly reduce industrial CO_2 emissions and contribute to a sustainable future.

References

Allan RP, Arias PA, Berger S et al (2023) Intergovernmental panel on climate change (IPCC). Summary for Policymakers. Climate change 2021: The physical science basis. Contribution of working group I to the sixth assessment report of the intergovernmental panel on climate change. Cambridge University Press

Anjos M, Fernandes BD, Vicente AA et al (2013) Optimization of CO2 bio-mitigation by *Chlorella vulgaris*. Bioresourc Technol 139:149–154. https://doi.org/10.1016/j.biortech.2013.04.032

Anwar M, Lou S, Chen L et al (2019) Recent advancement and strategy on bio-hydrogen production from photosynthetic microalgae. Bioresourc Technol 292:121972. https://doi.org/10.1016/j.biortech.2019.121972

Balraj A, Sekaran APC, Ramamurthy N et al (2022) Systematic review on sono-assisted CO2 stripping, solvent recovery and energy demand aspects in solvent-based post-combustion carbon dioxide capture process. Chem Eng Processing - Process Intensification 170:108723. https://doi.org/10.1016/j.cep.2021.108723

Bello A, Ivanova A, Cheremisin A (2023) A comprehensive review of the role of CO2 foam EOR in the reduction of carbon footprint in the petroleum industry. Energies [Online] 16. https://doi.org/10.3390/en16031167

Benson SM, Cole DR (2008) CO2 sequestration in deep sedimentary formations. Elements 4:325–331. https://doi.org/10.2113/gselements.4.5.325

Bhattacharya M, Goswami S (2020) Microalgae – a green multi-product biorefinery for future industrial prospects. Biocatal Agric Biotechnol 25:101580. https://doi.org/10.1016/j.bcab.2020.101580

Blomen E, Hendriks C, Neele F (2009) Capture technologies: improvements and promising developments. Energy Procedia 1:1505–1512. https://doi.org/10.1016/j.egypro.2009.01.197

Busby JW, Baker K, Bazilian MD et al (2021) Cascading risks: understanding the 2021 winter blackout in Texas. Energy Res Soc Sci 77:102106. https://doi.org/10.1016/j.erss.2021.102106

Cabral RP, Bui M, Mac Dowell N (2019) A synergistic approach for the simultaneous decarbonisation of power and industry via bioenergy with carbon capture and storage (BECCS). Int J Greenhouse Gas Control 87:221–237. https://doi.org/10.1016/j.ijggc.2019.05.020

Castellanos Diaz O, Katiyar A, Hassanzadeh A et al (2022) Evaluation of carbon footprint for a hydrocarbon foam EOR field pilot. SPE improved oil recovery conference. https://doi.org/10.2118/209366-MS

Chen Y, Xu C (2021) How to narrow the CO2 gap from growth-optimal to flue gas levels by using microalgae for carbon capture and sustainable biomass production. J Cleaner Prod:280

Chen S, Zhu M, Tang Y et al (2018) Molecular simulation and experimental investigation of CO2 capture in a polymetallic cation-exchanged 13X zeolite. J Mater Chem A 6:19570–19583

Chen S, Yu R, Soomro A et al (2019) Thermodynamic assessment and optimization of a pressurized fluidized bed oxy-fuel combustion power plant with CO2 capture. Energy 175:445–455. https://doi.org/10.1016/j.energy.2019.03.090

Cherubini F, Jungmeier G, Wellisch M et al (2009) Toward a common classification approach for biorefinery systems. Biofuels Bioproducts Biorefining 3:534–546. https://doi.org/10.1002/bbb.172

Chou H-H, Su H-Y, Song X-D et al (2019) Isolation and characterization of *Chlorella* sp. mutants with enhanced thermo- and CO2 tolerances for CO2 sequestration and utilization of flue gases. Biotechnol Biofuels 12:251. https://doi.org/10.1186/s13068-019-1590-9

Crippa M, Guizzardi D, Solazzo E et al (2021) GHG emissions of all world countries. Publications Office of the European Union. https://doi.org/10.2760/173513

Damrow R, Maldener I, Zilliges YJFIM (2016) The multiple functions of common microbial carbon polymers, glycogen and PHB, during stress responses in the non-diazotrophic cyanobacterium *Synechocystis* sp. PCC 6803. 7:966. https://doi.org/10.3389/fmicb.2016.00966

Daneshvar E, Wicker RJ, Show P-L et al (2022) Biologically-mediated carbon capture and utilization by microalgae towards sustainable CO2 biofixation and biomass valorization – review. Chem Eng J 427:130884. https://doi.org/10.1016/j.cej.2021.130884

Du J, Förster B, Rourke L et al (2014) Characterisation of cyanobacterial bicarbonate transporters in *E. coli* shows that SbtA homologs are functional in this heterologous expression system. PLoS One 9:e115905. https://doi.org/10.1371/journal.pone.0115905

Dziejarski B, Krzyżyńska R, Andersson K (2023) Current status of carbon capture, utilization, and storage technologies in the global economy: a survey of technical assessment. Fuel 342:127776. https://doi.org/10.1016/j.fuel.2023.127776

Elhenawy SEM, Khraisheh M, Almomani F et al (2020) Metal-organic frameworks as a platform for CO2 capture and chemical processes: adsorption, membrane separation, catalytic-conversion, and electrochemical reduction of CO_2. Catalysts 10:–1293. https://doi.org/10.3390/catal10111293

Estrada-Graf A, Hernández S, Morales M (2020) Biomitigation of CO2 from flue gas by *Scenedesmus obtusiusculus* AT-UAM using a hybrid photobioreactor coupled to a biomass recovery stage by electro-coagulation-flotation. Environ Sci Pollut Res 27:28561–28574. https://doi.org/10.1007/s11356-020-08240-2

Gerotto C, Norici A, Giordano M (2020) Toward enhanced fixation of CO2 in aquatic biomass: focus on microalgae. Front Energy Res 8. https://doi.org/10.3389/fenrg.2020.00213
Gibbins J, Chalmers H (2008) Carbon capture and storage. Energy Policy 36:4317–4322. https://doi.org/10.1016/j.enpol.2008.09.058
Güleç F, Meredith W, Snape CE (2020) Progress in the CO2 capture technologies for fluid catalytic cracking (FCC) units—a review. Front Energy Res 8. https://doi.org/10.3389/fenrg.2020.00062
Hüner NPA, Grodzinski B (2011) Photosynthesis and photoautotrophy
Iñiguez C, Capó-Bauçà S, Niinemets Ü et al (2020) Evolutionary trends in RuBisCO kinetics and their co-evolution with CO2 concentrating mechanisms. Plant J 101:897–918. https://doi.org/10.1111/tpj.14643
Irons R, Sekkapan G, Panesar R et al (2007) CO2 capture ready plants. IEA Greenhouse Gas Programme
Jin X, Gong S, Chen Z et al (2021) Potential microalgal strains for converting flue gas CO2 into biomass. J Appl Phycol 33:47–55. https://doi.org/10.1007/s10811-020-02147-8
Klanchui A, Cheevadhanarak S, Prommeenate P et al (2017) Exploring components of the CO2-concentrating mechanism in alkaliphilic cyanobacteria through genome-based analysis. Comput Struct Biotechnol J 15:340–350. https://doi.org/10.1016/j.csbj.2017.05.001
Kosheleva RI, Mitropoulos AC, Kyzas GZ (2019) Synthesis of activated carbon from food waste. Environ Chem Lett 17:429–438. https://doi.org/10.1007/s10311-018-0817-5
Kumar N, Kar S, Shukla P (2022) Role of regulatory pathways and multi-omics approaches for carbon capture and mitigation in cyanobacteria. Bioresourc Technol 366:128104. https://doi.org/10.1016/j.biortech.2022.128104
Laisk A, Eichelmann H, Oja V (2015) Oxidation of plastohydroquinone by photosystem II and by dioxygen in leaves Biochim Biophys Acta (BBA) - Bioenergetics 1847:565-575. https://doi.org/10.1016/j.bbabio.2015.03.003
Lau HC, Ramakrishna S, Zhang K et al (2021) The role of carbon capture and storage in the energy transition. Energy Fuels 35:7364–7386. https://doi.org/10.1021/acs.energyfuels.1c00032
Li J, Tang X, Pan K et al (2020) The regulating mechanisms of CO2 fixation and carbon allocations of two *Chlorella* sp. strains in response to high CO2 levels. Chemosphere 247:125814. https://doi.org/10.1016/j.chemosphere.2020.125814
Liu X-Y, Hou W-T, Wang L et al (2021) Structures of cyanobacterial bicarbonate transporter SbtA and its complex with PII-like SbtB. Cell Discov 7:–63. https://doi.org/10.1038/s41421-021-00287-w
Mabhaudhi T, Senzanje A, Modi AT et al (2022) Water-energy-food nexus narratives and resource securities: a global south perspective. Elsevier
Mangan NM, Brenner MP (2014) Systems analysis of the CO2 concentrating mechanism in cyanobacteria. eLife 3:e02043. https://doi.org/10.7554/eLife.02043
Masojídek J, Torzillo G, Koblížek M (2013) Photosynthesis in microalgae. In: Handbook of microalgal culture. https://doi.org/10.1002/9781118567166.ch2
Mata TM, Martins AA, Caetano NS (2010) Microalgae for biodiesel production and other applications: a review. Renew Sustainab Energy Rev 14:217–232. https://doi.org/10.1016/j.rser.2009.07.020
Middepogu A, Hou J, Gao X et al (2018) Effect and mechanism of TiO2 nanoparticles on the photosynthesis of *Chlorella pyrenoidosa*. Ecotoxicol Environ Safety 161:497–506. https://doi.org/10.1016/j.ecoenv.2018.06.027
Mohan SV, Modestra JA, Amulya K et al (2016) A circular bioeconomy with biobased products from CO2 sequestration. Trends Biotechnol 34:506–519. https://doi.org/10.1016/j.tibtech.2016.02.012
Montalvo GEB, Thomaz-Soccol V, Vandenberghe LPS et al (2019) *Arthrospira maxima* OF15 biomass cultivation at laboratory and pilot scale from sugarcane vinasse for potential biological new peptides production. Bioresourc Technol 273:103–113. https://doi.org/10.1016/j.biortech.2018.10.081

Olajire AA (2010) CO2 capture and separation technologies for end-of-pipe applications – a review. Energy 35:2610–2628. https://doi.org/10.1016/j.energy.2010.02.030

Omodolor IS, Otor HO, Andonegui JA et al (2020) Dual-function materials for CO2 capture and conversion: a review. Indus Eng Chem Res 59:17612–17631. https://doi.org/10.1021/acs.iecr.0c02218

Osman AI, Hefny M, Abdel Maksoud MIA et al (2021) Recent advances in carbon capture storage and utilisation technologies: a review. Environ Chem Lett 19:797–849. https://doi.org/10.1007/s10311-020-01133-3

Pang N, Gu X, Chen S et al (2019) Exploiting mixotrophy for improving productivities of biomass and co-products of microalgae. Renew Sustain Energy Rev 112:450–460. https://doi.org/10.1016/j.rser.2019.06.001

Peter AP, Khoo KS, Chew KW et al (2021) Microalgae for biofuels, wastewater treatment and environmental monitoring. Environ Chem Lett 19:2891–2904. https://doi.org/10.1007/s10311-021-01219-6

Rae Benjamin D, Long Benedict M, Badger Murray R et al (2013) Functions, compositions, and evolution of the two types of carboxysomes: polyhedral microcompartments that facilitate CO2 fixation in cyanobacteria and some proteobacteria. Microbiol Molecular Biol Rev 77:357–379. https://doi.org/10.1128/mmbr.00061-12

Rajabloo T, Valee J, Marenne Y et al (2023) Carbon capture and utilization for industrial applications. Energy Rep 9:111–116. https://doi.org/10.1016/j.egyr.2022.12.009

Rodas-Zuluaga LI, Castañeda-Hernández L, Castillo-Vacas EI et al (2021) Bio-capture and influence of CO2 on the growth rate and biomass composition of the microalgae *Botryococcus braunii* and *Scenedesmus* sp. J CO2 Util 43:101371. https://doi.org/10.1016/j.jcou.2020.101371

Rottet S, Förster B, Hee WY et al (2021) Engineered accumulation of bicarbonate in plant chloroplasts: known knowns and known unknowns. Front Plant Sci 12. https://doi.org/10.3389/fpls.2021.727118

Satya IA, Satya A, Chrismadha T et al (2024) Cultivation of *Spirulina Platensis* for carbon dioxide bio sequestration in hybrid photobioreactor with real-time monitoring system. J Environ Chem Eng 12:112396. https://doi.org/10.1016/j.jece.2024.112396

Su Y (2021) Revisiting carbon, nitrogen, and phosphorus metabolisms in microalgae for wastewater treatment. Sci Total Environ 762:144590. https://doi.org/10.1016/j.scitotenv.2020.144590

Talekar S, Jo BH, Dordick JS et al (2022) Carbonic anhydrase for CO2 capture, conversion and utilization. Curr Opin Biotechnol 74:230–240. https://doi.org/10.1016/j.copbio.2021.12.003

Theo WL, Lim JS, Hashim H et al (2016) Review of pre-combustion capture and ionic liquid in carbon capture and storage. Appl Energy 183:1633–1663. https://doi.org/10.1016/j.apenergy.2016.09.103

Vecchi V, Barera S, Bassi R et al (2020) Potential and challenges of improving photosynthesis in algae. Plants 9:67. https://doi.org/10.3390/plants9010067

Wang C, Sun B, Zhang X et al (2019) Structural mechanism of the active bicarbonate transporter from cyanobacteria. Nat Plants 5:1184–1193. https://doi.org/10.1038/s41477-019-0538-1

Wei X, Manovic V, Hanak DP (2020) Techno-economic assessment of coal- or biomass-fired oxy-combustion power plants with supercritical carbon dioxide cycle. Energy Convers Manage 221:113143. https://doi.org/10.1016/j.enconman.2020.113143

Wienchol P, Szlęk A, Ditaranto M (2020) Waste-to-energy technology integrated with carbon capture – challenges and opportunities. Energy 198:117352. https://doi.org/10.1016/j.energy.2020.117352

Xiong W, Lee T-C, Rommelfanger S et al (2015) Phosphoketolase pathway contributes to carbon metabolism in cyanobacteria. 2:1–8. https://doi.org/10.1038/nplants.2015.187

Yadav G, Meena DK, Sahoo AK et al (2020a) Effective valorization of microalgal biomass for the production of nutritional fish-feed supplements. J Cleaner Prod 243:118697. https://doi.org/10.1016/j.jclepro.2019.118697

Yadav G, Panda SP, Sen R (2020b) Strategies for the effective solid, liquid and gaseous waste valorization by microalgae: a circular bioeconomy perspective. J Environ Chem Eng 8:104518. https://doi.org/10.1016/j.jece.2020.104518

Yang B, Liu J, Ma X et al (2017) Genetic engineering of the Calvin cycle toward enhanced photosynthetic CO2 fixation in microalgae. Biotechnol Biofuels 10:229. https://doi.org/10.1186/s13068-017-0916-8

Yao J, Han H, Yang Y et al (2023) A review of recent progress of carbon capture, utilization, and storage (CCUS) in China. App Sci 13. https://doi.org/10.3390/app13021169

Chapter 9
Cyanobacterial Biomass as a Feedstock for Biochemicals

Antira Wichaphian, Nanthakrit Sriket, Nang Yee Moo Htoo, May Thu Zin, and Sirasit Srinuanpan

Abstract The ongoing rise in the global human population and the depletion of natural energy resources present substantial environmental threats. Effective addressing of these challenges necessitates the sustainable provision of food and energy devoid of any adverse environmental impact to meet the sustainable development goals outlined by the United Nations. Cyanobacteria have recently emerged as promising candidates capable of meeting the requirements mentioned above. Their proficiency in efficiently harnessing solar energy and converting it into high-value biomass, utilizing CO_2, water, and nutrients, positions them as noteworthy contributors to the circular bioeconomy framework. During this conversion of radiant energy into chemical energy, these biological systems concurrently generate oxygen as a byproduct. The cyanobacterial biomass includes a diverse array of biochemical compounds, including polysaccharides, proteins, lipids, vitamins, and elements, constituting a valuable feedstock for the production of biochemicals with applications ranging from food and energy to biofertilizers, secondary metabolites, cosmetics, and medicinal products. While cyanobacteria represent a feasible source

A. Wichaphian · N. Sriket · M. T. Zin
Department of Biology, Faculty of Science, Chiang Mai University, Chiang Mai, Thailand

Microbial Biorefinery and Biochemical Process Engineering Research Group, Chiang Mai University, Chiang Mai, Thailand

N. Y. M. Htoo
Department of Biology, Faculty of Science, Chiang Mai University, Chiang Mai, Thailand

S. Srinuanpan (✉)
Department of Biology, Faculty of Science, Chiang Mai University, Chiang Mai, Thailand

Biorefinery and Bioprocess Engineering Research Cluster, Chiang Mai University, Chiang Mai, Thailand

Center of Excellence in Microbial Diversity and Sustainable Utilization, Chiang Mai University, Chiang Mai, Thailand

Office of Research Administration, Office of the University, Chiang Mai University, Chiang Mai, Thailand
e-mail: sirasit.s@cmu.ac.th

M. A. Mehmood et al. (eds.), *Cyanobacteria Biotechnology*,
https://doi.org/10.1007/978-3-031-70698-1_9

of biochemical feedstock, their industrial applications are impeded by energy and cost challenges. Researchers have therefore expressed interest in addressing these challenges by applying the biorefinery approach. This chapter provides an in-depth exploration of cyanobacterial biomass as a biochemical feedstock for multiproduct biorefinery, elucidating its potential applications and contributing to the advancement of sustainable industrial practices.

Keywords Cyanobacteria · Biomass · Biochemicals · Feedstock · High value

9.1 Introduction

Algae, especially cyanobacteria, emerge as the predominant and ubiquitous organisms on Earth, displaying a propensity for flourishing in a myriad of environments, notably within marine and freshwater ecosystems. Marine waters stand out as particularly favorable, offering nutrient-rich habitats conducive to the thriving cultivation of cyanobacteria (Deviram et al. 2020). These organisms, characterized by their diminutive size and multicellular/unicellular nature, often aggregate to form extensive colonies. The taxonomic spectrum of cyanobacteria is notably diverse, showcasing an array of morphologies with various shapes and sizes, exemplified by the identification of over 150 genera to date. Noteworthy is their antiquity, with discernible characteristics traceable to ancient fossils dating back more than 3.5 billion years, underscoring their enduring presence throughout Earth's evolutionary history. Beyond their historical significance, cyanobacteria assume a pivotal role in shaping the contemporary oxygenic environment, thus contributing fundamentally to the evolutionary trajectory of the planet (Pekkoh et al. 2023).

Cyanobacteria, tracing their origins to the earliest life forms, have undergone extensive evolutionary processes and metabolic adaptations spanning centuries, endowing them with the capacity to withstand formidable environmental challenges. This inherent resilience positions algae as a paramount candidate for a myriad of industrial and consumer applications (Pathak et al. 2018). The utilization of cyanobacterial biomass includes a spectrum of applications (Fig. 9.1) ranging from energy production, including bioethanol, biodiesel, biogas, biohydrogen, and bio-oil, to a diverse array of bioproducts, including pigments, bioplastics, fatty acids, proteins, biofertilizers, biochar, vitamins, polyphenols, and phytosterols (Arora and Philippidis 2023). In the pursuit of economically viable commercialization of cyanobacterial-based biofuels and bioproducts, the industry has ardently embraced the biorefinery concept, underscored by the comprehensive valorization of the entire biomass, spanning lipids, carbohydrates, proteins, and other micro-components culminating in the production of marketable products (Pekkoh et al. 2022). As of 2017, the global market for algae-based bioproducts was conservatively estimated at USD 33 billion, with a prognosticated surge to USD 53 billion by 2026 (Rahman et al. 2020). Additionally, considering the escalating market interest and elevated consumer consciousness regarding natural and sustainable commodities, the industry is

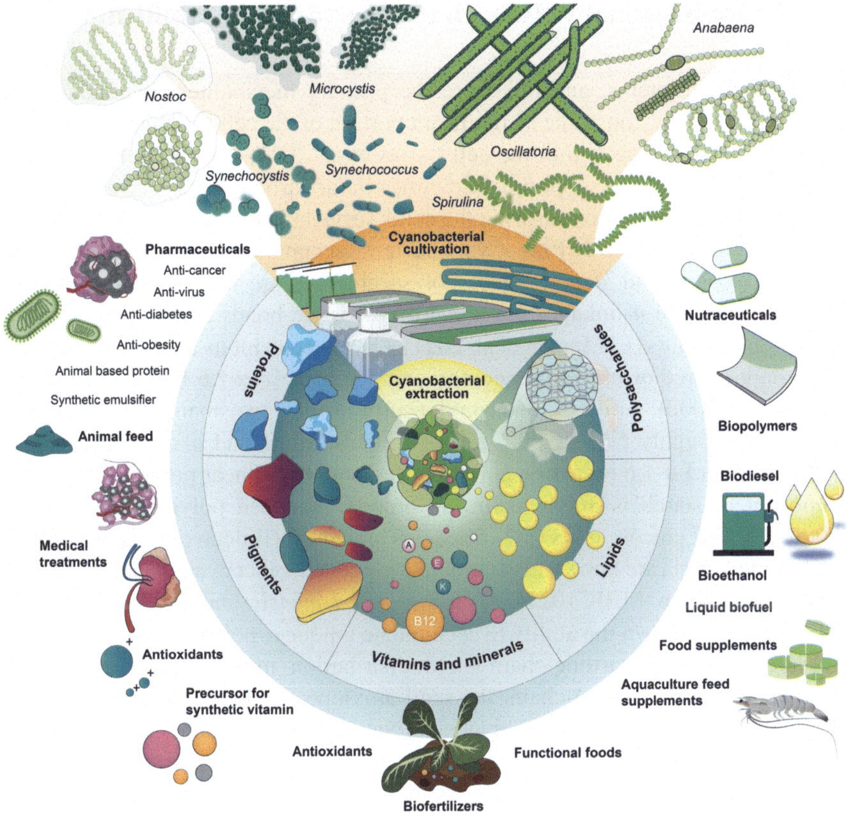

Fig. 9.1 Diverse applications of biochemicals from cyanobacteria

undergoing a discernible shift, gradually replacing synthetic compounds with their natural equivalents, distinguished by attributes of safety, renewability, and environmental friendliness. Particularly noteworthy is the food and cosmetics sector, which, with a substantial global market size of USD 278 billion in 2020, is poised to manifest a compound annual growth rate (CAGR) of 5% from 2021 to 2028 (Arora and Philippidis 2023), has experienced a significant transformation toward products sourced from botanical origins and microorganisms. In the strategic diversification of their product portfolio, cyanobacterial-derived biochemicals are increasingly garnering attention as value-added commodities, propelled by their robust market demand and versatile applications across the food, feed, cosmetic, chemical, and pharmaceutical industries.

This chapter thoroughly explores the application of cyanobacterial biomass as a biochemical feedstock within a multiproduct biorefinery framework, elucidating its potential applications and contributing to the progress of sustainable industrial practices.

9.2 Cyanobacterial Polysaccharides and Their Applications

Carbohydrates, proteins, and lipids are the primary metabolites accumulated by cyanobacteria. They constitute a significant portion of the cell wall and serve as the primary carbon storage within the cell (Arias et al. 2021). Extracellular carbohydrates provide structural support, while intracellular carbohydrates serve as energy sources or defense systems, ensuring the cell's survival in the face of environmental challenges (Singh et al. 2019a). Cyanobacteria accumulate carbohydrates in the form of glycogen and starch. These polysaccharides are made up of D-glucose polymerized by α-1,4-linkages using α-1,6-glucosidic bonds at different frequencies of branch points (El Mannai et al. 2021). Glycogen exhibits a uniform structure with a high degree of branching, comprising 7% to 10% α-1,6 branches (Shearer and Graham 2002). Amylose and amylopectin are the two main components of starch. With roughly 500–20,000 α-1,4-D-glucose units and a small percentage (less than 1%) of α-1,6 branches, amylose is essentially a linear polymer. The main component in starch is amylopectin, an orderly branched polymer made up of 10,000–100,000 glucose units (5%–6% α-1,6 branches) (Buléon et al. 1998). While most cyanobacteria are known to accumulate glycogen, other cyanobacteria, such as *Cyanobacterium* sp. MBIC10216, *Myxosarcina burmensis*, and *Synechococcus* sp. BG043511 have shown the ability to accumulate unique α-polyglucans identified as semi-amylopectin. Regarding chain length distribution, molecular size, and length of the most abundant α-1,4-chain, the semi-amylopectin fell between rice amylopectin and typical cyanobacterial glycogen (Nakamura et al. 2005).

Extracellular polysaccharides, or exopolysaccharides (EPS), represent another significant source of carbohydrates in cyanobacteria. To withstand adverse conditions such as variations in light intensity (Phélippé et al. 2019), nutrient scarcity (Marchus et al. 2018), and changes in humidity (Adessi et al. 2018), these compounds create microenvironments within the soil. Two distinctive properties of cyanobacterial exopolysaccharides distinguish them from other bacterial polysaccharides. The first notable characteristic is their anionic nature, attributed to the presence of uronic acid and sugars containing sulfate. Additionally, most cyanobacterial exopolysaccharides are complex heteropolysaccharides, with approximately 75% composed of six or more distinct types of monosaccharide units (Pereira et al. 2009). Depending on the species, soluble and insoluble fractions of monosaccharides make up most exopolysaccharides. Although glucose is the most prevalent component, numerous species also include xylose, arabinose, fucose, galactose, mannose, rhamnose, and glucosamine (De Philippis and Vincenzini 1998; Tiwari et al. 2020). EPS can be coupled with lipids (glycolipids) and proteins (proteoglycans) when they are produced in significant quantities (Pereira et al. 2009). There is currently no consensus concerning the classification of cyanobacterial EPS (Delattre et al. 2016). While many authors agree that they can generally be categorized into two main groups based on their location: (1) cell-bound polysaccharides (CPS), including slime, sheaths, and "capsular"; and (2) released polysaccharides (RPS) (Kamravamanesh et al. 2018).

Table 9.1 illustrates the applications of cyanobacterial carbohydrates or polysaccharides. Cyanobacterial carbohydrates have some advantages over other higher plants or green algae. For example, cyanobacteria contain peptidoglycan-based cell walls instead of cellulose ones, easily broken down by yeasts or other fermentative bacteria (Aikawa et al. 2015; Klanchui et al. 2018). These cell walls make cyanobacterial biomass easy to ferment to produce fermented products. Glycogen is also a better feedstock than starch for storing carbohydrates, particularly for dark fermentation. Numerous environmental conditions trigger the generation of carbohydrates in cyanobacteria and other microalgae species. In well-regulated cultures, cyanobacteria may accumulate more significant percentages of carbohydrates in their dry cell weight due to various conditions that can stress the cells (Arias et al. 2021). Cyanobacteria generate more glycogen when employing carbohydrate accumulation strategies such as the three-stage feeding strategy (growth phase, feast-famine phase, and feast phase). In the case of *Synechocystis* sp., the highest glycogen content was 69% (Rueda et al. 2020a). Wastewater-borne cyanobacteria cultivated under nitrogen limitation, exposed to alternating light-dark cycle (12 h of light and 12 h of darkness) and continuous light for 24 h, significantly accumulated 75% and 62% of carbohydrates, respectively (Arias et al. 2018). Moreover, nitrogen and phosphorus depletion boost the carbohydrate accumulation of water-borne cyanobacteria to a level of 69% (Rueda et al. 2020b). In the effort to remove nutrients and organic substances from industrial effluents and simultaneously generate biomass rich in carbohydrates, a culture of filamentous cyanobacteria *Geitlerinema* sp. was introduced into a laboratory-scale stirred tank photobioreactor using a varied hydraulic retention time (HRT). The findings reveal that changes in HRT impacted the inorganic carbon profiles and N:P ratios of the culture, subsequently influencing microbial changes and the amount of carbohydrates produced. The highest carbohydrate accumulation was achieved with an 8-day HRT, characterized by an N:P ratio of 22:1 and ample carbon availability from the effluent ($CaCO_3$ concentration: 109.63 ± 57.15 mg L^{-1}). Under these conditions, a steady 57% carbohydrate content was observed without any additional carbon source (Sánchez-Contreras et al. 2021).

The study of EPS in cyanobacteria has gained considerable attention, alongside investigating intracellular carbohydrates. The types and compositional monosaccharides of EPS vary among different cyanobacterial species. For instance, *Aphanothece sacrum*, *Nostoc flagelliforme*, and *Cyanothece* sp. 113 share the same EPS type, namely CPS. However, their monosaccharide compositions differ. *A. sacrum* has up to 8 types of monosaccharides, *N. flagelliforme* has five types, and *Cyanothece* sp. 113 EPS contains only one type, which is D-glucose (Ngatu et al. 2012; Chi et al. 2007; Kanekiyo et al. 2005). Numerous prior studies have consistently demonstrated that cyanobacteria exhibit a wide range of types proficiently producing EPS. *Spirulina platensis* (*Arthrospira platensis*), a filamentous cyanobacterium, was examined for the impact of light intensity and temperature on its growth and EPS production. Under optimal light intensity conditions at 180 μmol m^{-2} s^{-1} and a temperature of 33 °C, the cultivation resulted in an EPS production of 290 ± 5 mg L^{-1} (Trabelsi et al. 2009). *Limnothrix redekei* PUPCCC 116, another filamentous cyanobacterium, has been documented to produce 304 mg of EPS per

Table 9.1 Carbohydrates and polysaccharides produced by different cyanobacterial species and their applications

Species	Polysaccharide type	Applications	Efficiency	References
Spirulina platensis	Carbohydrates	Bioethanol production	6.5 g L^{-1}	Aikawa et al. (2013)
Synechococcus elongatus PCC7942	Carbohydrates	Bioethanol production	7.2 g L^{-1}	Chow et al. (2015)
Synechococcus sp. PCC 7002	Carbohydrates	Bioethanol production	30 g L^{-1}	Möllers et al. (2014)
Microcystis wesenbergii and *Microcystis aeruginosa*	Carbohydrates	Biohydrogen production	24.96 mL H_2 g^{-1} TVS	Cheng et al. (2019)
Anabaena siamensis TISTR8012	Carbohydrates	Biohydrogen production	8.68 mmol H_2 mg^{-1} chl a^{-1} h^{-1}	Khetkorn et al. (2010)
Nostoc punctiforme ATCC 29133	Carbohydrates	Biohydrogen production	20.7 mmol H_2 mg^{-1} chl a^{-1} h^{-1}	Khetkorn et al. (2010)
S. elongatus PCC 7942	Carbohydrates	Biobutanol production	14.5 mgL^{-1}	Lan and Liao (2011)
Synechocystis PCC 6803	Carbohydrates	Biobutanol production	35 mg g^{-1} biomass	Anfelt et al. (2015)
Aphanothece flocculosa	EPS	Hg^{2+} absorption	456 mg g^{-1}	Cain et al. (2008)
S. platensis	EPS	Co^{2+} absorption Cu^{2+} absorption Zn^{2+} absorption	181 mg g^{-1} 272.1 mg g^{-1} 250.3 mg g^{-1}	Vannela and Verma (2006)
Spirulina sp.	EPS	Cu^{2+} absorption Cr^{3+}, Cr^{6+} absorption Ni^{2+} absorption	100–389 mg g^{-1} 167–304 mg g^{-1} 515–1378 mg g^{-1}	Doshi et al. (2007)
Aphanothece sacrum	CPS (Sacran)	Anti-inflammatory and anti-allergic properties	Improving skin barrier function and reducing T_l cytokine production	Ngatu et al. (2012)
Nostoc flagelliforme	CPS (Nostoflan)	Anti-viral activity	Anti-herpes simplex virus type 1 (HSV-1) activity with a selectivity index of 13,000	Kanekiyo et al. (2005)
S. platensis	CPS (Calcium spirulan)	Anti-viral activity	Inhibit the replication of several enveloped viruses	Hayashi et al. (1996)

(continued)

Table 9.1 (continued)

Species	Polysaccharide type	Applications	Efficiency	References
Anabaena J-1	EPS	Bioflocculants (Cell-bound flocculant)	1.1 U μg^{-1} cellular protein	Bar-Or and Shilo (1988)
Anabaena circularis 6720	EPS	Bioflocculants (Cell-bound flocculant)	2.0 U μg^{-1} cellular protein	Bar-Or and Shilo (1988)
Cyanothece epiphytica	EPS	Bio-lubricants	Effective hydrophobic dispersant, outstanding emulsifying agent, and a proficient flocculant	Borah et al. (2018)
Phormidium valderianum BDU 30501	EPS	Bioremediation (Decolorization)	55% color reduction	Kalavathi et al. (2001)
Oscillatoria boryana BDU 92181	EPS	Bioremediation (Decolorization)	60% color reduction	Kalavathi et al. (2001)
Oscillatoria, Synechococcus, Nodularia, Nostoc, Anabaena, and *Microcystis*	EPS	Bioremediation (Biodegradation)	Degradation of lindane [g-hexachlorocyclohexane (g-HCH)]	El-Bestawy et al. (2007)

liter of culture over 21 days (Khattar et al. 2010). Some studies on cyanobacteria cultivation under salt stress conditions have indicated that elevated EPS production is linked to sodium chloride (NaCl) tolerance. Among the examined strains, *Synechocystis* sp. BASO444 exhibited the highest salt tolerance and EPS secretion at 500 mg L^{-1} (Ozturk and Aslim 2010). *Cyanothece* sp. 113 was subjected to EPS production testing in a modified medium incorporating seawater and sea mud extract. Under optimal conditions, this strain demonstrated the capability to produce a substantial quantity of CPS, reaching 18.4 g L^{-1} (Su et al. 2007).

Owing to their strong CO_2 consumption and adaptability as a feedstock for various biotechnological and biorefinery applications, these microorganisms have garnered increased attention from researchers. Compared to first- and second-generation feedstocks, cyanobacteria provide several benefits as a third-generation feedstock. Recent research has concentrated on biofuels that are based on carbohydrates, including bioethanol, biohydrogen, and biobutanol (Arias et al. 2021). To produce ethanol directly from *S. platensis*, the researchers used lysozyme with a recombinant yeast strain expressing amylase. This procedure removes the need for pretreatment of biomass or amylase hydrolysis. *S. platensis* was directly converted into ethanol, yielding 6.5 g L^{-1} (86% of the theoretical yield) (Aikawa et al. 2013). Indeed, many studies have employed pretreatment methods on cyanobacteria biomass before initiating ethanol fermentation processes. Pretreatments are often implemented to enhance the accessibility of cellulose and other biomass components, facilitating more efficient conversion of cyanobacterial biomass into ethanol during the fermentation stage. These pretreatment techniques aim to break down complex structures, reduce inhibitory substances, and improve overall substrate digestibility, ultimately optimizing ethanol production. *S. elongatus* PCC 7942, a genetically engineered cyanobacterial strain, demonstrates increased rates of photosynthesis, enhanced cell growth, and improved efficiency in carbohydrate production. The resulting cyanobacterial biomass undergoes effective hydrolysis with dilute acid, achieving an impressive 90% glucose recovery at a high biomass concentration of 80 g L^{-1}. Subsequently, bioethanol is produced by fermenting the acidic hydrolysate, resulting in a concentration of 7.2 g L^{-1} with a remarkable 91% theoretical yield (Chow et al. 2015)—the biomass derived from the rapidly growing photosynthetic marine cyanobacterium *Synechococcus* sp. PCC 7002 underwent enzymatic hydrolysis utilizing lysozyme and two alpha-glucanases. Subsequently, the enzymatic hydrolysate was subjected to ethanol fermentation by *Saccharomyces cerevisiae*. The highest achieved ethanol concentration was 30 g L^{-1}, with 90% of the glucose in the biomass successfully converted to ethanol (Möllers et al. 2014).

Within the spectrum of biohydrogen production techniques, harnessing cyanobacterial carbohydrates directly or indirectly for biohydrogen synthesis through dark fermentation or indirect biophotolysis is a notable pursuit. Both methodologies involve the participation of numerous heterotrophic bacteria or cyanobacterial species equipped with nitrogenase or hydrogenase enzymes (Bandyopadhyay et al. 2010; Mishra et al. 2019). The study shows hydrogen production by dark fermentation of carbohydrate-enriched cyanobacterial biomass. The biomass, originating from an algal bloom in Dianchi Lake, underwent hydrothermal acid pretreatment before undergoing sequential processes of dark hydrogen fermentation and anaerobic digestion. This led to achieving a maximum energy conversion efficiency of 44.1%, accompanied by the production of 25 mL H_2 g^{-1} total volatile solids (TVS) (Cheng et al. 2019). *Anabaena siamensis* TISTR 8012 and *Nostoc punctiforme* ATCC 29133 can produce biohydrogen through indirect biophotolysis. They exhibit H_2 production rates of 8.68 and 20.7 mmol H_2 mg^{-1} chl a^{-1} h^{-1}, respectively (Khetkorn et al. 2010).

Biobutanol, another type of fuel derived from cyanobacteria, can be produced through fermentative processing by *Clostridium* spp. Many investigations into one-stage biobutanol production through genetic engineering have focused on two cyanobacterial model strains: *Synechococcus elongatus* PCC 7942 and *Synechocystis* PCC 6803 (Arias et al. 2021). These studies found that *S. elongatus* PCC 7942 accumulated 14.5 mg L^{-1} of butanol, while *Synechocystis* sp. PCC 6803 achieved an n-butanol yield of 35 mg g^{-1} of biomass with a productivity of 2.7 mg g^{-1} biomass per day (Lan and Liao 2011; Anfelt et al. 2015).

Cyanobacterial EPS possess unique properties that make them valuable in various biotechnological applications. For heavy metal removal and wastewater treatment properties, Cyanobacterial EPS can form complex structures with heavy metals, aiding in their removal from water systems. The EPS acts as a biosorbent, helping to sequester and immobilize heavy metals, making them less harmful to the environment (Choi et al. 2022). *Aphanothece flocculosa* exhibits a significant capability to Hg^{2+}, with an absorption potential of 456 mg g^{-1} (Cain et al. 2008). The research indicates that *S. platensis* possesses the ability to absorb three different types of heavy metals, namely Co^{2+} (181 mg g^{-1}), Cu^{2+} (272.1 mg g^{-1}), and Zn^{2+} (250.3 mg g^{-1}) (Vannela and Verma 2006). Similarly, *Spirulina* sp. has been reported to exhibit a high absorption capacity for three distinct types of heavy metals, specifically Cu^{2+} (100–389 mg g^{-1}), $Cr^{3+,6+}$ (167–304 mg g^{-1}), and Ni^{2+} (515–1378 mg g^{-1}) (Doshi et al. 2007).

EPS from cyanobacteria can be utilized in the food and pharmaceutical industries for various purposes. Due to their gel-like consistency and water-absorbing properties, they may serve as thickening or gelling agents in food products (Singha 2012; Singh et al. 2019b). In the pharmaceutical industry, these substances are potentially utilized in drug delivery systems (Saha and Datta 2022). Moreover, a prior study indicates that *Aphanothece sacrum* CPS exhibits anti-inflammatory and anti-allergic properties (Ngatu et al. 2012). Additionally, both *Nostoc flagelliforme* and *S. platensis* CPS have been reported to display anti-viral activity (Kanekiyo et al. 2005; Hayashi et al. 1996).

Cyanobacterial EPS can act as bioflocculants, promoting the aggregation of suspended particles in water. This property is beneficial in processes such as water clarification, where the coagulated particles settle, leaving more transparent water. Species like *Anabaena*, including strains such as *Anabaena* species strain J-1 and *Anabaena circularis* 6720, are known to secrete extracellular flocculants into their surroundings. These substances have practical applications in clarifying turbid lakes and other water bodies. Extracellular flocculants aid in aggregating and settling suspended particles in water, contributing to the purification and clearing of turbidity in aquatic environments. This natural process can be harnessed for water treatment and environmental remediation (Bar-Or and Shilo 1988).

The unique properties of cyanobacterial EPS, such as their viscosity and lubricating abilities, make them suitable candidates for bio-lubricants. Bio-lubricants derived from cyanobacterial EPS can find applications in various industries where environmentally friendly lubricants are preferred. The analysis revealed that the EPS from *Cyanothece epiphytica* has potential as a bio-lubricant, exhibiting characteristics superior to conventional lubricants like "grease." This implies that the EPS could be a promising alternative in lubrication applications, offering improved performance and environmentally friendly attributes compared to traditional lubricants (Borah et al. 2018).

Cyanobacteria and their EPS can play a role in bioremediation processes, helping to break down and remove environmental pollutants. Their ability to thrive in diverse conditions makes them helpful in cleaning up contaminated sites (Parwani et al. 2021). A consortium composed of *Phormidium valderianum* BDU 30501 and *Oscillatoria boryana* BDU 92181 has been reported to remove phenol and melanoidin from distillery effluents effectively (Kalavathi et al. 2001). Additionally, cyanobacterial strains such as *Oscillatoria*, *Synechococcus*, *Nodularia*, *Nostoc*, *Anabaena*, and *Microcystis* have demonstrated the capability to break down lindane residues (El-Bestawy et al. 2007). Moreover, certain EPS-producing cyanobacterial strains, including *Anabaena* sp., *Lyngbya* sp., *Microcystis* sp., *Nostoc* sp., and *Spirulina* sp., have been identified for their ability to eliminate glyphosate herbicide from contaminated soil (Lipok et al. 2009).

9.3 Cyanobacteria Proteins and Their Applications

Proteins have captured the attention of numerous researchers, and cyanobacteria have been a subject of exploration since the mid-twentieth century due to their novel properties. Proteins, typically composed of 2–20 amino acids, offer insights into functionality and nutritional quality, forming a crucial link between protein intake and the ability to predict health outcomes (Akinbule et al. 2023). Furthermore, proteins play a vital role in every stage of the human life cycle, from early infancy to old age. Inadequate protein intake can have severe consequences, including growth failure and an increased susceptibility to infections, and this has become evident in many regions globally due to food scarcity.

In this context, it is imperative to acknowledge the potential role of natural products as dietary supplements, providing viable options in response to insufficient protein sources due to food scarcity. As a result, numerous researchers have sought alternative protein sources to reduce reliance on existing agricultural and food production systems, which may struggle to meet the increasing global food demand, including exploring the role of cyanobacteria (Béné et al. 2015; Rösch et al. 2019).

Among the 2000 species of modern cyanobacteria, *Spirulina* has been widely employed as a food supplement for numerous years due to its elevated nutritional value and safety for human consumption. Remarkably, its protein content in dry weight surpasses that of soybeans by 20 times, corn by 40 times, and beef by an astounding 400 times. For example, another cyanobacterium, *Phormidium* sp., has garnered attention for its versatility across various industries, including food, pharmaceuticals, and cosmetics. Similarly, phycoerythrins (PE) extracted from *Phormidium* sp. A27DM exhibits apoptotic potentials on A549 human lung carcinoma cells (Madamwar et al. 2015). Furthermore, an 11-kDa protein known as cyanovirin-N (CV-N), initially isolated from the cyanobacterium *Nostoc ellipsosporum*, demonstrates the ability to inactivate diverse strains of human immunodeficiency virus type 1 (HIV-1), HIV-2, simian immunodeficiency virus, and feline immunodeficiency virus (Mori and Boyd 2001).

However, the cell wall of cyanobacteria exhibits complex structures with carbohydrate-rich compositions, necessitating mechanical or non-mechanical pretreatment to facilitate cell wall disruption. The mechanical pretreatment can be achieved using a bead mill, high- or low-pressure homogenizer, high-speed homogenizer, ultrasonication, hydrodynamic and microfluidizer, while non-mechanical can further be divided into three methods: physical, chemical, and enzymatic (Amorim et al. 2021).

An alternative technique for cell wall disruption involves using polysaccharide-digesting enzymes, specifically cellulase, characterized as a food-grade approach to break down the cell wall. However, this method is contingent on factors such as the enzyme-to-substrate ratio, temperature, and pH for optimal protein recovery (Le Nguyen Doan et al. 2022). Additionally, a traditional method utilizes alkaline extraction with sodium hydroxide (NaOH) solution, which is widely employed in various food matrices (Naseri et al. 2020). This is followed by acidic extraction using hydrochloric (HCl), and the combination of these two steps ensures a pH shift from 10 to 3, ultimately maximizing the yield of extracted components.

From the standpoint of human health and safety profiles, protein hydrolysates are typically generated through enzymatic hydrolysis of whole protein sources using appropriate proteolytic enzymes under controlled conditions. This process aims to isolate desired bioactive peptides from a complex mixture of active and inactive peptides. Depending on their amino acid composition, sequences, hydrophobicity, and length, peptides released from food proteins may exhibit various biological activities, including antihypertensive, antioxidative, antithrombotic, hypoglycemic, hypocholesterolemic, and antibacterial activities (Nasri 2017).

Enzymatic hydrolysis is also employed to synthesize bioactive peptides from microalgal proteins for clinical applications, as it depletes harmful substances to yield desirable outcomes (Sathya et al. 2021). It is important to note that various protein extraction techniques can influence the bioactive metabolite content of cyanobacteria species. For instance, hydrolyzation yields bioactive peptides from spirulina, reported to exhibit anti-cancer properties and effectiveness against immunodeficiency diseases caused by HIV and other viral infections (Sathasivam et al. 2019). Protein hydrolysate is commonly achieved through chemical hydrolysis or the use of proteases. Chemical hydrolysis involves acids or alkalis at temperatures exceeding 40 °C. While this technique is simple and cost-effective, controlling the composition of the hydrolysate and resulting products is challenging (Branco et al. 2018). Strong organic acids such as HCl and trichloroacetic acid can destroy tryptophan, a crucial amino acid for the human body, making this method less favorable. In contrast, alkaline hydrolysis with NaOH can form unusual amino acid residues like lysinoalanine or lanthionine while reducing the amounts of cystine, arginine, threonine, serine, and isoleucine.

Consequently, many researchers have explored an alternative hydrolysis method known as enzymatic hydrolysis using protease enzymes. Different protease enzymes exhibit distinct preferred cleavage sites, leading to variations in bioactive properties. Utilizing enzymatic hydrolysis with gastrointestinal proteases can activate previously inactive peptide sequences in parent proteins. The use of proteases for protein hydrolysis gained popularity due to reports indicating that peptides obtained from hydrolysis are biologically active and enhance bioactive properties (Kim and Wijesekara 2010). Two major types of broad protease enzymes exist, exopeptidases and endopeptidases, although specific oligopeptides may work for particular peptide bonds. These proteases can be sourced from plants, animals, bacteria, and viruses. Widely used proteases for protein hydrolysis include trypsin, pepsin, papain, and α-chymotrypsin. Additionally, enzymatic hydrolysis will likely generate more specific peptides, resulting in a higher yield (Echave et al. 2021). For instance, immune peptides play a role in modulating lymphocyte proliferation and increasing the phagocytic activity of macrophages in humans (Ahn et al. 2008). Previous studies on the protein hydrolysate of *Spirulina* with trypsin, pepsin, and chymotrypsin under optimal conditions exhibited anti-cancer properties on five different cell lines (HepG-2, MCF-7, SGC-7901, A549, and HT-29) (Wang and Zhang 2017).

In the past, protein powder was encapsulated into edible tablets for therapeutic purposes; however, incorporating all compounds into these tablets could compromise the properties of individual components. Due to its capacity to produce diverse bioactive compounds and other valuable co-products, cyanobacteria have extensive applications across various fields, including antioxidants, biofertilizers, biodegradable polymers, nutrient supplements, and color substances. Table 9.2 illustrates proteins obtained from different species of cyanobacteria. For example, spirulina (*A. platensis*) is a notable source of antioxidant peptides, and its hydrolysate holds significant potential in the pharmaceutical and food industries. A pepsin hydrolysate of spirulina demonstrated a higher degree of hydrolysis and antioxidant activity. Moreover, it markedly reduced lipid oxidation in emulsions compared to the original protein (Mohammadi et al. 2022). Furthermore, wild halotolerant cyanobacteria, particularly the novel pigment-protein complex phycocyanin from *Phormidium versicolor* NCC 466, exhibited hepatoprotective effects by increasing antioxidant enzyme activities, inhibiting lipid peroxidation, and decreasing the levels of hepatic markers (Gammoudi et al. 2019).

Furthermore, *Spirulina* has emerged as a promising nutrient source in aquaculture, particularly as animal feed, owing to its rich protein content and antioxidant properties. These attributes contribute to the growth of fish, enhance feeding efficiency, and influence the physiological response to diseases. A previous study detailed the preparation of a pellet-form feed for fattening Tilapia, incorporating *Spirulina* sp. (20%), cassava leaf flour (50%), gliricidia leaf flour (20%), and rice husk flour (10%). This formulation exhibited notable characteristics, including low disintegration, high water absorption capacity, and enhanced flotage due to the expansion effect. The floatation analysis revealed that 60% of the pellets remained afloat for over 40 min, and 80% retained their shape for 4 h. In another experiment involving rainbow trout (*Oncorhynchus mykiss*), brook trout (*Salvelinus fontinalis*), brown trout (*Salmo trutta fario*), African catfish (*Clarias gariepinus*), and a diet supplemented with *Spirulina* showed improved digestibility. The 16S rRNA sequencing results indicated that the *Spirulina*-enriched diet did not significantly impact the overall microbial community structure. It may be advantageous to conduct a dedicated laboratory trial focusing on protein extraction or protein hydrolysate using proteases for the animal feed industry. Protein hydrolysate is considered milder and safer for consumption, supported by numerous studies demonstrating its reduced toxicity both in vitro and in vivo. Furthermore, peptides with lower molecular weight are generally less allergenic than their parent proteins (Nasri 2017).

Table 9.2 Cyanobacteria proteins and their applications

Strains	Extraction method	Yield	Applications	References
S. platensis	Ultrasound extraction (SPE)	Proteins66.6 (wt%)	Animal-based proteins, synthetic emulsifiers	Silva et al. (2022)
S. platensis	NaCl	Phycobiliprotein 102.4 mg g^{-1} DW Proteins	Antimicrobial agent	Wang et al. (2023)
S. platensis	Aqueous extraction and high pressure	Allophycocyanin 31.3%	Anti-viral property	Shih et al. (2003)
S. platensis	Cellulase-assisted protein extraction	Proteins (76.45%)	Plant-based meat products	Le Nguyen Doan et al. (2022)
S. platensis	Hydrolysis with nanoliposomes	NA	Accelerating wound healing via angiogenesis and collagen production	Ebrahimi et al. (2023)
S. platensis	Ultrasound coupled with subcritical water (USW) technology	NA	Anti-diabetes (type2)	Hu et al. (2019)
S. platensis	Enzymatic hydrolysis	NA	Inhibitory effect on muscle atrophy	Lee et al. (2022)
S. platensis	Enzymatic hydrolysis	NA	Anti-obesity	Zhao et al. (2019)
S. platensis	Enzymatic hydrolysis	NA	Anti-cancer properties	Wang and Zhang (2017)
Phormidium versicolor NCC466	Sodium buffer	Phycocyanin (C-PC)	Antioxidant properties	Gammoudi et al. (2019)
Phormidium tergestinum	Aqueous two-phase system Polyethylene glycol (PEG) and salts	Phycocyanin (C-PC)	NA	Tan et al. (2021)
Nostoc ellipsosporum	Water extraction	Cyanovirin	Anti-viral properties	Mori and Boyd (2001)
Microcystis aeruginosa	NA	Microvirin	Anti-viral properties	Mori and Boyd (2001)

NA = Not Available

9.4 Cyanobacteria Lipids and Their Applications

In congruence with bacteria, fungi, plants, and fish, cyanobacteria, as poikilothermic entities, possess the adaptive capability to modulate their membrane lipid composition in response to environmental fluctuations (Gaysina et al. 2019; Sato and Wada 2009; Siebenaller and Schneider 2023; Wada and Murata 1998). Cyanobacteria feature two distinctive membrane systems: the plasma and thylakoid membranes.

The plasma membrane serves the function of demarcating the cellular internal milieu from the external surroundings. In contrast, the thylakoid membranes constitute secluded internal structures housing the photosynthetic and respiratory electron-transfer chains (Boudière et al. 2014). Noteworthy is the observed similarity between the plasma and thylakoid membranes of cyanobacteria and the inner envelope and thylakoid membranes inherent to chloroplasts in higher plants (Nandagopal et al. 2021; Mikami and Murata 2003; Siebenaller and Schneider 2023). The lipid composition characterizing these membranes is distinctive to cyanobacteria and chloroplasts, primarily marked by the presence of glycolipids, specifically monogalactoyldiacylglycerol (MGDG), digalactosyldiacylglycerol (DGDG), and sulfoquinovosyldiacylglycerol (SQDG). Notably, the sole phospholipid identified in these membranes is phosphatidylglycerol (PG), constituting 10–20% of the total lipids in the membrane (Mikami and Murata 2003; Ernst et al. 2016). This distinctive composition distinguishes cyanobacteria from most bacteria, where phospholipids predominate as the principal glycerolipids. The distribution of glycerolipids within cyanobacterial membranes discloses MGDG accounting for approximately 50%, DGDG for 20%, and SQDG for 16% of the total glycerolipids. Furthermore, cyanobacteria host a minor glycerolipid, monoglucosyldiacylglycerol (MGluDG), absent in chloroplasts. However, the content of MGluDG constitutes approximately 1% of the total glycerolipids and exhibits variability contingent upon environmental conditions (Boudière et al. 2014).

Glycerolipids are systematically classified into four distinct categories: neutral glycolipids, anionic glycolipids, anionic phospholipids, and other glycerolipids. Within the neutral glycolipid classification, galactoglycerolipids, specifically MGDG and DGDG, overwhelmingly characterize the lipid composition of cyanobacterial membranes. MGDG, a pivotal constituent within this category, stands out for its rich content of polyunsaturated fatty acids. This includes 16- and 18-carbon omega-3 trienoic acids, such as cis-7,10,13-hexadecatrienoic acid (C16:3), and those with more than 20 carbon atoms and over three double bonds, exemplified by eicosapentaenoic acid. In contrast, saturated forms, particularly palmitic acid (C16:0), form the predominant constituents of DGDG. Anionic glycolipids, as exemplified by SQDG, feature a sulfur-containing anionic polar head. However, their presence constitutes the minority within the total glycerolipid composition. Some studies propose the potential substitution of SQDG function by another lipid. The SQDG level is intricately linked with PG, and the dynamic SQDG/PG ratio responds to fluctuations in phosphate availability. The anionic phospholipid PG, the exclusive phosphoglycerolipid identified in cyanobacterial membranes, possesses a negatively charged phosphate group in its polar head, akin to SQDG, resulting in an overall negative charge for the molecule. The anionic nature of phosphatidylglycerol plays a pivotal role in protein interactions. PG levels exhibit an inverse relationship with SQDG; in phosphate-deficient environments, PG levels decrease while sulfolipid levels increase. In addition to core lipid components such as MGDG, DGDG, SQDG, and PG, which form the foundational lipid composition in cyanobacteria, diverse lipid constituents may also be present in their membranes, as documented by Boudière et al. (2014). This particular lipid profile is susceptible to significant variations contingent upon environmental conditions. For instance, phosphatidylcholine (PC) has been discerned in *Chlorogloeopsis fritschii* (Sallal et al. 1987), N-acylphosphatidylethanolamines have been identified in *Calothrix* sp. (Řezanka et al. 2009), and diacylglyceryltrimethylhomoserines have been reported in *Nannochloropsis granulata* (Banskota et al. 2013).

The fatty acid composition of cyanobacterial glycerolipids diverges significantly from that of the majority of bacteria, primarily characterized by the synthesis of saturated and monounsaturated fatty acids. Kenyon (1972) and Kenyon et al. (1972) laid the foundation for a systematic classification system for cyanobacteria, emphasizing their fatty acid composition and delineating four distinct groups. The initial group encompasses both saturated and monounsaturated fatty acids. Subsequent groups, namely the second, third, and fourth, are characterized by the prevalence of α-18:3, γ-18:3, and 18:4, respectively. Prominent among the fatty acids identified in cyanobacteria are palmitic acid (C16:0), palmitoleic acid (C16:1n-7), 12-cis-hexadecadienoic acid (C16:2n-4,7), stearic acid (C18:0), oleic acid (cis-9 C18:1), linoleic acid (cis-9 cis-12 C18:2), γ-linolenic acid (C18:3n-6), and α-linolenic acid (C18:3n-3). Nevertheless, the specific types of fatty acids present are contingent upon environmental conditions and the unique attributes of cyanobacterial species (Sato and Wada 2009; Wada and Murata 1998). For instance, the prevalence of oleic acid is conspicuous in *Chroococcidiopsis supralittoralis* and *C. umbratilis* (Řezanka et al. 2003), while eicosapentaenoic acid dominates in *Phaeodactylum tricornutum* (Guedes et al. 2011a, b), and palmitic acid is notably abundant in *Scytonema* sp. (Dembitsky and Srebnik 2002). Cyanobacteria exhibit indispensable glycerolipids and fatty acids that hold diverse applications in fields such as biofuel production, pharmaceuticals, food and feed, and industrial chemicals (Abed et al. 2009; Jha and Zi-Rong 2004; da Costa et al. 2020). Table 9.3 presents the lipid accumulation in cyanobacteria, the types of fatty acids synthesized under various conditions, and the applications of these lipids.

In contemporary biofuel research, the focus extends beyond merely identifying materials for biofuel production, encompassing the imperative of ensuring environmentally sound and economically sustainable utilization of byproducts. Given the global reliance on fossil fuels and the urgent need for novel energy sources, biofuels are gaining prominence as a strategic avenue to achieve energy security and environmental sustainability. Cyanobacteria play a pivotal role in biofuel production by offering the potential to generate substantial amounts of renewable energy (Zahra et al. 2020). Notably, thermophilic cyanobacteria possess the capacity to accumulate significant lipid quantities. The pinnacle of lipid content, reaching 42.8%, was achieved by *Synechococcus* sp., accompanied by corresponding yields of 46.9% for C16 and C18 methyl esters. Conversely, *Cyanobacterium aponinum* demonstrated lipid contents of up to 45.0%, corresponding yields of 67.7% for C16 and C18 methyl esters. The highest yield of C16 and C18 methyl ester, at 90.6%, was observed in *Phormidium* sp., which exhibited the highest lipid content at 38.2%. In *Synechococcus* sp., *C. aponinum*, and *Phormidium* sp., the percentages of saturated compounds were 74.5%, 77.9%, and 84.7%, respectively. These findings suggest that biodiesel produced from these raw lipids would possess a high cetane number

Table 9.3 Cyanobacterial lipid accumulation, fatty acid types, and applications under varied conditions

Species	Culture conditions	Lipid content (%)	Predominant types of fatty acids	SFAs	MUFAs	PUFAs	Applications	References
Synechococcus sp.	BG-11 medium contains 0.25 g L^{-1} $NaNO_3$, 30 °C, pH 7, cool light at 12.5 wm^{-2} (2400 L), 15 days	42.8	Caprylic acid, myristic acid, palmitoleic acid, palmitic acid, stearic acid	74.5	22.9	2.6	Biodiesel feedstocks	Karatay and Dönmez (2011)
Cyanobacterium aponinum	BG-11 medium contains 0.25 g L^{-1} $NaNO_3$, 30 °C, pH 8, light lamps at 12.5 wm^{-2} (2400 L), 15 days	45.0	Stearic acid, palmitic acid, caprylic acid	77.9	17.2	4.8	Biodiesel feedstocks	Karatay and Dönmez (2011)
Phormidium sp.	BG-11 medium contains 0.25 g L^{-1} NaNO3, 30 °C, pH 9, light 12.5 wm^{-2} (2400 L), 15 days	38.2	Stearic acid, palmitic acid	84.7	15.3	–	Biodiesel feedstocks	Karatay and Dönmez (2011)
Synechocystis sp. NN	BG-11 medium, 25 °C, light at 1500 lux, 15–20 days	18.9	Palmitic acid, eicosenoic acid, erucic acid	–	–	–	Biodiesel feedstocks	Jawaharraj et al. (2016)
Nostoc calcicola MBDU 602	BG-11 N_0 medium, 28 ± 2 °C, light at 50 µmol photons m^{-2} s^{-1}, 24 days	18.92	Palmitic acid, stearic acid	87.66	6.80	5.54	Biodiesel feedstocks	Anahas and Muralitharan (2018)
Nostoc sp. MCC41	BG-11 medium, 25 °C, light at 54 µmol photons m^{-2} s^{-1}, 0.5% glucose	15.69	Palmitic acid, palmitoleic acid, linoleic acid	46.9	37.0	16.3	Biodiesel feedstocks	Nagappan et al. (2020)

Aliinostoc vietnamicum VP225-stationary phase	–	–	Linoleic acid, palmitoleic acid, palmitic acid	46.97	14.62	38.41	Biodiesel feedstocks	Maltseva et al. (2022)
Nostoc sp. AARL C008	Modified BG-11 medium, 25 ± 1 °C, pH 7.4, fluorescent lamp at 40 µmol photons m^{-2} s^{-1}, 30 days	10.21	Palmitic acid, palmitoleic acid, oleic acid, elaidic acid, linoleic acid, stearic acid	51.87	40.34	7.78	Biodiesel feedstocks	Pekkoh et al. (2022)
Leptolyngbya sp. KC45	Castenhölz's D medium, 35 °C, pH 8.2, light emitting diode at 10 µmol photons m^{-2} s^{-1}, 0.04%–40% of CO_2 concentration, 10 days	2.6–11.5	Palmitic acid, stearic acid, elaidic acid, linoleic acid	61.57–65.21	21.41–23.22	11.72–16.19	Biodiesel feedstocks	Pekkoh et al. (2023)
Nostoc linckia HA-46	40.54 µmol photons m^{-2} s^{-1}, 25 ± 1 °C	–	–	–	–	–	Biohydrogen producer	Mona and Kaushik (2015)
Myxosarcina spectabilis HP-43	40.54 µmol photons m^{-2} s^{-1}, 25 ± 1 °C	–	–	–	–	–	Biohydrogen producer	Mona and Kaushik (2015)
Gloeocapsa calcarea HP-45	40.54 µmol photons m^{-2} s^{-1}, 25 ± 1 °C	–	–	–	–	–	Biohydrogen producer	Mona and Kaushik (2015)

(continued)

Table 9.3 (continued)

Species	Culture conditions	Lipid content (%)	Predominant types of fatty acids	SFAs	MUFAs	PUFAs	Applications	References
Anabaena variabilis ATCC 29413	70 μmol photons $m^{-2} s^{-1}$, 30 °C	–	–	–	–	–	Biohydrogen producer	Salleh et al. (2016)
Geitlerinema sp. RMK-SH10	ASN III medium, 30 μmol photons $m^{-2} s^{-1}$, 30 °C	–	–	–	–	–	Biohydrogen producer	Tinpranee et al. (2018)
Anabaena sp. UTEX 1448	82.14 μmol photons $m^{-2} s^{-1}$	–	–	–	–	–	Biohydrogen producer	Vargas et al. (2018)
S. platensis	–	9.36–11.52	Palmitic acid, palmitoleic acid, gamma-linoleic acid	–	–	–	Antioxidative, anti-inflammatory, anti-allergic activities	Yang et al. (2019)
Gloeothece sp. ATCC 27152	BG-11 medium, pH 8, 250 μmol photons $m^{-2} s^{-1}$, 14 days	12.3	Palmitic acid, linoleic acid, oleic acid	–	–	–	Antioxidant, anti-inflammatory activity	Da Costa et al. (2020)
S. platensis	–	36.39	Palmitic acid, gamma-linoleic acid	48.83	5.14	46.02	Food supplement	Ramírez-Rodrigues et al. (2021)
Aliinostoc vietnamicum VP225-exponential phase	–	–	Linoleic acid, palmitoleic acid, palmitic acid	27.02	20.95	52.03	Supplement in agriculture, aquaculture feeds, food supplement	Maltseva et al. (2022)

– : not determined, SFAs: saturated fatty acids, MUFAs: monounsaturated fatty acids, PUFAs: polyunsaturated fatty acids and outstanding oxidative stability, facilitating extended biodiesel storage (Karatay and Dönmez 2011). Furthermore, utilizing biological hydrogen as a fuel presents an alternative method involving hydrogen in fuel cells to produce electricity from chemical energy (Mona et al. 2020). The generation of biological hydrogen offers numerous advantages, including environmentally friendly characteristics, efficiency, renewability, and the absence of carbon dioxide emissions throughout production and utilization (Abed et al. 2009). Cyanobacteria, as microorganism-based options, can generate hydrogen as a byproduct of nitrogen fixation. In a study by Kossalbayev et al. (2022), thirteen cyanobacterial strains from Kazakhstan's rice paddies were investigated, revealing that *Synechocystis* sp. S-1 exhibited the highest H_2 production activity (2.35 μmol H_2 mg^{-1} Chl a^{-1} h^{-1}) in the light. Conversely, the wild-type cyanobacterium *Anabaena variabilis* A-1 demonstrated a more substantial capacity for hydrogen production in the dark (8.67 μmol H_2 mg^{-1} Chl a^{-1} h^{-1}). Additionally, Bozieva et al. (2023) explored a novel strain of cyanobacteria with hydrogen-producing capabilities. According to their research, the *Dolichospermum* sp. IPPAS B-1213 culture showed the highest rate of H_2 photoproduction at 4.24 μmol H_2 mg^{-1} Chl a^{-1} h^{-1} after being treated with 20 μM DCMU (3-(3,4-dichlorophenyl)-1,1-dimethylurea). Therefore, cyanobacteria hold significant potential in addressing global energy and environmental challenges.

Glycerolipids, specifically MGDGs, DGDGs, and SQDGs, comprise the predominant constituents within cyanobacteria cells. As members of the metabolite class, these glycerolipids manifest diverse biological characteristics, positioning them as promising candidates for various biotechnological applications. Notably, they exhibit anti-inflammatory, anti-tumor, and antimicrobial properties. The assessment of the anti-inflammatory potential of sulfolipids and galactolipids employed the thermophilic blue-green alga ETS-05 to isolate MGDG, DGDG, and SQDG. Results indicated that SQDG exhibited significantly lower anti-inflammatory activity than MGDG and DGDG, which demonstrated biological activity (Da Costa et al. 2016). Furthermore, cyanobacteria containing sulfolipids and galactolipids exhibited anti-viral effects against HSV-1 by inhibiting polymerase activity. The hypothesis suggests that the sulfonate group of SQDG may interact with the positively charged DNA polymerase, influencing polymerase inhibition (Da Costa et al. 2016). In addition to their fundamental features, glycerolipids are acknowledged as a rich source of polyunsaturated fatty acids (PUFAs), specifically those with diverse biological applications, including α-linolenic (ALA), eicosapentaenoic (EPA), and docosahexaenoic (DHA) fatty acids (Da Costa et al. 2016). These essential PUFAs play pivotal roles in sustaining human health, contributing to the development of the nervous system, maintaining cardiovascular health, and regulating inflammation. Significantly, cyanobacteria are recognized as abundant sources of γ-linolenic acid (GLA), a specific type of omega-6 fatty acid. GLA can undergo conversion into arachidonic acid and prostaglandin E2 in the human body, influencing various aspects of lipid metabolism (Abed et al. 2009). Research indicates that the consumption of PUFAs derived from cyanobacteria holds promise for potential health benefits, with ongoing studies exploring their role in preventing chronic diseases and promoting overall well-being. A comprehensive understanding of the unique composition of PUFAs in cyanobacteria provides valuable insights into their potential as functional food sources and nutritional supplements.

9.5 Cyanobacteria Pigments and Their Applications

Pigments are compounds capable of absorbing light across various segments of the visible spectrum. They possess intricate conjugated structures that allow them to capture and energize electrons at varying energy levels. Natural pigments are crucial in sustaining the life and energy production of these organisms that rely on them and contribute to the environment as primary producers of oxygen and biomass (Richmond 2004). Additionally, these pigments have been used by society for millennia as colorants, highlighting their historical and cultural significance. Cyanobacteria contain various bio-pigments, primarily chlorophyll, carotenoids, and phycobiliproteins (PBP). PBP and orange carotenoids are located out of the membrane, whereas chlorophyll and the majority of the carotenoid compounds are positioned within the photosystem.

9.5.1 Photosynthetic Pigment

Chlorophyll stands as the fundamental pigment, playing a crucial role in oxygenic photosynthesis, and is commonly found in all photosynthetic microorganisms. Cyanobacteria naturally possess six distinct types of chlorophyll, Chl *a* being the prevalent chlorophyll (Gan and Bryant 2015). Chlorophyll has applications as a natural food colorant (Timberlake and Henry 1986) and a deodorant (da Silva Ferreira and Sant'Anna 2017). Additionally, it can be utilized as an ingredient in toothpaste and skincare cosmetics manufacturing. It was demonstrated that chlorophyll a is clinically beneficial in diminishing facial acne lesions and severity when used in conjunction with photodynamic therapy, with minimal side effects. Chlorophyll has shown effectiveness in the external treatment of certain inflammatory skin conditions, including hemostatic eczema and ulcers (Zeligman 1949). Chlorophyll is suggested as a color agent in pharmaceuticals, food packaging, and cosmetics. Specifically, in the cosmetic industry, chlorophyll is employed as a substance to treat wavy hair.

9.5.2 Secondary Pigments

Carotenoids serve both primary pigment and secondary pigment, with their composition varying among species and growth conditions. They are commonly recognized as bioactive compounds with various beneficial properties such as antioxidant, anti-inflammatory, antimicrobial, and anti-tumoral. They find applications in animal feed as color enhancers and cosmetics for their roles as antioxidants and anti-aging agents (Guedes et al. 2011a, b; Honda 2020; Morone et al. 2019; Yusoff et al. 2020). Notably, commercially significant carotenoids include astaxanthin, lutein, and β-carotene (Mandal et al. 2020). β-carotene is identified as a plentiful vitamin source and recognized as a pro-vitamin. It is employed as a supplementary diet for vitamin intake. Astaxanthin is capable of coloring different aquatic animals like shrimp, lobsters, and some other crustaceans. Canthaxanthin contributes to preventing blood disorders; however, it has been demonstrated that its consumption in excessive amounts can pose health risks. The carotenoid concentration in cyanobacteria is markedly less than in microalgal species or higher plants (Hirschberg and Chamovitz 1994). In cyanobacteria, carotenoids may exist within the reaction center alongside chlorophyll or within a protein complex that includes a singular carotenoid. This molecule plays a crucial role in photoprotection. Carotenoids primarily dissipate energy and prevent oxidative damage (Zhang et al. 2014). Methanolic pigment-rich extracts in which β-carotene is one of the main identified carotenoids obtained from *Spirulina*, *Pseudanabaena*, and *Lyngbya* could potentially act as preventive agents against anti-urolithiasis (kidney stone disease) based on in vitro assays (Paliwal et al. 2015a, b). Lopes et al. (2020) conducted a screening of cyanobacteria, namely *Alkalinema aff. pantanalense*, *Nodosilinea* (*Leptolyngbya*) *antarctica*, *Cyanobium gracile*, *Cuspidothrix issatschenkoi*, and *Leptolyngbya*-like species, to identify potential carotenoid-rich extracts with anti-inflammatory properties for the treatment of psoriasis. Regarding the ability to inhibit tumor growth, Schwartz and Shklar (1989) proposed that carotenoid-rich extract from *S. platensis* and *Dunaliella salina* (microalgae) could serve as an anti-tumoral substance. Numerous studies have highlighted the potential of *A. platensis* as a feed supplement to enhance carotenoid accumulation in animals, including fish species such as *Tilapia nilotica, Cyprinus carpio, T. mossambica,* and *Plecoglossus altivelis*, as well as shrimps like *Penaeus monodon* by adding raw cyanobacterium into the diet of poultry, up to a concentration of 10%, has been associated with higher carotenoid levels in poultry eggs (Pagels et al. 2021).

9.5.3 Phycobilins

A unique class of pigments called PBP, exclusive to red and blue-green algae (cyanobacteria), serves as primary light absorbers. Among these, phycocyanin (blue), phycoerythrin (red), and allophycocyanin (bluish green) are the most renowned phycobiliproteins. This PBP serves as an accessory pigment for capturing light, absorbing wavelengths that chlorophyll cannot absorb (Ghosh and Mishra 2020). Regarding quantity, phycocyanin stands out as the predominant phycobiliprotein in cyanobacteria, with phycoerythrin and allophycocyanin following closely behind, as noted by Wildman and Bowen in 1974. In cyanobacteria, phycobiliproteins can account for more than 50–60% of the overall content of soluble proteins in some cyanobacteria, such as *A. platensis* and *A. maxima* (Pagels et al. 2019). The effectiveness of cyanobacterial phycobilin is similar to that of other phytochemical compounds employed in the prevention of conditions, for example, cardiovascular diseases, inflammation, diabetes, cancers, degenerative diseases, ischemia, and anemia (Hirata et al. 2000).

Phycocyanin (PC) can operate through the non-enzymatic antioxidant mechanism, typically by scavenging reactive oxygen species (ROS), neutralizing these free radicals, and reducing oxidative levels. Both phycocyanobilin and phycocyanin have shown effectiveness in scavenging peroxynitrite and preventing DNA damage (Bhat and Madyastha 2001). PC is investigated for its potential anti-cancer properties. An advantage of phycocyanin lies in its ability to be harmful to cancer cells while sparing normal cells from toxicity (Jiang et al. 2017; Ravi et al. 2015). PC supports both anti-proliferative and pro-apoptotic effects, while the characteristics of anti-cancer drugs are the inhibition of cell proliferation and/or inducing tumor cell apoptosis. Furthermore, the light-absorbing property of phycocyanin can be applied as a photosensitizer in the delivery of drugs. Zinc phthalocyanine and phycocyanin have shown promising results regarding photodynamic activity and targeted effects on the tumor cells (Wan et al. 2017). The factor contributing to the anti-inflammation effects of phycobiliproteins is primarily associated with the impact of the molecules on various mechanisms. These mechanisms include the regulation of enzyme expression and activation and the modulation of macrophage function. PBP exerts its anti-inflammatory properties by inhibiting signals that promote inflammation. PC has the capability to inhibit inflammatory reactions and apoptosis by suppressing cofactors. PC can be employed to alleviate symptoms in diverse inflammatory processes or as a potential treatment for various diseases (González et al. 1999). PBP possesses other bioactivities. PC effectively reduced the production of reactive oxygen species, thereby creating a more favorable cellular state (McCarty et al. 2021). As a result, the potential therapeutic application of phycocyanobilin in treating pulmonary diseases could be significant. PC is recognized for its anti-obesity effects, possibly attributed to inhibiting pancreatic lipase activity (Han et al. 2006). PC reduces serum cholesterol, total cholesterol, triglycerides, and low-density lipoproteins (Chakdar and Pabbi 2017). Furthermore, phycocyanin has been recognized as a neuroprotective agent. Fibrils are distinctive structures in neurodegenerative conditions like Parkinson's and Alzheimer's.

Phycoerythrins (PE) exhibit a vibrant pink hue. Regarding energy transfer, PE serves as the initial phycobiliprotein that captures light and transfers the energy to PC and APC (allophycocyanin) (Bryant et al. 1979). PE is typically categorized into three types based on distinct spectral properties and the photosynthetic organisms responsible for their production. These types are C-phycoerythrin, B-phycoerythrin, and R-phycoerythrin. C-PE of cyanobacteria has received comparatively less attention due to public apprehensions related to cyanotoxins. PE possesses antioxidant properties due to its performance in neutralizing free radicals (Punampalam et al. 2018). It enables them to prevent the occurrence of potentially life-threatening conditions like heart disease, cancer, and liver damage. The potent antioxidant action supports human well-being and enhances the immune system's ability to combat infections (Hemlata et al. 2018). PE has gained recognition in pharmaceuticals, colorants, nutraceuticals, and biologically active compounds. The phytopigments found in *Spirulina* sp. are effective as antioxidants, anti-wrinkle, anti-melanogenic, and anti-aging agents. Additionally, they serve as natural, non-toxic colorants in products such as eye shadows, eyeliners, and lipsticks (Nowruzi et al. 2020). Phycoerythrin is identified by its unique spectral characteristics, including fluorescence emission (Onay 2021). The distinct fluorescent characteristics of PE have expanded its uses as a fluorescent marker in various applications, including immunoassays, cell biology, flow cytometry, and fluorescence microscopy. This versatility makes PE valuable in biomedical research and diagnostics (Bailey et al. 1987; Glazer 1994; Wilson et al. 1991). PE can also be employed as a marker in electrofocusing or gel electrophoresis and a photosensitizer in cancer therapy (Aráoz et al. 1998; Wilson et al. 1991). R-phycoerythrin purified from *Colaconema formosanum* exhibits anti-aging and anti-allergic characteristics without causing toxicity in various mammal cell lines and epidermal tissues. This suggests that the compound holds promise for use in cosmetics (Lee et al. 2021). Compared to phycocyanin and allophycocyanin, the distinctive fluorescence features of PE have rendered it a valuable tool in the examination and treatment of tumors (Chen et al. 2022). PE, a natural colorant, has emerged as a substitute for synthetic dyes due to its properties (Sonani et al. 2016). The pink pigment produced by PE has been used as a dye in dairy items like yogurt and milkshakes. This addition enhances the appeal of processed products, imparting color to otherwise bland or colorless foods. PE can potentially improve the immune response of whiteleg shrimp in both vivo and in vitro laboratory experiments. This suggests its possible use as an immunomodulator in shrimp production (Liu et al. 2009). Adding PE to the diet can alter the composition of gut microbiota, leading to enhanced nutrition of the intestine and increased resistance to diseases in organisms. This is achieved by promoting beneficial bacteria growth while decreasing harmful ones in the intestine (Qi et al. 2019).

Scytonemin, an extracellular pigment found in blue-green algae, is being researched due to its potential use as a photoprotectant (Gao and Garcia-Pichel 2011). It captures the majority of UV radiation before it reaches the photosystem, preventing damage from high UV levels (Hernando et al. 2018). This characteristic can be employed to create UV sunscreen by blending it with ascorbic acid. The ability of scytonemin to protect against sunlight enables cyanobacteria to thrive in challenging, rocky, and intensely radiated environments (Pathak et al. 2017). Table 9.4 shows different types of pigments found in most cyanobacteria together with the yield and application.

9.6 Cyanobacterial Vitamins and Their Applications

Vitamins play pivotal roles within diverse metabolic pathways, functioning as indispensable coenzymes with multifaceted contributions, including antioxidative activities and the precise regulation of cellular processes and tissue proliferation. Living organisms rely on distinct categories of vitamins, namely lipid-soluble vitamins

Table 9.4 Cyanobacterial pigments and their application

Pigments	Color	Species	Yield	Applications	References
Chlorophyll *a*	Green	*Leptolyngbya* sp., *Phormidium* sp., *Rhabdoderma* sp., *Nostoc* sp.	Up to 15 μg m L^{-1}	Colorant in cosmetics, pharmaceuticals, and food packaging	Baracho and Lombardi (2023)
Carotenoids	Yellow	*Synechocystis* sp., *S. platensis*, *Prochlorococcus* sp.	3.34–7.99 mg g^{-1} biomass	A color enhancer in animal feed, antioxidant and anti-aging components in cosmetical application	Amaro et al. (2018), D'Alessandro et al. (2019), Erdoğan et al. (2016), Montero et al. (2005), Paliwal et al. (2015a, b), Tavanandi et al. (2019)
β-carotene (Carotenoid)	Orange	*Gloeothece* sp., *Synechococcus* sp.	0.5–1.5 mg g^{-1} biomass		
Astaxanthin (Carotenoid)	Red	*Geitlerinema amphibium*	2.74 mg g^{-1} biomass		
Lutein (Carotenoid)	Yellowish red	*Gloeothece* sp.	2.9 mg g^{-1} biomass		
Phycobiliproteins		*S. platensis*	92 mg g^{-1} biomass	Natural colorants for food and cosmetics, fluorescent agent	Chittapun et al. (2020), Martínez et al. (2017) Silveira et al. (2007), Tavanandi et al. (2019), Pekkoh et al. (2023)
Phycocyanin (Phycobiliproteins)	Blue	*Nostoc commune, S. platensis*	29.66–100 mg g^{-1} biomass		
Allophycocyanin (Phycobiliproteins)	Bluish green	*S. platensis*	32.27 mg g^{-1} biomass		
Phycoerythrin (Phycobiliproteins)	Red	*Phormidium rubidium, Gloeobacter violaceus, Halomicronema* sp., *Leptolyngbya* sp.	Up to 400 mg g^{-1} biomass		

such as A, D, E, and K, and water-soluble vitamins, which include B_1, B_2, B_3, B_5, B_6, B_7, B_9, B_{12}, and C (Żymańczyk-Duda et al. 2022). Within the botanical realm, algae demonstrate a remarkable capacity to synthesize and accumulate an extensive array of vitamins. Noteworthy is the capacity of cyanobacteria to harbor essential vitamins, including B_{12}, K, or D, which are conventionally absent in higher plants. To illustrate, vitamin D exhibits a heightened concentration in consumable sea organisms, such as fishes, procuring it through algal-based diets due to their incapacity for endogenous synthesis. Moreover, cyanobacteria, such as *Spirulina*, have garnered recognition for their substantial accumulation of vitamins B_2, B_{12}, B_9, and B_3, in addition to vitamins C and E. Furthermore, reports indicate a substantial content of vitamin C in cyanobacteria like *Anabaena cylindrica* (Del Mondo et al. 2020; Żymańczyk-Duda et al. 2022).

Cyanobacteria, as elucidated by Asensi-Fabado and Munné-Bosch (2010) and Helliwell et al. (2016), manifest the capability to synthesize a spectrum of vitamins, namely A, B, C, E, and K, in response to environmental stressors. The pervasive global deficiency of vitamins within the human population is a formidable threat, necessitating urgent attention. In the context of bioactive metabolite generation, cyanobacteria emerge as highly promising entities with versatile applications in biotechnology. Their role facilitates the advancement of eco-sustainable methodologies for producing natural bioactive metabolites. It is significant to note that the biosynthesis of all vitamins is not uniformly distributed among plants, and specific vitamins, including B, D, and K, are notably limited in various plant species (Asensi-Fabado and Munné-Bosch 2010; Del Mondo et al. 2020). Consequently, cyanobacteria have been empirically demonstrated to produce these essential yet scarce vitamins, addressing deficiencies observed in numerous vascular plants (Del Mondo et al. 2020). Although individuals typically derive a substantial proportion of their vitamin intake from a varied diet comprising fruits, vegetables, dairy products, meat, fish, and eggs, those grappling with specific vitamin deficiencies can undergo treatment with artificially or synthetically derived vitamins administered through oral or intramuscular routes. Exemplars of such therapies include B_{12} (both oral and intramuscular routes), D_1 (oral ergocalciferol), C (oral), and K_1 (intramuscular) (Żymańczyk-Duda et al. 2022).

Spirulina is a pivotal source of vitamin B_{12} for individuals adhering to vegan and vegetarian lifestyles. To align with the European Food Safety Authority's (EFSA) recommended daily consumption of 4 μg day^{-1} for cobalamin (vitamin B12), a suggested daily *Spirulina* intake of 1.6–3.2 g is proposed (Grosshagauer et al. 2020; Żymańczyk-Duda et al. 2022). However, suppliers advocate a regular intake of 3–9 g day^{-1} to meet vitamin B_{12} requirements effectively (Muys et al. 2019). *Spirulina* has been introduced as a nutritional supplement for human consumption, explicitly addressing vitamin B_{12} deficiency and contributing to enhancements in bone strength, stiffness, and ulcer prevention (Żymańczyk-Duda et al. 2022).

Furthermore, *Spirulina* has been recognized as a noteworthy source of vitamin E and β-carotene. In a previous investigation by Carcea et al. (2015), the generation of vitamin E from *Spirulina* demonstrated content ranging from 2.8 to 12.5 mg per 100 g in improved techniques, with higher levels observed in traditionally dried

Spirulina. Traditional drying methods recorded a vitamin E content of 30.9 mg per 100 g. Additionally, *Spirulina* exhibits a substantial amount of β-carotene, ranging between 33.5 and 231.6 mg per 100 g. Elevated levels of β-carotene are also identified in *Synechoccus* sp., *Anabaena cylindrica*, and various fruits such as oranges, potatoes, broccoli, and carrots (Asensi-Fabado and Munné-Bosch 2010). Noteworthy is the exceptional status of *A. cylindrica* among non-vascular plants, serving as a remarkable source of vitamin C—an antioxidant that offers a defense mechanism against oxidants in cyanobacteria (Aaronson et al. 1977). In response to UV radiation stress and the protection of cell membranes, cyanobacteria, including *Spirulina* demonstrate the ability to synthesize a modest amount of vitamin D (Fais et al. 2022). Tarento et al. (2018) highlighted the experimental examination of a significant quantity of vitamin K1 in the *A. cylindrica*, surpassing the content found in parsley and spinach.

9.7 Conclusions and Prospects

The exploration of cyanobacterial biomass as a feedstock for biochemicals constitutes a promising frontier within the domain of sustainable and environmentally conscious resource management. The unique capacity of cyanobacteria to efficiently synthesize a diverse spectrum of vital compounds, including vitamins, minerals, essential amino acids, and proteins, positions them as compelling candidates for bio-based applications. This emphasis on cyanobacteria, focusing on prominent strains such as *Spirulina*, highlights their pivotal role as versatile reservoirs of essential nutrients, including but not limited to vitamin B_{12}, E, β-carotene, and even vitamin K_1. The bioactive metabolites engendered by cyanobacteria not only address nutritional deficiencies but also harbor the potential to substantially contribute to diverse industrial sectors, including pharmaceuticals, nutraceuticals, and the burgeoning field of bioplastics. This accentuates the intricate biochemical prowess inherent to cyanobacteria, thereby underscoring their integral role in propelling sustainable practices and mitigating reliance on conventional resources. In essence, the inquiry into the viability of cyanobacterial biomass as a feedstock for biochemicals not only signifies a novel avenue for bio-based industries but also presents a nuanced perspective that aligns with broader environmental and societal imperatives. Integrating ecological stewardship and socio-economic benefits signifies a concerted step toward a more sustainable future.

References

Aaronson S, Dhawale SW, Patni NJ, De Angelis B, Frank O, Baker H (1977) The cell content and secretion of water-soluble vitamins by several freshwater algae. Arch Microbiol 112:57–59. https://doi.org/10.1007/BF00446654

Abed RM, Dobretsov S, Sudesh K (2009) Applications of cyanobacteria in biotechnology. J Appl Microbiol 106(1):1–12. https://doi.org/10.1111/j.1365-2672.2008.03918.x

Adessi A, de Carvalho RC, De Philippis R, Branquinho C, da Silva JM (2018) Microbial extracellular polymeric substances improve water retention in dryland biological soil crusts. Soil Biol Biochem 116:67–69. https://doi.org/10.1016/j.soilbio.2017.10.002

Ahn G, Hwang I, Park E, Kim J, Jeon YJ, Lee J, Park JW, Jee Y (2008) Immunomodulatory effects of an enzymatic extract from *Ecklonia cava* on murine splenocytes. Mar Biotechnol 10:278–289. https://doi.org/10.1007/s10126-007-9062-9

Aikawa S, Joseph A, Yamada R, Izumi Y, Yamagishi T, Matsuda F, Kawai H, Chang J-S, Hasunuma T, Kondo A (2013) Direct conversion of *Spirulina* to ethanol without pretreatment or enzymatic hydrolysis processes. Energy Environ Sci 6(6):1844–1849. https://doi.org/10.1039/C3EE40305J

Aikawa S, Ho SH, Nakanishi A, Chang JS, Hasunuma T, Kondo A (2015) Improving polyglucan production in cyanobacteria and microalgae via cultivation design and metabolic engineering. Biotechnol J 10(6):886–898. https://doi.org/10.1002/biot.201400344

Akinbule OO, Onabanjo OO, Sanni SA, Adegunwa MO, Akinbule AS, Sosanya SK, Afolabi I (2023) Amino acid composition and protein quality of commonly consumed cooked foods in Nigeria. J Food Compos Anal 119:105295. https://doi.org/10.1016/j.jfca.2023.105295

Amaro HM, Guedes AC, Preto MAC, Sousa-Pinto I, Malcata FX (2018) *Gloeothece* sp. as a nutraceutical source—an improved extraction method of carotenoids and fatty acids. Mar Drugs 16(9):327. https://doi.org/10.3390/md16090327

Amorim ML, Soares J, Coimbra JSDR, Leite MO, Albino LFT, Martins MA (2021) Microalgae proteins: production, separation, isolation, quantification, and application in food and feed. Crit Rev Food Sci Nutr 61(12):1976–2002. https://doi.org/10.1080/10408398.2020.1768046

Anahas AMP, Muralitharan G (2018) Characterization of heterocystous cyanobacterial strains for biodiesel production based on fatty acid content analysis and hydrocarbon production. Energy Conv Manag 157:423–437. https://doi.org/10.1016/j.enconman.2017.12.012

Anfelt J, Kaczmarzyk D, Shabestary K, Renberg B, Rockberg J, Nielsen J, Uhlén M, Hudson EP (2015) Genetic and nutrient modulation of acetyl-CoA levels in *Synechocystis* for n-butanol production. Microb Cell Factories 14:167. https://doi.org/10.1186/s12934-015-0355-9

Aráoz R, Lebert M, Häder DP (1998) Electrophoretic applications of phycobiliproteins. Electrophoresis 19(2):215–219. https://doi.org/10.1002/elps.1150190213

Arias DM, Uggetti E, García-Galán MJ, García J (2018) Production of polyhydroxybutyrates and carbohydrates in a mixed cyanobacterial culture: effect of nutrients limitation and photoperiods. New Biotechnol 42:1–11. https://doi.org/10.1016/j.nbt.2018.01.001

Arias DM, Ortíz-Sánchez E, Okoye PU, Rodríguez-Rangel H, Balbuena Ortega A, Longoria A, Domínguez-Espíndola R, Sebastian PJ (2021) A review on cyanobacteria cultivation for carbohydrate-based biofuels: cultivation aspects, polysaccharides accumulation strategies, and biofuels production scenarios. Sci Total Environ 794:148636. https://doi.org/10.1016/j.scitotenv.2021.148636

Arora N, Philippidis GP (2023) The prospects of algae-derived vitamins and their precursors for sustainable cosmeceuticals. PRO 11(2):587. https://doi.org/10.3390/pr11020587

Asensi-Fabado MA, Munné-Bosch S (2010) Vitamins in plants: occurrence, biosynthesis and antioxidant function. Trends Plant Sci 15(10):582–592. https://doi.org/10.1016/j.tplants.2010.07.003

Bailey MP, Rocks BF, Riley C (1987) On the use of fluorescent labels in immunoassay. J Pharm Biomed Anal 5(7):649–658. https://doi.org/10.1016/0731-7085(87)80077-8

Bandyopadhyay A, Stöckel J, Min H, Sherman LA, Pakrasi HB (2010) High rates of photobiological H_2 production by a cyanobacterium under aerobic conditions. Nat Commun 1(1):139. https://doi.org/10.1038/ncomms1139

Banskota AH, Stefanova R, Sperker S, McGinn PJ (2013) New diacylglyceryltrimethylhomoserines from the marine microalga *Nannochloropsis granulata* and their nitric oxide inhibitory activity. J Appl Phycol 25:1513–1521. https://doi.org/10.1007/s10811-012-9967-1

Baracho DH, Lombardi AT (2023) Study of the growth and biochemical composition of 20 species of cyanobacteria cultured in cylindrical photobioreactors. Microb Cell Factories 22(1):36. https://doi.org/10.1186/s12934-023-02035-z

Bar-Or Y, Shilo M (1988) The role of cell-bound flocculants in coflocculation of benthic cyanobacteria with clay particles. FEMS Microbiol Ecol 4(3–4):169–174. https://doi.org/10.1016/0378-1097(88)90439-9

Béné C, Barange M, Subasinghe R, Pinstrup-Andersen P, Merino G, Hemre G-I, Williams M (2015) Feeding 9 billion by 2050 – putting fish back on the menu. Food Secur 7(2):261–274. https://doi.org/10.1007/s12571-015-0427-z

Bhat VB, Madyastha KM (2001) Scavenging of peroxynitrite by phycocyanin and phycocyanobilin from *Spirulina platensis*: protection against oxidative damage to DNA. Biochem Biophys Res Commun 285(2):262–266. https://doi.org/10.1006/bbrc.2001.5195

Borah D, Nainamalai S, Gopalakrishnan S, Rout J, Alharbi NS, Alharbi SA, Nooruddin T (2018) Biolubricant potential of exopolysaccharides from the cyanobacterium *Cyanothece epiphytica*. Appl Microbiol Biotechnol 102(8):3635–3647. https://doi.org/10.1007/s00253-018-8892-x

Boudière L, Michaud M, Petroutsos D, Rébeillé F, Falconet D, Bastien O, Roy S, Finazzi G, Rolland N, Jouhet J, Block MA, Maréchal E (2014) Glycerolipids in photosynthesis: composition, synthesis and trafficking. Biochim Biophys Acta-Bioenerg 1837(4):470–480. https://doi.org/10.1016/j.bbabio.2013.09.007

Bozieva AM, Khasimov MK, Voloshin RA, Sinetova MA, Kupriyanova EV, Zharmukhamedov SK, Dunikov DO, Tsygankov AA, Tomo T, Allakhverdiev SI (2023) New cyanobacterial strains for biohydrogen production. Int J Hydrog Energy 48(21):7569–7581. https://doi.org/10.1016/j.ijhydene.2022.11.198

Branco RHR, Serafim LS, Xavier AMRB (2018) Second generation bioethanol production: on the use of pulp and paper industry wastes as feedstock. Fermentation 5(1):4. https://doi.org/10.3390/fermentation5010004

Bryant DA, Guglielmi G, de Marsac NT, Castets A-M, Cohen-Bazire G (1979) The structure of cyanobacterial phycobilisomes: a model. Arch Microbiol 123:113–127. https://doi.org/10.1007/BF00446810

Buléon A, Colonna P, Planchot V, Ball S (1998) Starch granules: structure and biosynthesis. Int J Biol Macromol 23(2):85–112. https://doi.org/10.1016/S0141-8130(98)00040-3

Cain A, Vannela R, Woo LK (2008) Cyanobacteria as a biosorbent for mercuric ion. Bioresour Technol 99(14):6578–6586. https://doi.org/10.1016/j.biortech.2007.11.034

Carcea M, Sorto M, Batello C, Narducci V, Aguzzi A, Azzini E, Fantauzzi P, Finotti E, Gabrielli P, Galli V, Gambelli L, Maintha KM, Namba F, Ruggeri S, Turfani V (2015) Nutritional characterization of traditional and improved dihé, alimentary blue-green algae from the lake Chad region in Africa. LWT-Food Sci Technol 62(1):753–763. https://doi.org/10.1016/j.lwt.2014.10.039

Chakdar H, Pabbi S (2017) Algal pigments for human health and cosmeceuticals. In: Algal green chemistry. Elsevier, pp 171–188. https://doi.org/10.1016/B978-0-444-63784-0.00009-6

Chen H, Qi H, Xiong P (2022) Phycobiliproteins—a family of algae-derived biliproteins: productions, characterization and pharmaceutical potentials. Mar Drugs 20(7):450. https://doi.org/10.3390/md20070450

Cheng J, Yue L, Ding L, Li Y-Y, Ye Q, Zhou J, Cen K, Lin R (2019) Improving fermentative hydrogen and methane production from an algal bloom through hydrothermal/steam acid pretreatment. Int J Hydrog Energy 44(12):5812–5820. https://doi.org/10.1016/j.ijhydene.2019.01.046

Chi Z, Su CD, Lu WD (2007) A new exopolysaccharide produced by marine *Cyanothece* sp. 113. Bioresour Technol 98(6):1329–1332. https://doi.org/10.1016/j.biortech.2006.05.001

Chittapun S, Jonjaroen V, Khumrangsee K, Charoenrat T (2020) C-phycocyanin extraction from two freshwater cyanobacteria by freeze thaw and pulsed electric field techniques to improve extraction efficiency and purity. Algal Res 46:101789. https://doi.org/10.1016/j.algal.2020.101789

Choi JS, Park YH, Kim S, Son J, Park J, Choi Y-E (2022) Strategies to control the growth of cyanobacteria and recovery using adsorption and desorption. Bioresour Technol 365:128133. https://doi.org/10.1016/j.biortech.2022.128133

Chow T-J, Su H-Y, Tsai T-Y, Chou H-H, Lee T-M, Chang J-S (2015) Using recombinant cyanobacterium (*Synechococcus elongatus*) with increased carbohydrate productivity as feedstock for bioethanol production via separate hydrolysis and fermentation process. Bioresour Technol 184:33–41. https://doi.org/10.1016/j.biortech.2014.10.065

D'Alessandro EB, Soares AT, de Oliveira D'Alessandro NC, Antoniosi Filho NR (2019) Potential use of a thermal water cyanobacterium as raw material to produce biodiesel and pigments. Bioprocess Biosyst Eng 42(12):2015–2022. https://doi.org/10.1007/s00449-019-02196-5

Da Costa E, Silva J, Mendonça SH, Abreu MH, Domingues MR (2016) Lipidomic approaches towards deciphering glycolipids from microalgae as a reservoir of bioactive lipids. Mar Drugs 14(5):101. https://doi.org/10.3390/md14050101

da Costa E, Amaro HM, Melo T, Guedes AC, Domingues MR (2020) Screening for polar lipids, antioxidant, and anti-inflammatory activities of *Gloeothece* sp. lipid extracts pursuing new phytochemicals from cyanobacteria. J Appl Phycol 32:3015–3030. https://doi.org/10.1007/s10811-020-02173-6

da Silva Ferreira V, Sant'Anna C (2017) Impact of culture conditions on the chlorophyll content of microalgae for biotechnological applications. World J Microbiol Biotechnol 33(1):20. https://doi.org/10.1007/s11274-016-2181-6

De Philippis R, Vincenzini M (1998) Exocellular polysaccharides from cyanobacteria and their possible applications. FEMS Microbiol Rev 22(3):151–175. https://doi.org/10.1111/j.1574-6976.1998.tb00365.x

Del Mondo A, Smerilli A, Sané E, Sansone C, Brunet C (2020) Challenging microalgal vitamins for human health. Microb Cell Factories 19(1):1–23. https://doi.org/10.1186/s12934-020-01459-1

Delattre C, Pierre G, Laroche C, Michaud P (2016) Production, extraction and characterization of microalgal and cyanobacterial exopolysaccharides. Biotechnol Adv 34(7):1159–1179. https://doi.org/10.1016/j.biotechadv.2016.08.001

Dembitsky VM, Srebnik M (2002) Variability of hydrocarbon and fatty acid components in cultures of the filamentous cyanobacterium *Scytonema* sp. isolated from microbial community "black cover" of limestone walls in Jerusalem. Biochem-Moscow 67:1276–1282. https://doi.org/10.1023/A:1021309623541

Deviram G, Mathimani T, Anto S, Ahamed TS, Ananth DA, Pugazhendhi A (2020) Applications of microalgal and cyanobacterial biomass on a way to safe, cleaner and a sustainable environment. J Clean Prod 253:119770. https://doi.org/10.1016/j.jclepro.2019.119770

Doan DLN, Le TTH, Nguyen TXQ, Nguyen DN (2022) Protein extraction from *Spirulina platensis* with the cellulase enzyme assistance. J Technol Educ 70B:25–32. https://doi.org/10.54644/jte.70B.2022.1213

Doshi H, Ray A, Kothari IL (2007) Bioremediation potential of live and dead *Spirulina*: spectroscopic, kinetics and SEM studies. Biotechnol Bioeng 96(6):1051–1063. https://doi.org/10.1002/bit.21190

Ebrahimi A, Farahpour MR, Amjadi S, Mohammadi M, Hamishehkar H (2023) Nanoliposomal peptides derived from *Spirulina platensis* protein accelerate full-thickness wound healing. Int J Pharm 630:122457. https://doi.org/10.1016/j.ijpharm.2022.122457

Echave J, Fraga-Corral M, Garcia-Perez P, Popović-Djordjević J, Avdović EH, Radulović M, Xiao J, Prieto MA, Simal-Gandara J (2021) Seaweed protein hydrolysates and bioactive peptides: extraction, purification, and applications. Mar Drugs 19(9):500. https://doi.org/10.3390/md19090500

El Mannai Y, Deto R, Kuroki M, Suzuki R, Suzuki E (2021) Cyanobacterial branching enzymes bind to α-glucan via surface binding sites. Arch Biochem Biophys 702:108821. https://doi.org/10.1016/j.abb.2021.108821

El-Bestawy EA, Abd El-Salam AZ, Mansy AERH (2007) Potential use of environmental cyanobacterial species in bioremediation of lindane-contaminated effluents. Int Biodeterior Biodegrad 59(3):180–192. https://doi.org/10.1016/j.ibiod.2006.12.005

Erdoğan A, Demirel Z, Eroğlu AE, Dalay MC (2016) Carotenoid profile in *Prochlorococcus* sp. and enrichment of lutein using different nitrogen sources. J Appl Phycol 28:3251–3257. https://doi.org/10.1007/s10811-016-0861-0

Ernst R, Ejsing CS, Antonny B (2016) Homeoviscous adaptation and the regulation of membrane lipids. J Mol Biol 428(24):4776–4791. https://doi.org/10.1016/j.jmb.2016.08.013

Fais G, Manca A, Bolognesi F, Borselli M, Concas A, Busutti M, Broggi G, Sanna P, Castillo-Aleman YM, Rivero-Jiménez RA, Bencomo-Hernandez AA, Ventura-Carmenate Y, Altea M, Pantaleo A, Gabrielli G, Biglioli F, Cao G, Giannaccare G (2022) Wide range applications of *Spirulina*: from earth to space missions. Mar Drugs 20(5):299. https://doi.org/10.3390/md20050299

Gammoudi S, Athmouni K, Nasri A, Diwani N, Grati I, Belhaj D, Bouaziz-Ketata H, Fki L, El Feki A, Ayadi H (2019) Optimization, isolation, characterization and hepatoprotective effect of a novel pigment-protein complex (phycocyanin) producing microalga: *Phormidium versicolor*NCC-466 using response surface methodology. Int J Biol Macromol 137:647–656. https://doi.org/10.1016/j.ijbiomac.2019.06.237

Gan F, Bryant DA (2015) Adaptive and acclimative responses of cyanobacteria to far-red light. Environ Microbiol 17(10):3450–3465. https://doi.org/10.1111/1462-2920.12992

Gao Q, Garcia-Pichel F (2011) Microbial ultraviolet sunscreens. Nat Rev Microbiol 9(11):791–802. https://doi.org/10.1038/nrmicro2649

Gaysina LA, Saraf A, Singh P (2019) Cyanobacteria in diverse habitats. In: Cyanobacteria. Academic Press, pp 1–28. https://doi.org/10.1016/B978-0-12-814667-5.00001-5

Ghosh T, Mishra S (2020) Studies on extraction and stability of C-phycoerythrin from a marine cyanobacterium. Front Sustain Food Syst 4:561714. https://doi.org/10.3389/fsufs.2020.00102

Glazer AN (1994) Phycobiliproteins – a family of valuable, widely used fluorophores. J Appl Phycol 6:105–112. https://doi.org/10.1007/BF02186064

Gonzalez R, Rodriguez S, Romay CHEYLA, González A, Armesto J, Remirez D, Merino N (1999) Anti-inflammatory activity of phycocyanin extract in acetic acid-induced colitis in rats. Pharmacol Res 39(1):55–59. https://doi.org/10.1006/phrs.1998.0409

Grosshagauer S, Kraemer K, Somoza V (2020) The true value of *Spirulina*. J Agric Food Chem 68(14):4109–4115. https://doi.org/10.1021/acs.jafc.9b08251

Guedes AC, Amaro HM, Barbosa CR, Pereira RD, Malcata FX (2011a) Fatty acid composition of several wild microalgae and cyanobacteria, with a focus on eicosapentaenoic, docosahexaenoic and α-linolenic acids for eventual dietary uses. Food Res Int 44(9):2721–2729. https://doi.org/10.1016/j.foodres.2011.05.020

Guedes AC, Amaro HM, Malcata FX (2011b) Microalgae as sources of carotenoids. Mar Drugs 9(4):625–644. https://doi.org/10.3390/md9040625

Han LK, Li DX, Xiang L, Gong XJ, Kondo Y, Suzuki I, Okuda H (2006) Isolation of pancreatic lipase activity-inhibitory component of *Spirulina platensis* and it reduce postprandial triacylglycerolemia. J Pharm Soc Jpn 126(1):43–49. https://doi.org/10.1248/yakushi.126.43

Hayashi T, Hayashi K, Maeda M, Kojima I (1996) Calcium spirulan, an inhibitor of enveloped virus replication, from a blue-green alga *Spirulina platensis*. J Nat Prod 59(1):83–87. https://doi.org/10.1021/np960017o

Helliwell KE, Lawrence AD, Holzer A, Kudahl UJ, Sasso S, Kräutler B, Scanlan DJ, Warren MJ, Smith AG (2016) Cyanobacteria and eukaryotic algae use different chemical variants of vitamin B12. Curr Biol 26(8):999–1008. https://doi.org/10.1016/j.cub.2016.02.041

Hemlata, Afreen S, Fatma T (2018) Extraction, purification and characterization of phycoerythrin from *Michrochaete* and its biological activities. Biocatal Agric Biotechnol 13:84–89. https://doi.org/10.1016/j.bcab.2017.11.012

Hernando M, Minaglia MCC, Malanga G, Houghton C, Andrinolo D, Sedan D, Rosso L, Giannuzzi L (2018) Physiological responses and toxin production of *Microcystis aeruginosa* in short-term exposure to solar UV radiation. Photochem Photobiol Sci 17:69–80. https://doi.org/10.1039/c7pp00265c

Hirata T, Tanaka M, Ooike M, Tsunomura T, Sakaguchi M (2000) Antioxidant activities of phycocyanobilin prepared from *Spirulina platensis*. J Appl Phycol 12:435–439. https://doi.org/10.1023/A:1008175217194

Hirschberg J, Chamovitz D (1994) Carotenoids in cyanobacteria. In: The molecular biology of cyanobacteria. Springer, pp 1–28. https://doi.org/10.1007/978-94-011-0227-8_18

Honda M (2020) Nutraceutical and pharmaceutical applications of carotenoids. In: Pigments from microalgae handbook. Springer, pp 1–28. https://doi.org/10.1007/978-3-030-50971-2_18

Hu S, Fan X, Qi P, Zhang X (2019) Identification of anti-diabetes peptides from *Spirulina platensis*. J Funct Food 56:333–341. https://doi.org/10.1016/j.jff.2019.03.024

Jawaharraj K, Karpagam R, Ashokkumar B, Pratheeba CN, Varalakshmi P (2016) Enhancement of biodiesel potential in cyanobacteria: using agro-industrial wastes for fuel production, properties and acetyl CoA carboxylase D (*acc*D) gene expression of *Synechocystis* sp.NN. Renew Energy 98:72–77. https://doi.org/10.1016/j.renene.2016.02.038

Jha RK, Zi-Rong X (2004) Biomedical compounds from marine organisms. Mar Drugs 2(3):123–146. https://doi.org/10.3390/md203123

Jiang L, Wang Y, Yin Q, Liu G, Liu H, Huang Y, Li B (2017) Phycocyanin: a potential drug for cancer treatment. J Cancer 8(17):3416–3429. https://doi.org/10.7150/jca.21058

Kalavathi DF, Uma L, Subramanian G (2001) Degradation and metabolization of the pigment—melanoidin in distillery effluent by the marine cyanobacterium *Oscillatoria boryana* BDU 92181. Enzym Microb Technol 29(4–5):246–251. https://doi.org/10.1016/S0141-0229(01)00383-0

Kamravamanesh D, Lackner M, Herwig C (2018) Bioprocess engineering aspects of sustainable polyhydroxyalkanoate production in cyanobacteria. Bioengineering-Basel 5(4):111. https://doi.org/10.3390/bioengineering5040111

Kanekiyo K, Lee J-B, Hayashi K, Takenaka H, Hayakawa Y, Endo S, Hayashi T (2005) Isolation of an antiviral polysaccharide, nostoflan, from a terrestrial cyanobacterium, *Nostoc flagelliforme*. J Nat Prod 68(7):1037–1041. https://doi.org/10.1021/np050056c

Karatay SE, Dönmez G (2011) Microbial oil production from thermophile cyanobacteria for biodiesel production. Appl Energy 88(11):3632–3635. https://doi.org/10.1016/j.apenergy.2011.04.010

Kenyon C (1972) Fatty acid composition of unicellular strains of blue-green algae. J Bacteriol 109(2):827–834. https://doi.org/10.1128/jb.109.2.827-834.1972

Kenyon CN, Rippka R, Stanier RY (1972) Fatty acid composition and physiological properties of some filamentous blue-green algae. Arch Mikrobiol 83:216–236. https://doi.org/10.1007/BF00645123

Khattar JIS, Singh DP, Jindal N, Kaur N, Singh Y, Rahi P, Gulati A (2010) Isolation and characterization of exopolysaccharides produced by the cyanobacterium *Limnothrix redekei* PUPCCC 116. Appl Biochem Biotechnol 162:1327–1338. https://doi.org/10.1007/s12010-010-8922-3

Khetkorn W, Lindblad P, Incharoensakdi A (2010) Enhanced biohydrogen production by the N_2-fixing cyanobacterium *Anabaena siamensis* strain TISTR 8012. Int J Hydrog Energy 35(23):12767–12776. https://doi.org/10.1016/j.ijhydene.2010.08.135

Kim S-K, Wijesekara I (2010) Development and biological activities of marine-derived bioactive peptides: a review. J Funct Food 2(1):1–9. https://doi.org/10.1016/j.jff.2010.01.003

Klanchui A, Dulsawat S, Chaloemngam K, Cheevadhanarak S, Prommeenate P, Meechai A (2018) An improved genome-scale metabolic model of *Arthrospira platensis* C1 (iAK888) and its application in glycogen overproduction. Meta 8(4):84. https://doi.org/10.3390/metabo8040084

Kossalbayev BD, Kakimova AB, Bolatkhan K, Zayadan BK, Sandybayeva SK, Bozieva AM, Sadvakasova AK, Alwasel S, Allakhverdiev SI (2022) Biohydrogen production by novel cyanobacterial strains isolated from rice paddies in Kazakhstan. Int J Hydrog Energy 47(37):16440–16453. https://doi.org/10.1016/j.ijhydene.2022.03.126

Lan EI, Liao JC (2011) Metabolic engineering of cyanobacteria for 1-butanol production from carbon dioxide. Metab Eng 13(4):353–363. https://doi.org/10.1016/j.ymben.2011.04.004

Lee P-T, Yeh H-Y, Lung W-Q-C, Huang J, Chen Y-J, Chen B, Nan F-H, Lee M-C (2021) R-phycoerythrin from *Colaconema formosanum* (Rhodophyta), an anti-allergic and collagen promoting material for cosmeceuticals. Appl Sci 11(20):9425. https://doi.org/10.3390/app11209425

Lee CW, Chang YB, Park CW, Han SH, Suh HJ, Ahn Y (2022) Protein hydrolysate from *Spirulina platensis* prevents dexamethasone-induced muscle atrophy via Akt/Foxo3 signaling in C2C12 myotubes. Mar Drugs 20(6):365. https://doi.org/10.3390/md20060365

Lipok J, Wieczorek D, Jewgiński M, Kafarski P (2009) Prospects of in vivo 31P NMR method in glyphosate degradation studies in whole cell system. Enzym Microb Technol 44(1):11–16. https://doi.org/10.1016/j.enzmictec.2008.09.011

Liu L-N, Su H-N, Yan S-G, Shao S-M, Xie B-B, Chen X-L, Zhang X-Y, Zhou B-C, Zhang Y-Z (2009) Probing the pH sensitivity of R-phycoerythrin: investigations of active conformational and functional variation. Biochim Biophys Acta-Bioenerg 1787(7):939–946. https://doi.org/10.1016/j.bbabio.2009.02.018

Lopes G, Clarinha D, Vasconcelos V (2020) Carotenoids from cyanobacteria: a biotechnological approach for the topical treatment of psoriasis. Microorganisms 8(2):302. https://doi.org/10.3390/microorganisms8020302

Madamwar D, Patel DK, Desai SN, Upadhyay KK, Devkar RV (2015) Apoptotic potential of C-phycoerythrin from *Phormidium* sp. A27DM and *Halomicronema* sp. A32DM on human lung carcinoma cells. EXCLI J 14:527–539. https://doi.org/10.17179/excli2014-696

Maltseva S, Kezlya E, Krivova Z, Gusev E, Kulikovskiy M, Maltsev Y (2022) Phylogeny and fatty acid profiles of *Aliinostoc vietnamicum* sp. nov. (cyanobacteria) from the soils of Vietnam. J Phycol 58(6):789–803. https://doi.org/10.1111/jpy.13283

Mandal MK, Chanu NK, Chaurasia N (2020) Cyanobacterial pigments and their fluorescence characteristics: applications in research and industry. In: Advances in cyanobacterial biology. Academic Press, pp 55–72. https://doi.org/10.1016/B978-0-12-819311-2.00005-X

Marchus KA, Blankinship JC, Schimel JP (2018) Environmental controls on extracellular polysaccharide accumulation in a California grassland soil. Soil Biol Biochem 125:86–92. https://doi.org/10.1016/j.soilbio.2018.07.009

Martínez JM, Luengo E, Saldaña G, Álvarez I, Raso J (2017) C-phycocyanin extraction assisted by pulsed electric field from *Arthrosphira platensis*. Food Res Int 99:1042–1047. https://doi.org/10.1016/j.foodres.2016.09.029

McCarty MF, DiNicolantonio JJ, Lerner A (2021) Review–nutraceuticals can target asthmatic bronchoconstriction: NADPH oxidase-dependent oxidative stress, RhoA and calcium dynamics. J Asthma Allergy 14:685–701. https://doi.org/10.2147/JAA.S307549

Mikami K, Murata N (2003) Membrane fluidity and the perception of environmental signals in cyanobacteria and plants. Prog Lipid Res 42(6):527–543. https://doi.org/10.1016/S0163-7827(03)00036-5

Mishra AK, Kaushik MS, Tiwari DN (2019) Nitrogenase and hydrogenase: enzymes for nitrogen fixation and hydrogen production in cyanobacteria. In: Cyanobacteria. Academic Press, pp 173–191. https://doi.org/10.1016/B978-0-12-814667-5.00008-8

Mohammadi M, Soltanzadeh M, Ebrahimi AR, Hamishehkar H (2022) *Spirulina platensis protein* hydrolysates: techno-functional, nutritional and antioxidant properties. Algal Res 65:102739. https://doi.org/10.1016/j.algal.2022.102739

Möllers KB, Cannella D, Jørgensen H, Frigaard N-U (2014) Cyanobacterial biomass as carbohydrate and nutrient feedstock for bioethanol production by yeast fermentation. Biotechnol Biofuels 7(1):64. https://doi.org/10.1186/1754-6834-7-64

Mona S, Kaushik A (2015) Screening metal-dye-tolerant photoautotrophic microbes from textile wastewaters for biohydrogen production. J Appl Phycol 27:1185–1194. https://doi.org/10.1007/s10811-014-0396-1

Mona S, Kumar SS, Kumar V, Parveen K, Saini N, Deepak B, Pugazhendhi A (2020) Green technology for sustainable biohydrogen production (waste to energy): a review. Sci Total Environ 728:138481. https://doi.org/10.1016/j.scitotenv.2020.138481

Montero O, Macías-Sánchez MD, Lama CM, Lubián LM, Mantell C, Rodríguez M, de la Ossa EM (2005) Supercritical CO_2 extraction of β-carotene from a marine strain of the cyanobacterium *Synechococcus* species. J Agric Food Chem 53(25):9701–9707. https://doi.org/10.1021/jf051283n

Mori T, Boyd MR (2001) Cyanovirin-N, a potent human immunodeficiency virus-inactivating protein, blocks both CD4-dependent and CD4-independent binding of soluble gp120 (sgp120) to target cells, inhibits sCD4-induced binding of sgp120 to cell-associated CXCR4, and dissociates bound sgp120 from target cells. Antimicrob Agents Chemother 45(3):664–672. https://doi.org/10.1128/AAC.45.3.664-672.2001

Morone J, Alfeus A, Vasconcelos V, Martins R (2019) Revealing the potential of cyanobacteria in cosmetics and cosmeceuticals – a new bioactive approach. Algal Res 41:101541. https://doi.org/10.1016/j.algal.2019.101541

Muys M, Sui Y, Schwaiger B, Lesueur C, Vandenheuvel D, Vermeir P, Vlaeminck SE (2019) High variability in nutritional value and safety of commercially available Chlorella and Spirulina biomass indicates the need for smart production strategies. Bioresour Technol 275:247–257. https://doi.org/10.1016/j.biortech.2018.12.059

Nagappan S, Bhosale R, Nguyen DD, Pugazhendhi A, Tsai PC, Chang SW, Ponnusamy VK, Kumar G (2020) Nitrogen-fixing cyanobacteria as a potential resource for efficient biodiesel production. Fuel 279:118440. https://doi.org/10.1016/j.fuel.2020.118440

Nakamura Y, Takahashi J, Sakurai A, Inaba Y, Suzuki E, Nihei S, Fujiwara S, Tsuzuki M, Miyashita H, Ikemoto H, Kawachi M, Sekiguchi H, Kurano N (2005) Some cyanobacteria synthesize semi-amylopectin type alpha-polyglucans instead of glycogen. Plant Cell Physiol 46(3):539–545. https://doi.org/10.1093/pcp/pci045

Nandagopal P, Steven AN, Chan LW, Rahmat Z, Jamaluddin H, Mohd Noh NI (2021) Bioactive metabolites produced by cyanobacteria for growth adaptation and their pharmacological properties. Biology 10(10):1061. https://doi.org/10.3390/biology10101061

Naseri A, Marinho GS, Holdt SL, Bartela JM, Jacobsen C (2020) Enzyme-assisted extraction and characterization of protein from red seaweed *Palmaria palmata*. Algal Res 47:101849. https://doi.org/10.1016/j.algal.2020.101849

Nasri M (2017) Protein hydrolysates and biopeptides: production, biological activities, and applications in foods and health benefits. A review. Adv Food Nutr Res 81:109–159. https://doi.org/10.1016/bs.afnr.2016.10.003

Ngatu NR, Okajima MK, Yokogawa M, Hirota R, Eitoku M, Muzembo BA, Dumavibhat N, Takaishi M, Sano S, Kaneko T, Tanaka T, Nakamura H, Suganuma N (2012) Anti-inflammatory effects of sacran, a novel polysaccharide from *Aphanothece sacrum*, on 2,4,6-trinitrochlorobenzene–induced allergic dermatitis in vivo. Ann Allergy Asthma Immunol 108(2):117–122.e2. https://doi.org/10.1016/j.anai.2011.10.013

Nowruzi B, Sarvari G, Blanco S (2020) The cosmetic application of cyanobacterial secondary metabolites. Algal Res 49:101959. https://doi.org/10.1016/j.algal.2020.101959

Onay M (2021) Enhancing phycoerythrin and phycocyanin production from *Porphyridium cruentum* CCALA 415 in synthetic wastewater: the application of theoretical methods on microalgae. J Nat Appl Sci 25(3):499–512. https://doi.org/10.19113/sdufenbed.846985

Ozturk S, Aslim B (2010) Modification of exopolysaccharide composition and production by three cyanobacterial isolates under salt stress. Environ Sci Pollut Res 17:595–602. https://doi.org/10.1007/s11356-009-0233-2

Pagels F, Guedes AC, Amaro HM, Kijjoa A, Vasconcelos V (2019) Phycobiliproteins from cyanobacteria: chemistry and biotechnological applications. Biotechnol Adv 37(3):422–443. https://doi.org/10.1016/j.biotechadv.2019.02.010

Pagels F, Vasconcelos V, Guedes AC (2021) Carotenoids from cyanobacteria: biotechnological potential and optimization strategies. Biomolecules 11(5):735. https://doi.org/10.3390/biom11050735

Paliwal C, Ghosh T, Bhayani K, Maurya R, Mishra S (2015a) Antioxidant, anti-nephrolithe activities and in vitro digestibility studies of three different cyanobacterial pigment extracts. Mar Drugs 13(8):5384–5401. https://doi.org/10.3390/md13085384

Paliwal C, Pancha I, Ghosh T, Maurya R, Chokshi K, Vamsi Bharadwaj SV, Ram S, Mishra S (2015b) Selective carotenoid accumulation by varying nutrient media and salinity in *Synechocystis* sp. CCNM 2501. Bioresour Technol 197:363–368. https://doi.org/10.1016/j.biortech.2015.08.122

Parwani L, Bhatt M, Singh J (2021) Potential biotechnological applications of cyanobacterial exopolysaccharides. Braz Arch Biol Technol 64:e21200401. https://doi.org/10.1590/1678-4324-2021200401

Pathak J, Sonker AS, Rajneesh R, Kannaujiya VK, Singh V, Ahmed H, Sinha RP (2017) Screening and partial purification of photoprotective pigment scytonemin from cyanobacterial crusts dwelling on the historical monuments in and around Varanasi, India. Microbiol Res 8(1):6559. https://doi.org/10.4081/mr.2017.6559

Pathak J, Rajneesh MPK, Singh SP, Haeder DP, Sinha RP (2018) Cyanobacterial farming for environment friendly sustainable agriculture practices: innovations and perspectives. Front Environ Sci 6:7. https://doi.org/10.3389/fenvs.2018.00007

Pekkoh J, Lomakool S, Chankham J, Duangjan K, Thurakit T, Phinyo K, Ruangrit K, Tragoolpua Y, Pumas C, Pathom-aree W, Srinuanpan S (2022) Maximizing biomass productivity of cyanobacterium *Nostoc* sp. through high-throughput bioprocess optimization and application in multiproduct biorefinery towards a holistic zero waste. Biomass Convers Biorefinery 1–21. https://doi.org/10.1007/s13399-021-02285-0

Pekkoh J, Duangjan K, Phinyo K, Kaewkod T, Ruangrit K, Thurakit T, Pumas C, Pathom-aree W, Cheirsilp B, Gu W, Wang G, Chaichana C, Srinuanpan S (2023) Turning waste CO_2 into value-added biorefinery co-products using cyanobacterium *Leptolyngbya* sp. KC45 as a highly efficient living photocatalyst. Chem Eng J 460:141765. https://doi.org/10.1016/j.cej.2023.141765

Pereira S, Zille A, Micheletti E, Moradas-Ferreira P, De Philippis R, Tamagnini P (2009) Complexity of cyanobacterial exopolysaccharides: composition, structures, inducing factors and putative genes involved in their biosynthesis and assembly. FEMS Microbiol Rev 33(5):917–941. https://doi.org/10.1111/j.1574-6976.2009.00183.x

Phélippé M, Gonçalves O, Thouand G, Cogne G, Laroche C (2019) Characterization of the polysaccharides chemical diversity of the cyanobacteria *Arthrospira platensis*. Algal Res 38:101426. https://doi.org/10.1016/j.algal.2019.101426

Punampalam R, Khoo KS, Sit NW (2018) Evaluation of antioxidant properties of phycobiliproteins and phenolic compounds extracted from *Bangia atropurpurea*. Malays J Fundam Appl Sci 14:289–297. https://doi.org/10.11113/mjfas.v14n2.1096

Qi H, Liu Y, Qi X, Liang H, Chen H, Jiang P, Wang D (2019) Dietary recombinant phycoerythrin modulates the gut microbiota of H22 tumor-bearing mice. Mar Drugs 17(12):665. https://doi.org/10.3390/md17120665

Rahman A, Agrawal S, Nawaz T, Pan S, Selvaratnam T (2020) A review of algae-based produced water treatment for biomass and biofuel production. Water 12(9):2351. https://doi.org/10.3390/w12092351

Ramírez-Rodrigues MM, Estrada-Beristain C, Metri-Ojeda J, Pérez-Alva A, Baigts-Allende DK (2021) *Spirulina platensis* protein as sustainable ingredient for nutritional food products development. Sustain For 13(12):6849. https://doi.org/10.3390/su13126849

Ravi M, Tentu S, Baskar G, Rohan Prasad S, Raghavan S, Jayaprakash P, Jeyakanthan J, Rayala SK, Venkatraman G (2015) Molecular mechanism of anti-cancer activity of phycocyanin in triple-negative breast cancer cells. BMC Cancer 15:768. https://doi.org/10.1186/s12885-015-1784-x

Řezanka T, Víden I, Go JV, Dor I, Dembitsky VM (2003) Polar lipids and fatty acids of three wild cyanobacterial strains of the genus *Chroococcidiopsis*. Folia Microbiol 48:781–786. https://doi.org/10.1007/BF02931514

Řezanka T, Nedbalová L, Elster J, Cajthaml T, Sigler K (2009) Very-long-chain iso and anteiso branched fatty acids in N-acylphosphatidylethanolamines from a natural cyanobacterial mat of *Calothrix* sp. Phytochemistry 70(5):655–663. https://doi.org/10.1016/j.phytochem.2009.02.003

Richmond A (ed) (2004) Handbook of microalgal culture: biotechnology and applied phycology. Blackwell science, Oxford

Rösch C, Roßmann M, Weickert S (2019) Microalgae for integrated food and fuel production. GCB Bioenergy 11(1):326–334. https://doi.org/10.1111/gcbb.12579

Rueda E, García-Galán MJ, Díez-Montero R, Vila J, Grifoll M, García J (2020a) Polyhydroxybutyrate and glycogen production in photobioreactors inoculated with wastewater borne cyanobacteria monocultures. Bioresour Technol 295:122233. https://doi.org/10.1016/j.biortech.2019.122233

Rueda E, García-Galán MJ, Ortiz A, Uggetti E, Carretero J, García J, Díez-Montero R (2020b) Bioremediation of agricultural runoff and biopolymers production from cyanobacteria cultured in demonstrative full-scale photobioreactors. Process Saf Environ Protect 139:241–250. https://doi.org/10.1016/j.psep.2020.03.035

Saha I, Datta S (2022) Bacterial exopolysaccharides in drug delivery applications. J Drug Deliv Sci Technol 74:103557. https://doi.org/10.1016/j.jddst.2022.103557

Sallal AK, Ghannoum MA, Al-Hasan RH, Nimer NA, Radwan SS (1987) Lanosterol and diacylglycerophosphocholines in lipids from whole cells and thylakoids of the cyanobacterium *Chlorogloeopsis fritschii*. Arch Microbiol 148:1–7. https://doi.org/10.1007/BF00429638

Salleh SF, Kamaruddin A, Uzir MH, Karim KA, Mohamed AR (2016) Investigation of the links between heterocyst and biohydrogen production by diazotrophic cyanobacterium *A. variabilis* ATCC 29413. Arch Microbiol 198:101–113. https://doi.org/10.1007/s00203-015-1164-6

Sánchez-Contreras MI, Morales-Arrieta S, Okoye PU, Guillén-Garcés RA, Sebastian PJ, Arias DM (2021) Recycling industrial wastewater for improved carbohydrate-rich biomass production in a semi-continuous photobioreactor: effect of hydraulic retention time. J Environ Manag 284:112065. https://doi.org/10.1016/j.jenvman.2021.112065

Sathasivam R, Radhakrishnan R, Hashem A, Abd-Allah EF (2019) Microalgae metabolites: a rich source for food and medicine. Saudi J Biol Sci 26(4):709–722. https://doi.org/10.1016/j.sjbs.2017.11.003

Sathya R, MubarakAli D, MohamedSaalis J, Kim J-W (2021) A systemic review on microalgal peptides: bioprocess and sustainable applications. Sustain For 13(6):3262. https://doi.org/10.3390/su13063262

Sato N, Wada H (2009) Lipid biosynthesis and its regulation in cyanobacteria. In: Lipids in photosynthesis. Springer, pp 8–33. https://doi.org/10.1007/978-90-481-2863-1_8

Schwartz JL, Shklar G (1989) A cyanobacteria extract and β-carotene stimulate an antitumor host response against an oral cancer cell line. Phytother Res 3(6):243–248. https://doi.org/10.1002/ptr.2650030605

Shearer J, Graham TE (2002) New perspectives on the storage and organization of muscle glycogen. Can J Appl Physiol 27(2):179–203. https://doi.org/10.1139/h02-012

Shih S-R, Tsai K-N, Li Y-S, Chueh C-C, Chan E-C (2003) Inhibition of enterovirus 71-induced apoptosis by allophycocyanin isolated from a blue-green alga *Spirulina platensis*. J Med Virol 70(1):119–125. https://doi.org/10.1002/jmv.10363

Siebenaller C, Schneider D (2023) Cyanobacterial membrane dynamics in the light of eukaryotic principles. Biosci Rep 43(2):BSR20221269. https://doi.org/10.1042/BSR20221269

Silva SC, Almeida T, Colucci G, Santamaria-Echart A, Manrique YA, Dias MM, Barros L, Fernandes Â, Colla E, Barreiro MF (2022) Spirulina (*Arthrospira platensis*) protein-rich extract as a natural emulsifier for oil-in-water emulsions: optimization through a sequential experimental design strategy. Colloid Surf A-Physicochem Eng Asp 648:129264. https://doi.org/10.1016/j.colsurfa.2022.129264

Silveira ST, Burkert JFM, Costa JAV, Burkert CAV, Kalil SJ (2007) Optimization of phycocyanin extraction from *Spirulina platensis* using factorial design. Bioresour Technol 98(8):1629–1634. https://doi.org/10.1016/j.biortech.2006.05.050

Singh H, Varanasi JL, Banerjee S, Das D (2019a) Production of carbohydrate enrich microalgal biomass as a bioenergy feedstock. Energy 188:116039. https://doi.org/10.1016/j.energy.2019.116039

Singh S, Kant C, Yadav RK, Reddy YP, Abraham G (2019b) Cyanobacterial exopolysaccharides: composition, biosynthesis, and biotechnological applications. In: Cyanobacteria. Academic Press, pp 347–358. https://doi.org/10.1016/B978-0-12-814667-5.00017-9

Singha TK (2012) Microbial extracellular polymeric substances: production, isolation and applications. IOSR J Pharm 2(2):271–281. https://doi.org/10.9790/3013-0220276281

Sonani RR, Rastogi RP, Patel R, Madamwar D (2016) Recent advances in production, purification and applications of phycobiliproteins. World J Biol Chem 7(1):100–109. https://doi.org/10.4331/wjbc.v7.i1.100

Su C, Chi Z, Lu W (2007) Optimization of medium and cultivation conditions for enhanced exopolysaccharide yield by marine *Cyanothece* sp. 113. Chin J Oceanol Limnol 25:411–417. https://doi.org/10.1007/s00343-007-0411-3

Tan JS, Abbasiliasi S, Lalung J, Tam YJ, Murugan P, Lee CK (2021) Bioseparation of phycocyanin from *Phormidium tergestinum* using an aqueous two-phase system. Prep Biochem Biotechnol 51(3):260–266. https://doi.org/10.1080/10826068.2020.1808793

Tarento TD, McClure DD, Vasiljevski E, Schindeler A, Dehghani F, Kavanagh JM (2018) Microalgae as a source of vitamin K_1. Algal Res 36:77–87. https://doi.org/10.1016/j.algal.2018.10.008

Tavanandi HA, Vanjari P, Raghavarao KSMS (2019) Synergistic method for extraction of high purity Allophycocyanin from dry biomass of *Arthrospira platensis* and utilization of spent biomass for recovery of carotenoids. Sep Purif Technol 225:97–111. https://doi.org/10.1016/j.seppur.2019.05.064

Timberlake CF, Henry BS (1986) Plant pigments as natural food colours. Endeavour 10(1):31–36. https://doi.org/10.1016/0160-9327(86)90048-7

Tinpranee N, Incharoensakdi A, Phunpruch S (2018) Screening cyanobacteria from marine coastal waters of Thailand for biohydrogen production. J Appl Phycol 30:3471–3481. https://doi.org/10.1007/s10811-018-1490-6

Tiwari ON, Muthuraj M, Bhunia B, Bandyopadhyay TK, Annapurna K, Sahu M, Indrama T (2020) Biosynthesis, purification and structure-property relationships of new cyanobacterial exopolysaccharides. Polym Test 89:106592. https://doi.org/10.1016/j.polymertesting.2020.106592

Trabelsi L, Ben Ouada H, Bacha H, Ghoul M (2009) Combined effect of temperature and light intensity on growth and extracellular polymeric substance production by the cyanobacterium *Arthrospira platensis*. J Appl Phycol 21:405–412. https://doi.org/10.1007/s10811-008-9383-8

Vannela R, Verma SK (2006) Co^{2+}, Cu^{2+}, and Zn^{2+} accumulation by cyanobacterium *Spirulina platensis*. Biotechnol Prog 22(5):1282–1293. https://doi.org/10.1021/bp060075s

Vargas SR, dos Santos PV, Zaiat M, do Carmo Calijuri M (2018) Optimization of biomass and hydrogen production by *Anabaena* sp. (UTEX 1448) in nitrogen-deprived cultures. Biomass Bioenergy 111:70–76. https://doi.org/10.1016/j.biombioe.2018.01.022

Wada H, Murata N (1998) Membrane lipids in cyanobacteria. In: Lipids in photosynthesis: structure, function and genetics. Springer, pp 53–66. https://doi.org/10.1007/0-306-48087-5_4

Wan D-H, Zheng B-Y, Ke M-R, Duan J-Y, Zheng Y-Q, Yeh C-K, Huang J-D (2017) C-Phycocyanin as a tumour-associated macrophage-targeted photosensitiser and a vehicle of phthalocyanine for enhanced photodynamic therapy. Chem Commun 53(29):4112–4115. https://doi.org/10.1039/C6CC09541K

Wang Z, Zhang X (2017) Isolation and identification of anti-proliferative peptides from *Spirulina platensis* using three-step hydrolysis. J Sci Food Agric 97(3):918–922. https://doi.org/10.1002/jsfa.7815

Wang F, Yu X, Cui Y, Xu L, Huo S, Ding Z, Hu Q, Xie W, Xiao H, Zhang D (2023) Efficient extraction of phycobiliproteins from dry biomass of *Spirulina platensis* using sodium chloride as extraction enhancer. Food Chem 406:135005. https://doi.org/10.1016/j.foodchem.2022.135005

Wilson MR, Crowley S, Odgers GA, Shaw L (1991) Immunofluorescent labeling using covalently linked anti-phycoerythrin antibodies and phycoerythrin polymers. Cytometry A 12(4):373–377. https://doi.org/10.1002/cyto.990120413

Yang X, Li Y, Li Y, Ye D, Yuan L, Sun Y, Han D, Hu Q (2019) Solid matrix-supported supercritical CO_2 enhances extraction of γ-linolenic acid from the cyanobacterium *Arthrospira* (*Spirulina*) *platensis* and bioactivity evaluation of the molecule in zebrafish. Mar Drugs 17(4):203. https://doi.org/10.3390/md17040203

Yusoff FM, Banerjee S, Nagao N, Imaizumi Y, Shariff M, Toda T (2020) Use of microalgae pigments in aquaculture. In: Pigments from microalgae handbook. Springer, pp 123–134. https://doi.org/10.1007/978-3-030-50971-2_19

Zahra Z, Choo DH, Lee H, Parveen A (2020) Cyanobacteria: review of current potentials and applications. Environments 7(2):13. https://doi.org/10.3390/environments7020013

Zeligman I (1949) Topical chlorophyll therapy in the dermatoses. J Invest Dermatol 13(3):111–113. https://doi.org/10.1038/jid.1949.75

Zhang H, Liu H, Niedzwiedzki DM, Prado M, Jiang J, Gross ML, Blankenship RE (2014) Molecular mechanism of photoactivation and structural location of the cyanobacterial orange carotenoid protein. Biochemistry 53(1):13–19. https://doi.org/10.1021/bi401539w

Zhao B, Cui Y, Fan X, Qi P, Liu C, Zhou X, Zhang X (2019) Anti-obesity effects of *Spirulina platensis* protein hydrolysate by modulating brain-liver axis in high-fat diet fed mice. PLoS One 14(6):e0218543. https://doi.org/10.1371/journal.pone.0218543

Żymańczyk-Duda E, Samson SO, Brzezińska-Rodak M, Klimek-Ochab M (2022) Versatile applications of cyanobacteria in biotechnology. Microorganisms 10(12):2318. https://doi.org/10.3390/microorganisms10122318

Chapter 10
Bioprocessing of the Wastewater-Grown Cyanobacterial Biomass to Non-food Bioproducts

Muhammad Usman, Ayesha Shahid, Zhihao Fan, Wenlong Xiong, Yongkun Lv, Shen Zhang, Anqi Zhao, Md. Asraful Alam, and Jingliang Xu

Abstract Industrial production of bio-commodities through cyanobacteria is an emerging and attractive arena, but economic challenges pose significant constraints to this industry. Alternatively, escalated anthropogenic pollution and the need for a sustainable planet compel waste recycling into ecologically viable products. Wastewater cultivation of cyanobacteria is quintessential for pollutant removal and obtaining carbon-neutral valuable biomass in an environment-friendly manner. Although utilizing wastewater-cultivated biomass has been challenged by social acceptability, their distinctive characteristics intrigued scientists to explore this biomass for non-food products of agricultural and industrial importance. This book chapter is a recent compilation of the wide-scope prospect of wastewater-cultivated cyanobacterial biomass inclusion into the booming market of ecologically viable products. In the agricultural realm, exploration of cyanobacterial biomass as biofertilizer and biostimulant indicated its suitability to regulate aeration that augments soil's nutrient content and chemical/physical properties, leading to elevated crop productivity. Lipid and carbohydrate extraction and subsequent conversion into bioenergy compounds, biopolymers, and bioflocculant epitomize a promising waste-to-wealth approach for environmental sustainability. However, technological advancements, substantial research and investment are essential to smooth the lab-to-market transition of these products.

Keywords Cyanobacteria · Wastewater · Resource recovery · Bioproducts · Value addition

---

M. Usman · A. Shahid · Z. Fan · W. Xiong · Y. Lv · S. Zhang · M. A. Alam (✉) · J. Xu (✉)
School of Chemical Engineering, Zhengzhou University,
Zhengzhou, People's Republic of China
e-mail: alam@zzu.edu.cn; xujl@zzu.edu.cn

A. Zhao
School of Life Sciences, Zhengzhou University, Zhengzhou, People's Republic of China

M. A. Mehmood et al. (eds.), *Cyanobacteria Biotechnology*,
https://doi.org/10.1007/978-3-031-70698-1_10

10.1 Introduction

The ongoing surge in anthropogenic environmental pollution has led to multiple initiatives to achieve sustainability. Solutions must be economically viable, sustainable, and practical to address environmental challenges such as energy crises, food security, disease outbreaks, and global warming (Huisman et al. 2018). Over the last three billion years, cyanobacteria have helped increase atmospheric oxygen levels to secure their position as the planet's first oxygenic photosynthetic microbes (Fay 1992). These photoautotrophic organisms, also known as blue-green algae, primarily dominate freshwater, wastewater, marine, terrestrial, and desert ecosystems (Gaysina et al. 2019). They can be found as individual cells, colonies, or filaments. Even though cyanobacteria are tiny, they become visible when agglomerate in crusts. The cyanobacterial cellular mechanism allows them to adapt to environmental changes and multiply in dense populations. Their rapid expansion relies on fluctuations in nutrient levels, biotic/abiotic factors, climate change, or global warming (Fais et al. 2022). Moreover, they can thrive at an accelerated pace and offer more significant yield potential per acre compared to conventional food crops by utilizing non-productive and non-arable sites.

Industrial production of bio-commodities through cyanobacteria is an attractive arena (Wang et al. 2024), but economic challenges related to nutrient supply, high cultivation cost, and license to market wastewater-cultivated cyanobacterial products pose significant pressure on this industry. Currently, the production of 1 kg cyanobacterial biomass costs 1.6 USD kg^{-1} in open ponds and 12.4 USD kg^{-1} in photobioreactors (Xu et al. 2023), and water footprint is a significant cost component of this process, accounting for 2% of direct cost (Osman et al. 2024). Moreover, 2.4–6.8 m^3 of water requirement for 1 kg of dry biomass production has also been a concern in the present age of water scarcity. Fortunately, it could be reduced to as low as 400 L for open ponds and to 219 L for photobioreactors cultivation by using wastewater as nutrition and aquatic source as opposed to the 1546 L and 372 L freshwater requirement for open pond and photobioreactors, respectively (Shahid et al. 2019). Additionally, wastewater cultivation viably reduced the global warming potential (GWP) of cyanobacterial plants from 0.33 to −52 kg CO_2 eq. kg^{-1} biomass (Xu et al. 2023). The ability of cyanobacteria to harness wastewater as a low-cost nutrient source offers an economical and sustainable way to extract various high-value compounds. Implementation of this scenario could reduce the associated environmental impacts by five times (Arashiro et al. 2022). Researchers and scientists are intrigued by the distinctive characteristics of wastewater-cultivated cyanobacteria. Considering their widespread industrial potential, wastewater-cultivated cyanobacterial metabolites are explored for non-food products, including biofertilizers, soil conditioners, and bioenergy products.

In the future, wastewater as cultivation media is the most viable option for large-scale cultivation of cyanobacteria for commercial purposes. Life cycle assessment (LCA) of various cyanobacterial cultivation plants indicated that replacement of traditional nutrients with waste sources and traditional energy with bioenergy is the way forward to reduce the global warming potential of cyanobacteria cultivation as it reduces the water footprint, lowers energy consumption and the use of conventional cultivation chemicals (Table 10.1). The present study is the compilation of recent advances and potential applications of wastewater-cultivated cyanobacteria, focusing on life cycle assessment to evaluate its industrial suitability in terms of environmental burden and economics.

10.2 Cyanobacterial Metabolites

10.2.1 Phenolic Acids

Phenolic acids comprise a carboxyl group and one or more hydroxyl groups connected to the aromatic ring. These secondary metabolites belong to a vast collection of phenolic compounds that are derivatives from the shikimate pathway (Kumar and Goel 2019; Teodoro et al. 2015). Phenolic acids, generated by photosynthetic organisms, are crucial in preventing oxidative damage triggered by reactive oxygen species (ROS) and the hydroxyl radical (OH), along with exhibiting antimicrobial (Monroe et al. 2018) and antiviral properties (Li et al. 2018). They play a pivotal role in cyanobacteria, enabling them to withstand and adapt to environmental stresses. Still, these stresses can build up harmful free radicals in cells and cause chemical damage to deoxyribose and DNA. It is worth mentioning that the presence of gallic acid, caffeic acid, chlorogenic acid, ferulic acid, and vanillic acid was observed in the cultures of *Plectonema boryanum, Hapalosiphon intricatus, Anabaena doliolum*, and *Oscillatoria acuta* when a high concentration of NaCI was added. This indicates that these phenolic acids may have a role in neutralizing free radicals when the cyanobacteria are exposed to salt stress. Cold and hot fluctuations stimulated the abundance of phenolic compounds, leading to the antioxidant capacity of halotolerant *Halothece* sp. PCC 7418 (Patipong et al. 2019). The wastewater-grown cyanobacterial biomass has multiple applications in the packaging and textile industries (Fig. 10.1).

10.2.2 Vitamins

Photosynthetic organisms commonly synthesize vitamins such as water-soluble compounds include B_1 (thiamin), B_2 (riboflavin), B_3 (niacin), B_5 (pantothenic acid), B_6 (pyridoxine), B_7 (biotin), B_9 (folic acid), B_{12} (cobalamin), and C and lipid-soluble compounds include vitamin A, D, E, and K. Plants synthesize a variety of vitamins in their different organs to help mitigate the impact of environmental stresses (Asensi-Fabado and Munné-Bosch 2010). Nevertheless, it is worth noting that not all plants contain every vitamin. Usually, higher plants produce vitamin D, vitamin

Table 10.1 Lifecycle and environmental impact assessment of wastewater-cultivated algae to produce biofertilizers, bioenergy, and high-value bio-commodities

Wastewater source	Bioreactor	End product	Environmental burden	Economic assessment	References
Liquid anaerobic digestate	Biofilm system	Biogas and biofertilizers	9.3 g CO_2-Eq. MJ^{-1}, referring to least environmental impact	0.9 USD kWh^{-1} production cost, 0.7 NER	Xu et al. (2024)
Urban	HRAP followed by PBR	Biofertilizer, biogas, and pigments	1.2-fold to 2.4-fold higher environmental impact as compared to industrial wastewater	–	Arashiro et al. (2022)
Food industry	UASB reactor followed by HRAP	Biofertilizer, biogas, and pigments	High viability due to 34% biogas generation that compensate overall environmental impacts	–	Arashiro et al. (2022)
Agro-industry	Column bioreactor	Biofertilizers	−60.74 mPt, 2.2 kg CO_2 eq.	1.09 NER* (net energy ratio)	Magalhães et al. (2022)
	High-rate algal pond (HRAP)	Biofertilizers	−105.90 mPt, 0.45 kg CO_2 eq.	14.23 NER	Magalhães et al. (2022)
Agro-industry	HRAP	Biofertilizers	1.7 Pt, 51% lower environmental impact	36,094 USD/annum	Castro et al. (2023)
Undefined	Pilot-scale open ponds	Agriculture and aquaculture products	0.33 kg CO_2 eq., 77% lower environmental impact as compared to freshwater	–	Herrera et al. (2021)
Agro-industry	HRAP	Biofuels	1.7 Pt, 71% higher environmental impact due to electricity	39,710 USD/annum	Castro et al. (2023)
Anaerobic digestate	Pilot-scale pond	Biogas	−131 Pt, less environmentally viable	14.1% internal rate of return in 20 years	Romagnoli et al. (2024)
Manure and industrial	Commercial open pond	Bioethanol	87 m^3 GJ^{-1} water footprint	2.25 million USD profit/annum	Hossain et al. (2019)
Manure and industrial	Commercial PBR	Bioethanol	02 m^3 GJ^{-1} water footprint, 1.33 kg O_2 production kg^{-1} biomass	2.0 million USD profit/annum	Hossain et al. (2019)

(continued)

Table 10.1 (continued)

Wastewater source	Bioreactor	End product	Environmental burden	Economic assessment	References
Raw sewage	Open raceway pond	High-value biomass (40% water)	150 mPt, 2.69E + 00 kg CO_2 eq., −7.74 E-02 kg 1,4-DB eq. freshwater ecotoxicity	–	Schneider et al. (2018)
Domestic	HRAP	Pigments (extracted by supercritical solvent)	10 kPt, 2.2–6.2 times lower environmental impact	0.75 million US kg^{-1} production cost	Ferreira et al. (2024)

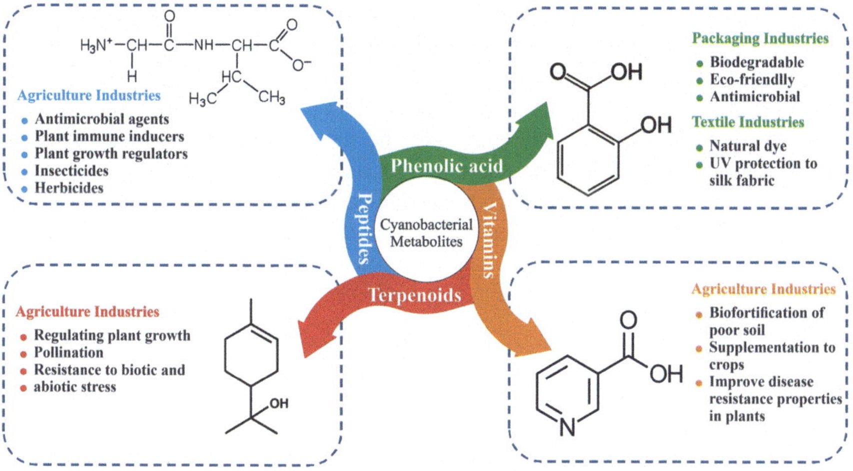

Fig. 10.1 Potential usage of wastewater-grown cyanobacterial biomass in various industries

K, and certain types of vitamin B in limited quantities. In contrast, cyanobacteria can produce all these vitamins D, K, and B_{12} (Del Mondo et al. 2020). *Arthrospira maxima, Anabaena cylindrica,* and *Synechococcus* sp. showed considerable amounts of β-carotene (pro-vitamin A) (Aaronson et al. 1977; Santiago-Morales et al. 2018) that is considerably higher when compared to certain fruits like carrots and oranges (Del Mondo et al. 2020). Like in plants, this compound produced by the cyanobacteria is highly effective at combating ROS, providing crucial protection against photooxidation. It is important to note that the effectiveness of $O_2^{\bullet -}$ scavenging by β-carotene surpasses that of vitamins E and C (Del Mondo et al. 2020).

Cyanobacteria play a crucial role in marine and wastewater ecosystems by being the primary providers of B vitamins to various aquatic organisms (Helliwell et al. 2016). Freshwater *A. cylindrica* was found to have a high content of B2, B5, and B6, while the marine *A. cylindrica* produced a considerable quantity of B12. Chromatographic analysis revealed a significant presence of B2, B3, B9, and B12 in the dried biomass of commercial *A. maxima* and *Arthrospira platensis* (Edelmann et al. 2019). It is essential to highlight that vitamin B significantly affects cyanobacterial adaptation to environmental stresses. *Nodularia spumigena* accumulates B1 in response to salinity and temperature stresses that help this planktonic cyanobacterium tolerate intense UV radiation (Sylvander et al. 2013).

Several cyanobacteria species, including *Anabaena variabilis, Mastigocladus laminosus, Nostoc muscorum, Prochlorococcus* sp., *Anacystis nidulans,* and *Synechocystis* sp. PCC 6803 are the cell factories of Phylloquinone commonly known as vitamin K1 (Collins and Jones 1981; Joliot et al. 2006). Phylloquinone serves a dual role in the photosystem I (PSI) complex. It functions as an electron carrier at the A1 binding site and also offers protection against growth damage under intense light conditions (Widhalm et al. 2009). However, its vitamin K-producing ability is highly species-dependent as marine *Anabaena cylindrica* may have a higher amount of vitamin K1 compared to spinach and parsley (Tarento et al. 2018) while *Spirulina* sp. CS-785 produces a small quantity of K1. In *Gloeobacter violaceus* and *Synechococcus* sp. PCC 7002, menaquinone (vitamin K2) functions as a secondary electron acceptor of PS1, just like vitamin K1 (Mimuro et al. 2005). Phylloquinone and menaquinone have similar structures. However, *Synechococcus* sp. PCC 7002 does not contain menaquinone, and *Gloeobacter violaceus* lacks two enzymes necessary for its biosynthesis.

10.2.3 Peptides

Marine natural products research has seen an enormous spike in interest due to the discovery of bioactive constituents from marine cyanobacteria. The diverse range of marine peptides derived from peptides-polyketides biosynthesis has garnered significant interest from pharmaceutical industries due to their high therapeutic potential (Gerwick et al. 2012). Research on these marine cyanobacteria has led to the discovery of numerous biologically active secondary metabolites, with peptides and peptide-derived compounds being the most common. Various structural classes of peptides, including linear peptides, linear depsipeptides, linear lipopeptides, cyclic peptides, cyclic depsipeptides, and cyclic lipopeptides have been previously identified in cyanobacteria (Singh et al. 2017).

10.2.4 Terpenoids

Terpenoids found in halophilic *Cylindrospermum muscicola* HPUSD12 and *Phormidium* sp. HPUSD13 are believed to have a protective effect against oxidative damage caused by the high-salinity environment of the Drang salt mine in India (Devi et al. 2020). 2-Methylhopanoids (2-MeBHPs) were found in large quantities in laboratory cyanobacterial cultures and natural cyanobacterium-dominated microbial mats. 2-MeBHPs demonstrate pentacyclic triterpenoids, serving as biomarkers for contemporary cyanobacteria in specific environmental conditions (Jahnke et al. 2004). In addition, 2-MeBHP enhances the ability of cyanobacteria to deal with osmotic, pH stress, and freezing/thawing, enabling their survival in various extreme environments such as desert soil crusts, hot springs, hypersaline lakes, Antarctic water, and Arctic soil (Garby et al. 2017).

Recent studies have highlighted cyanobacterial carotenoids' biological activities in treating various diseases. Various cyanobacteria strains, including *Alkalinema aff. pantanalense* LEGE15481, *Cyanobium gracile* LEGE12431, *Cuspidothrix issatschenkoi* LEGE03282, *Nodosilinea (Leptolyngbya) antarctica* LEGE13457, and *Leptolyngbya* sp. LEGE13412 have been studied for their significant carotenoid content (Lopes et al. 2020). The extracts from cyanobacteria exhibited significant antiageing properties by effectively inhibiting the activity of hyaluronidase, an enzyme that promotes the breakdown of hyaluronic acid in the presence of oxidative stress (Morone et al. 2020). Additionally, an antimicrobial sesterterpene, inhibits the growth of several harmful bacteria and fungi, including *Bacillus anthracis, Staphylococcus aureus, Escherichia coli, Candida albicans, Mycobacterium tuberculosis*, thus is a valuable antimicrobial agent. Furthermore, the presence of cybastacines A and B in *Nostoc* sp. BEA-0956 exhibited antibacterial properties against various clinical pathogenic bacteria (Cabanillas et al. 2018).

10.3 Cyanobacteria-Based Agricultural Products

In modern times, heavy reliance of agricultural practices on synthetic fertilizers and pesticides, along with intensive tillage and irrigation, pose adverse effects on soil fertility and overall production cost. Sustainable agriculture practices, such as eco-friendly and low-cost farming systems that utilize native cyanobacteria (blue-green algae), can enhance the resilience and self-regulation of the agricultural ecosystem while maintaining productivity and profitability (Table 10.2). Cyanobacterial biofertilizers, as opposed to chemical biofertilizers, offer various benefits as they can enhance nutrient utilization in the soil, boost crop stress tolerance, improve agricultural product quality and yield, and reduce the harmful effects of plant pathogens in the soil (Fig. 10.2) (Guo et al. 2020). Live algae biomass offers a promising alternative by reducing CO_2 emissions through carbon capture in the atmosphere and carbon sequestration, which helps advance green, sustainable agriculture (Renuka et al. 2018).

Table 10.2 Evaluation of wastewater-cultivated cyanobacteria for agriculture products

Agricultural product	Species	Waste stream	Product description/ application	Crop	Effects	References
Biostimulant	*Nostoc* sp. LS04	Municipal wastewater	Foliar application of 60% cellular extract prepared from deoiled biomass	Lettuce	• Augment root, shoot, and leaves • Enhance carotenoid content in leaves • Improve macronutrients and biochemicals in leaves	Silambarasan et al. (2021)
Biofertilizer	*Nostoc* sp.	Diluted kitchen wastewater	Wet inoculum equivalent to 1 g/kg dry mass was applied, foliar spray of 20% cellular extract from sonicated cyanobacteria as liquid fertilizer	*Oryza sativa*	• Enhance the weight of the plant and, especially shoots • Improve grain yield as compared to control.	Ruwanpathirana et al. (2023)
Biofertilizer	*Spirulina*	Aquaculture wastewater	2–10 g/L concentrate for seed germination, 5 g conc./pot for soil enrichment	Leafy vegetables (arugula, Baymen red, Chinese cabbage)	• 58.3% and 71.8% increase in the height of Baymen red and arugula, respectively • 110% higher fresh weight and 155.8% higher dry weight • 100% improvement in seed germination by 6–10% *Spirulina* conc.	Wuang et al. (2016)
Biostimulant	*Spirulina*	Commercial fertilizer filtrate	Liquid residue after protein extraction	Watercress, soybean, cucumber	• 300% higher root development • 15% higher weight • Exhibit potent auxin-like effect	Villaró et al. (2023)

Biofertilizer	*Chlamydomonas reinhardtii*	–	Hydroponic application of nutrient solution enriched with cyanobacteria at ~33 μM of total N content conc.	Corn (*Zea mays*)	• Increased micronutrient accumulation at roots and shoots	Martini et al. (2021)
Biostimulant	*Synechocystis* sp. and *Phormidium* sp.	Agriculture runoff and domestic wastewater	Foliar application of primary and secondary metabolites containing cellular extract	Tomato plantlets	• Provide essential phytohormones • Increase in plant height • Improvement in root length	Álvarez-González et al. (2023)
Nitrogen recovery	*Spirulina*	Agro-industry effluents	Stripping and acid absorption technology, 50 °C and 10–12 pH, citric acid/sulfuric acid solution, semi-continuous production	–	• Recovery of 70% ammonium sulfate • 72% ammonium citrate recovery	Attene et al. (2023)
Phosphorous recovery	*Chlamydomonas reinhardtii*	Sewage discharge, agriculture runoff	Genetically modified strain, overexpression of PSPR1 transcription factor for P overplus homeostasis	–	• Provision of phosphorous recirculation mechanism from wastewater to agriculture • Elevate PSR1 expression at starvation and induce luxury P-uptake at P-resupply	Slocombe et al. (2023)

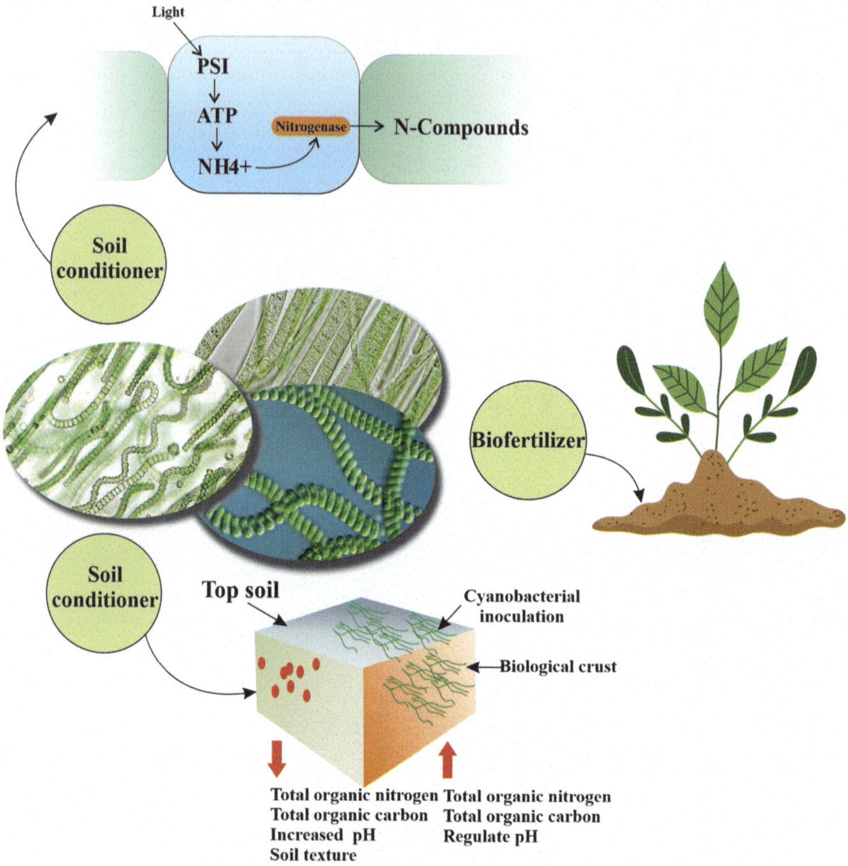

Fig. 10.2 Role of wastewater-grown cyanobacterial biomass in agriculture

10.3.1 Nitrogen Fixation

Nitrogen is an essential nutrient for plant growth and is crucial in lowland agricultural systems. Biological nitrogen fixation primarily contributes to soil nitrogen, typically carried out by microorganisms. Cyanobacteria play a vital role in forming surface crusts in different types of soils (Mahanty et al. 2017). These small photosynthetic units possess the capacity to convert atmospheric dinitrogen (N_2) into usable nitrogen, a process known as nitrogen fixation. Using cyanobacterial biomass for three to four consecutive crop seasons leads to the permanent establishment of cyanobacteria in the field. Cyanobacteria typically perform nitrogen fixation through free-living or symbiotic associations or by utilizing specialized thick-walled cells called "heterocysts." These cells are known for their importance in nitrogen fixation because of the nitrogenase enzyme. For example, when cyanobacteria (*Nostoc commune, Nostoc carneum*) were used in conjunction with chemical fertilizer, the growth parameters of *Oryza sativa* were significantly improved. An increase in root length ($p = 0.41$) and yield, including total number, grains per spike, and weight of 100 grains ($p < 0.001$) was observed. Combining cyanobacteria (*Anabaena*) with *Mesorhizobium ciceri* increased desi and kabuli chickpea (BG256, BG1003, and BG1108) productivity. Applying *Anabaena* sp. and *Microcystis aeruginosa* to *Salix viminalis L.* (willow) through biofertilization (triple foliar application) enhanced physiological performance. These improvements included increased net photosynthesis, chlorophyll content, stomatal conductance, and decreased intracellular CO_2 concentration.

10.3.2 Soil Conditioner

The connection between the earth, humans, and food sources highlights the essential function of soil as the basis of agriculture (Mahanty et al. 2017). Soil possesses distinct qualities that play an essential part in determining its texture. These include hydraulic conductivity, nutrient management, organic matter content, and citation exchange capacity. The soil composition consists primarily of soil, silt, and clay. There are 12 different types of soils found worldwide, each requiring specific care and attention. Therefore, the farmer's expertise and ability to manage the farm significantly impact the environment's well-being. Cyanobacteria are organisms that can be found worldwide and are recognized for their ability to improve soil quality. Using cyanobacteria in soil can improve soil stabilization, nutrient levels, and moisture retention (Rossi et al. 2017). Multiple reports have extensively studied cyanobacteria and soil, revealing the significant positive impact of these microorganisms on crust formation and the enhancement of soil health and quality. In a study conducted on the landslide area of Loess Plateau, it was found that certain cyanobacteria, specifically *Phormidium tenue* and *Oscillatoria* sp., played a significant role in enhancing carbon fixation and nutrient availability, particularly nitrogen and phosphorus (Ji et al. 2020).

10.3.3 Biofertilizers

Photosynthesis efficiently captures CO_2 (Singh et al. 2016), and introducing live cyanobacteria to the soil can lead to increased microbial biomass and improved soil health, including better aggregate stability and water retention (Alvarez et al. 2021). Furthermore, using cyanobacteria in soil as fertilizer may lead to improved CO_2 sequestration in soil and reduced CO_2 emissions. As discussed, the nitrogenase enzyme facilitates the N fixation process in cyanobacterial cells, which produces amino acids such as glutamine and glutamic acid. Plants typically absorb nitrate and ammonium, so the amino acids must be broken down through mineralization to be converted into these inorganic forms. A study compared the nitrogen mineralization of compost and solid cyanobacterial biofertilizer over 140 days in a controlled laboratory environment. The availability of nitrogen (in the form of nitrate and ammonium) from cyanobacterial biofertilizer was 31.6% at the end of the incubation period, whereas compost only showed 15.5%. The subsequent greenhouse study found that using cyanobacterial biofertilizer led to a notable increase in the *Brassica oleracea* yield, surpassing the results obtained from compost application (Jama et al. 2023). In addition, cyanobacteria have demonstrated the ability to dissolve inorganic phosphates, such as hydroxyapatite, in various environments. Some bacteria, such as *Anabaena*, *Nostoc,* and *Westiellopsis* sp., have been found to solubilize tricalcium phosphate and hydroxyapatite. Multiple mechanisms are being considered to explain this phenomenon. Phosphorus (P) from organic P fertilizers, like rock phosphate and bone meal, can be more bioavailable when used with *Anabaena* sp., according to findings of Afkairin et al. (2021). In addition to their intracellular phytohormones, cyanobacteria secrete gibberellins, auxin, cytokinin, and abscisic acid. Plants respond to cyanobacterial phytohormone synthesis and release by increasing their growth and performance (Singh et al. 2016). Commonly observed effects on plants include enhancements in germination, root and shoot length and weight, leaf number, or flower parameters. The synthesis of cyanobacterial fertilizer could be controlled to maximize phytohormone levels for various crops in the future. Certain algal compounds have been discovered to enhance plant growth by interacting with soil microbes, leading to increased nutrient availability through biomineralization or plant–microbe symbiosis. To confirm this theory, cyanobacteria were used to cultivate *Lupinus termis*. Germination rates were much higher when cyanobacterial filtrate and bacterial suspension were added to seeds.

Numerous studies have shown the effectiveness of fertilizers derived from *Spirulina* and *Chlorella* in promoting plant growth. In a survey conducted by Wuang et al. (2016) the potential of *S. platensis* biomass as agricultural fertilizer for various leafy vegetables was evaluated. The vegetables included Chinese Cabbage (*Brassica rapa* ssp. *chinensis*), Arugula (*Eruca sativa*), Pak Choy (*Brassica rapa* ssp. *chinensis*), Bayam Red (*Amaranthus gangeticus*), Kai Lan (*Brassica oleracea alboglabra*), and White Crown (*Brassica rapa* ssp. *chinensis*, F1 hybrid). These studies showed that biofertilizers based on *spirulina* have the ability to improve germination, plant height, root length, weight, and leaf number, among other biometric characteristics. Supplementing leaf vegetable crops with *Spirulina platensis* led to significantly higher plant growth rates than the control group. The number of leaves for pak choy increased from 10.33 to 13.00, while the root length increased from 2.33 to 7.00 cm. The performance of the *Spirulina*-based fertilizer was similar to chemical fertilizer in most plant growth parameters. In a study conducted by Dineshkumar et al. (2018), the impact of *Chlorella vulgaris* and *Spirulina platensis* had a significant positive effect on the critical growth parameters of the plants. This includes plant height, number of leaves, leaf area, and fresh and dry weight. As a result, reducing the recommended nitrogen fertilizer dosage by up to 50 or 75% was possible. In addition, the seed yield results showed key improvements in rice yield parameters using *Chlorella vulgaris*. The number of seeds/pods varied from 1.65 to 2.02, while the weight of 100 seeds ranged from 3.43 g to 4.03 g. The seed weight/

plant ranged from 2.07 g to 4.24 g, and the seed yield/pod was observed to be between 14.38 g and 19.33 g. On the other hand, *Spirulina platensis* was quite promising. The number of seeds per pod ranged from 2.22 to 2.51, the weight of 100 seeds varied from 5.94 g to 6.42 g, the seed weight per plant ranged from 7.12 g to 11.57 g, and the seed yield per pod was 15.56 g to 21.78 g. When algae biomass was applied to the soil, it led to an increase in nitrogenase and dehydrogenase activity. This, in turn, improved the availability of macronutrients in the soil. Additionally, it resulted in a decrease in electric conductivity and soil pH.

10.4 Cyanobacteria-Based Industrial Products

10.4.1 Bioflocculants

The most common cyanobacterial bioflocculant are exopolysaccharides EPS. The structural and physicochemical characteristics of EPS produced by cyanobacteria make them suitable for industrial and environmental applications. Using these biopolymers for biosurfactants and heavy metal biosorbents represents an influential advancement in environmental biotechnology. These methods are cost-effective and environmentally friendly ways to decrease pollution levels (Gondi et al. 2022). EPS play a vital role in developing biological soil crust (biofilm). This application inhibits water penetration into the soil by sealing its surface and blocking its pores. Resultantly, there has been a rise in the nutrients accessible and enhancement in the soil's aggregate stability (Tiwari et al. 2019). The production of algal EPS is influenced by various factors, including the algal species, cultivation conditions, and nutrient limitations like nitrogen and phosphorus deprivation (Costa et al. 2021). In addition, polysaccharides from algae can be categorized by their functions: (1) cell wall structure, (2) intracellular storage, and (3) extracellular or exopolysaccharides that are released into the environment. The most significant families of algae involved in EPS production include *Desmidiaceae, Chlamydomonadaceae, Chlorellaceae, Porphyridiaceae,* and *Glaucosphaeraceae* (Gaignard et al. 2019).

10.4.2 Biopolymers

Biopolymers and biodegradable polymers have distinct conceptual differences. Although biopolymers are renewable and are considered natural, they may or may not break down naturally. Biodegradable polymers, on the other hand, are specifically designed to be able to biodegrade. Cellulose, the most abundant polymer on earth, is commonly derived from crop residues, algae, and bacteria. Biopolymers derived from biological sources encompass many compounds, including proteins, polysaccharides, lipids, polyesters, and secondary metabolites like polyphenols

(Devadas et al. 2021). However, the biodegradability of biopolymers can be affected by various factors, including the choice of raw materials and the temperature conditions during the production process (Rai et al. 2021).

Bioplastics are substances that originate from biological sources and have the advantage of being degradable and eco-friendly. They can also undergo mineralization (Ray and Bousmina 2005). Bioplastics are a comparatively recent and promising development due to their ability to be used and broken down by microorganisms, such as bacteria. Biodegradation can be categorized into four groups: photodegradable, semi-biodegradable, chemically synthesized, and polyhydroxyalkanoates (PHAs). Microbially synthesized PHAs, including cyanobacterial PHA thermoplastics, have demonstrated significant potential in various applications in the industrial and agriculture sectors. PHAs biopolymers are primarily linear, thermoplastic materials with hydroxyalkanoic acid monomer units. These monomers are linked through ester bonds, with the carboxyl group of one unit connecting to the hydroxyl group of the next unit (Madison et al. 1999). It is interesting to note that cyanobacteria display three different modes of cell growth: photoautotrophic, mixotrophic, and heterotrophic. When sunlight is around, photoautotrophs use CO_2 as a raw material to grow and rely on light for energy. On the other hand, heterotrophs get all their energy and carbon from organic sources. However, in mixotrophic conditions, organisms can use both inorganic and organic carbon in the presence of light. Therefore, mixotrophs are advantageous over photoautotrophs and heterotrophs (Beardall and Raven 2016). A report published by Rippka et al. (1971) was the first to document the photoautotrophic accumulation of PHA in cyanobacteria. PHA accumulation was stimulated when organic compounds like acetate, glucose, citrate, and propionate were supplemented (Bhati and Mallick 2015). The cyanobacterial PHA bioplastics present the potential to utilize CO_2 and offset the emission of greenhouse gases from various industries. These properties have led individuals to consider cyanobacteria potential hosts for PHA production. For many years, scientists have been attracted by cyanobacteria's production of PHB polyester. Various thermoplastic processing methods were presented for producing thermoplastic algae articles, which can be used in plastic applications. A considerable amount of renewable algal biomass was used in polymer blends to create biodegradable plastics (Shi et al. 2012). The use of defatted algal biomass at various mass proportions with polyurethane (PU) resulted in the production of composite films. Compared to the neat PU-polyethylene glycol film, these films have shown enhanced antioxidant activity and increased bulk hydrophilicity when the biomass content reaches 70% (Saha et al. 2021). Another analysis conducted on the physicochemical properties of corn starch biocomposites derived from *Nannochloropsis*, *Spirulina*, and *Scenedesmus* showed a 50% reduction in water permeability, matrix rigidity, and improved oxygen permeability (Fabra et al. 2018). Items derived from a thermoplastic blend of algae and starch have various practical uses, including containers and automobile parts (Shi et al. 2012).

10.4.3 Colouring Agents

Cyanobacterial pigments, a natural source of colour, can enhance the appeal of industrial products. Natural colouring compounds are used in several industries, including food, cosmetics, and textiles, to improve product appeal and provide health benefits to customers. Colourants are classified as pigments in the pharmaceutical, ink, and cosmetic industries and dyes in the textile and food industries (Dikshit and Tallapragada 2018). Most colouring agents utilized in these businesses are chemically generated, posing risks to human health and the environment (Saini et al. 2018). Cyanobacteria are a primary source of pigments, including chlorophylls, carotenoids, phycobiliproteins, and scytonemin (Mourelle et al. 2017). Chlorophyll (Chl) is the primary green pigment found in cyanobacteria. Chlorophylls are pigments that have non-polar properties and consist of porphyrin or hydroporphyrin rings with magnesium atoms at their centres. They are present in all autotrophic organisms as they transform light into biological energy (Gan and Bryant 2015).

Cyanobacteria can produce six different types of chlorophyll, including chlorophylls a, b, d, f, and divinyl chlorophylls a and b. Carotenoids are the following light-harvesting molecules in cyanobacteria and are primarily associated with Chls. In addition to being non-polar pigments, carotenoids (CARs) are essential for photoprotection. They deactivate reactive oxygen species (ROS) generated when exposed to light and absorb light in the visible spectrum that differs from the absorption spectrum of Chl molecules (Nisar et al. 2015). From a biological perspective, these compounds fall into a class of pigments known for their highly conjugated polyene chains. These chains give CARs their vibrant colours, ranging from violet and red to orange and yellow. There are two groups of CARs: carotenes, which are unsaturated hydrocarbons like β-carotene and lycopene, and xanthophylls, which are similar to carotenes but have at least one oxygenated functional group like lutein, zeaxanthin, and fucoxanthin (Guedes et al. 2011). Previous research has primarily examined that algae tend to accumulate specific types of compounds, such as β-carotene in *Dunaliella salina*, astaxanthin in *Haematococcus pluvialis*, and lutein in *Scenedesmus almeriensis* and *Muriellopsis* sp. (Xu et al. 2018). However, recent studies have revealed that cyanobacterial strains also possess a diverse range and significant number of CARs.

Recent research investigated six strains of cyanobacteria from extreme habitats, specifically from the genera *Desertifilum* and *Haloleptolyngbya*. The findings revealed that *Desertifilum tharense* is a highly promising source of CARs (Singh et al. 2019). Phycobiliproteins (PBPs) are a group of water-soluble fluorescent compounds composed of proteins covalently linked to linear tetrapyrroles, known as phycobilins (Urtubia et al. 2016). These proteins capture light energy within the visible spectrum and facilitate energy transfer for photosynthesis. They also protect cyanobacteria from photolysis under intense light conditions (Sonani et al. 2019). PBPs are frequently observed in cyanobacteria, particularly in the genera Rhodophyta and Cryptophyta (Guskov et al. 2009). The pigment's colour determines whether it is phycoerythrin (PE) or phycocyanin (PC). At first, the pigment proteins were distinguished by their colours. The blue PBPs from cyanobacteria and the red-coloured PBPs from red algae were given the names PC and PE, respectively. Later, it was shown that cyanobacteria and red algae also contain PC and PE. As a result, these pigments differ from those previously found. Pigments and dyes derived from algae have various applications, including use in paints and the paper industry (Ariede et al. 2017). The utilization of algal pigments in the textile sector is currently minimal. This is primarily due to the lack of available reports on exploring algal-based colourants in textile processing and the insufficient information on extracting dyes from algal sources (Azeem et al. 2019). The results showed that *Sargassum muticum* had a higher colour strength on cotton fabrics when dyed at a temperature of 75 °C and a basic medium pH. The dyeing process lasted for 30 min and used 4.0 g per 100 mL NaCl as an exhausting agent. The cotton fabric's colour strength was higher when dyed with *I. stellata* extract at 75 °C, pH 5.0, using a NaCl concentration of 4.0 g per 100 mL and a time interval of 75 min. The colour strength achieved by *C. sinuosa* was found to be highest at pH 11, with a dyeing time of 60 min, a temperature of 60 °C, and a NaCl concentration of 8.0 g per 100 mL. The cotton fabric achieved optimal colour strength when dyed with *L. obtusa* dye at a dyeing temperature of 30 °C, a dyeing time of 45 min, a pH of 7.0, and a NaCl concentration of 6.0 g per 100 mL.

10.5 Cyanobacteria-Based Biofuels

Considering the quest of scientists for environmentally friendly energy sources in response to growing pollution concerns related to fossil fuels, wastewater-cultivated cyanobacteria show great potential as an energy source due to low cultivation cost and high pollutant removal potential (Agarwal et al. 2022). They are widely acknowledged as an excellent choice for producing third-generation biofuels (Sitther et al. 2020) as phototrophic cyanobacteria can produce nearly 30 times more energy per acre in comparison to certain first-generation fuel source plants (Sadvakasova et al. 2021). Their primary benefit over plants, such as being an abundant and uncontested food source and cultivation on non-food competing lands, makes them a promising bioenergy feedstock. Modern catalytic methods offer vast possibilities for converting cyanobacteria biomass into liquid fuels like biodiesel, biobutanol, bioethanol, bio-oil, and gaseous fuels like methane and biohydrogen.

10.5.1 Biodiesel

Biochemically, biodiesel comprises long-chain fatty acids of mono-alkyl ester, the best alternative to petroleum-derived diesel. Essential elements of biodiesel, such as glycerol and fatty acid methyl esters (FAMEs), can be derived from methanol and fatty acids (FAs) [81,82,83]. This metabolic process's end product is excellent lipids, as lipids store more energy than carbohydrates (Anahas and Muralitharan 2015). Additionally, producing esters from cellular lipids to create biodiesel is straightforward. Biofuel can be produced by combining cyanobacterial biomass with alcohol and a heterogeneous catalyst at high temperatures, directly extracting lipids. During this process, cyanobacteria cells react with methanol in the presence of a strong acid catalyst, such as sulfuric acid. This leads to the transesterification of fatty acids, followed by the separation of chloroform and methanol phases (Singh et al. 2023). The extraction of essential biofuel components, such as methyl ester from algae, which is compatible with traditional diesel engines, has recently been the subject of some research. Most cyanobacterial species contain around 40% fat by mass, according to the chemical compositions of the various algal species (Luque et al. 2010); the concentrations of lipids and FA in cyanobacteria can vary with the growth conditions. It has been reported that algal oil contains both saturated and monounsaturated fatty acids, including oleic (18:1), palmitic (16:0), stearic (18:0), linoleic (18:2), and caproic acids (C6:0). Cyanobacteria can store a significant amount of lipids, typically in the form of triacylglycerides (de Oliveira et al. 2022).

10.5.2 Bioethanol

Bioethanol is produced through the fermentation of carbohydrates obtained from various sources, including algae and plants like corn, sugarcane, wheat, and lignocellulosic biomass. Traditional bioethanol production involves using food crops like wheat, corn, potatoes, beets, and sugarcane as feedstocks for alcoholic fermentation. These crops are ideal for fermentation as they possess high levels of starch and sugar and are readily accessible in the agricultural sector. Nevertheless, with the growing human population, the pressure on our limited agricultural land becomes more pronounced, leading to significant worries about generating fuel from food crops. As part of sustainable bioethanol production at the commercial level, scientists are exploring various sources, including non-edible crops like lignocellulosic materials and algae, as potential feedstocks. Thus, bioethanol production can be achieved using feedstocks derived from non-food crops. Yet, certain strains of cyanobacteria that produce complex carbohydrates can also generate synthetic gas and bioethanol, just like non-edible crops. The carbohydrates obtained from cyanobacteria biomass can be converted into bioethanol through a series of steps involving hydrolysis and fermentation. An enzymatic process combined with yeast fermentation dramatically increases ethanol production when hydrolyzing *Synechococcus* sp. biomass. Due to its compact storage volume within cells, glycogen is the preferred form of carbohydrates for bioethanol production (Singh et al. 2023). Around 86% of ethanol production can be achieved by fermenting *Synechococcus* sp. with the assistance of yeast *Saccharomyces cerevisiae*. Cyanobacteria can undergo chemical hydrolysis to extract lipids, thereby recovering fermented carbohydrates and fats from the biomass. Various algal species, including *Chlamydomonas*, *Spirulina*, *Euglena*, *Chlorella*, *Scenedesmus*, and *Dunaliella*, have been extensively studied for their potential in bioethanol production (Fig. 10.3). Algae can be used to produce various valuable substances such as biobutanol, biomethane, biogas, and syngas, thanks to its sugar content (Hwang et al. 2016; Wang et al. 2020). However, certain nations can generate bioethanol using food crops as their feedstock. Brazil relies heavily on sugarcane for bioethanol production to meet a significant portion of its needs.

10.5.3 Biobutanol and Biohydrogen

In the past few decades, the direct production of short-chain fuels such as butanol has grown considerably. Commercial biobutanol production involves the utilization of cyanobacteria like *Oscillatoria obscura* and *Lyngbya limnetica*. These cyanobacteria are combined with *Clostridium beijerinckii* ATCC 35,702, a fermenting microorganism, and glucose to facilitate the process. The productivity of biobutanol from these cyanobacteria is reported to be 1.565 g/L, achieved by adding glucose in the batch mode condition (de Oliveira et al. 2022). In addition, it has been observed that the genetically modified cyanobacteria *Synechocystis* PCC6803 sp. emit fewer greenhouse gases (3.1 kg CO_2 eq. kg^{-1} biobutanol) into the environment. On the other hand, scientists are currently conducting experiments to test the viability of genetically engineered microorganisms for producing sustainable biobutanol (Singh et al. 2023). Another sustainable fuel source, biohydrogen, mainly produces water as a byproduct of burning. Studies have shown that *Anabaena* spp. can produce considerable quantities of H_2 (Villacreses-Freire et al. 2021). This biofuel is widely recognized for its clean, eco-friendly production methods and exceptional energy density. Cyanobacteria utilize sunlight to perform biophotolysis of water molecules, resulting in the production of biohydrogen.

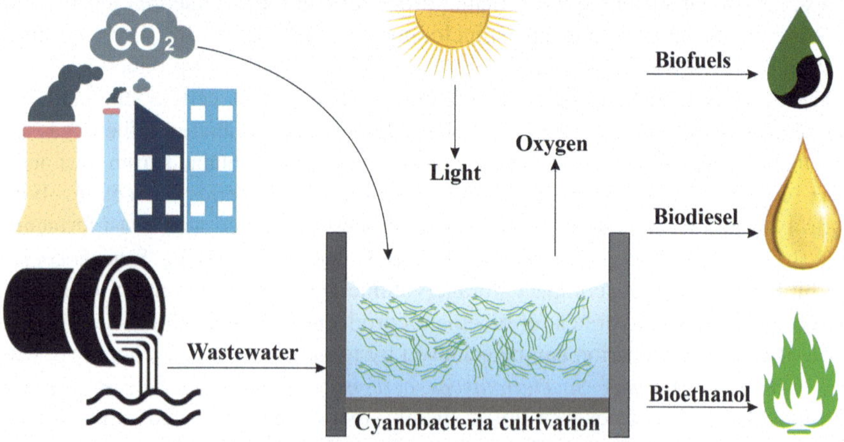

Fig. 10.3 Role of wastewater-grown cyanobacterial biomass in bioenergy

10.6 Conclusion and Prospects

The nascent development of ecologically viable products at the crossroads of escalated water pollution and the need for a sustainable planet spurred the "waste-to-wealth" movement. Cyanobacteria-mediated phytoremediation has been envisioned as a pragmatic approach to deal with the atrocious effects of wastewater as it reduces the associated environmental impacts by five times, reduces water footprint from 1546 to 219 L, and minimizes global warming potential to −52 kg CO_2 eq.kg^{-1} biomass. However, scrutiny of a technically feasible, environment-friendly, financially sustainable, and socially accepted biomass processing value chain is imperative to leverage the full potential of wastewater-cultivated cyanobacteria. Biofertilizers, biostimulants, nutrient recovery agents, bioenergy (biodiesel, bioethanol, biohydrogen, biobutanol), bioflocculants, and biopolymers are some intriguing applications of wastewater-cultivated biomass. However, the following are some considerations for the adroit transfer of these products from the lab scale to the market scale:

- Exploration of superior strains (especially of indigenous origin) with high nutrient removal, metabolite accumulation, and stress-tolerant potential.
- Optimization of production and processing routes for targeted bio-compounds acquisition. Production and processing of these products should be managed and controlled to avoid the accumulation of toxin metabolites.
- Implementation of safety assessment analyses to ensure the biomass quality and to increase customer acceptability.
- Including energy-efficient technologies, focus on byproduct valorization, adopting sustainable chemical alternatives, and designing simplified processing routes are necessary to develop an ecologically viable "cradle-to-gate" value chain.
- Execution of ecological and environmental risk analyses and life cycle assessments of pilot-scale and industrial-scale production plants to avoid environmental and economic implications.
- The focus of future LCA studies should be on circular or semi-circular processing routes.
- Coordinated networking among various stakeholders, including algal researchers, environmental protection agencies, government officials, industrialists, and consumers, to bridge market gaps.
- Investment and subsidies for cyanobacterial scale-up, infrastructure development, and market inclusion are mandatory.

References

Aaronson S, Dhawale S, Patni N, DeAngelis B, Frank O, Baker HJAom (1977) The cell content and secretion of water-soluble vitamins by several freshwater algae. Archives of Microbiology. 112:57–59. https://doi.org/10.1007/BF00446654

Afkairin A, Ippolito JA, Stromberger M, Davis JGJASE (2021) Solubilization of organic phosphorus sources by cyanobacteria and a commercially available bacterial consortium. Applied Soil Ecology. 162:103900. https://doi.org/10.1016/j.apsoil.2021.103900

Agarwal P, Soni R, Kaur P, Madan A, Mishra R, Pandey J et al (2022) Cyanobacteria as a promising alternative for sustainable environment: synthesis of biofuel and biodegradable plastics. Frontiers in Microbiology. 13:939347. https://doi.org/10.3389/fmicb.2022.939347

Alvarez AL, Weyers SL, Goemann HM, Peyton BM, Gardner RDJAR (2021) Microalgae, soil and plants: a critical review of microalgae as renewable resources for agriculture. Algal Research. 54:102200. https://doi.org/10.1016/j.algal.2021.102200

Álvarez-González A, Greque de Morais E, Planas-Carbonell A, Uggetti E (2023) Enhancing sustainability through microalgae cultivation in urban wastewater for biostimulant production and nutrient recovery. Science of The Total Environment 904:166878. https://doi.org/10.1016/j.scitotenv.2023.166878

Anahas AMP, Muralitharan GJBT (2015) Isolation and screening of heterocystous cyanobacterial strains for biodiesel production by evaluating the fuel properties from fatty acid methyl ester (FAME) profiles. Bioresource Technology. 184:9–17. https://doi.org/10.1016/j.biortech.2014.11.003

Arashiro LT, Josa I, Ferrer I, Van Hulle SWH, Rousseau DPL, Garfí M (2022) Life cycle assessment of microalgae systems for wastewater treatment and bioproducts recovery: natural pigments, biofertilizer and biogas. Science of The Total Environment. 847:157615. https://doi.org/10.1016/j.scitotenv.2022.157615

Ariede MB, Candido TM, Jacome ALM, Velasco MVR, de Carvalho JCM, Baby ARJAr (2017) Cosmetic attributes of algae – a review. Algal Research. 25:483–487. https://doi.org/10.1016/j.algal.2017.05.019

Asensi-Fabado M.A., Munné-Bosch SJTips (2010) Vitamins in plants: occurrence, biosynthesis and antioxidant function. Trends in Plant Science. 15:582–592. https://doi.org/10.1016/j.tplants.2010.07.003

Attene L, Deiana A, Carucci A, De Gioannis G, Asunis F, Ledda C (2023) Efficient nitrogen recovery from agro-energy effluents for cyanobacteria cultivation (*Spirulina*). Sustainability. https://doi.org/10.3390/su15010675

Azeem M, Iqbal N, Mir RA, Adeel S, Batool F, Khan AA et al (2019) Harnessing natural colorants from algal species for fabric dyeing: a sustainable eco-friendly approach for textile processing. Journal of Applied Phycology. 31:3941–3948. https://doi.org/10.1007/s10811-019-01848-z

Beardall J, Raven JAJTpom (2016) Carbon acquisition by microalgae. The Physiology of Microalgae. 89–99. https://doi.org/10.1007/978-3-319-24945-2_4

Bhati R, Mallick NJAR (2015) Poly (3-hydroxybutyrate-co-3-hydroxyvalerate) copolymer production by the diazotrophic cyanobacterium *Nostoc muscorum Agardh*: Process optimization and polymer characterization. Algal Research. 7:78–85. https://doi.org/10.1016/j.algal.2014.12.003

Cabanillas A.H., Tena Pérez Vc, Maderuelo Corral S, Rosero Valencia DF, Martel Quintana A, Ortega Doménech M et al (2018) Cybastacines A and B: antibiotic sesterterpenes from a Nostoc sp. cyanobacterium. Journal of Natural Products. 81:410–413. https://doi.org/10.1021/acs.jnatprod.7b00638

Castro JS, Ferreira J, Magalhães IB, Jesus Junior MM, Marangon BB, Pereira ASAP et al (2023) Life cycle assessment and techno-economic analysis for biofuel and biofertilizer recovery as by-products from microalgae. Renewable and Sustainable Energy Reviews. 187:113781. https://doi.org/10.1016/j.rser.2023.113781

Collins MD, Jones DJMr (1981) Distribution of isoprenoid quinone structural types in bacteria and their taxonomic implication.Microbiology and Molecular Biology Reviews. 45:316–354. https://doi.org/10.1128%2Fmr.45.2.316-354.1981

Costa JAV, Lucas BF, Alvarenga AGP, Moreira JB, de Morais MGJP (2021) Microalgae polysaccharides: an overview of production, characterization, and potential applications. Polysaccharides. 2:759–772. https://doi.org/10.3390/polysaccharides2040046

de Oliveira AdN, Ferreira IM, Jimenez DEQ, da Silva LS, da Costa AAF, Lima ETL et al (2022) Mining waste valorisation to catalytically active mesoporous materials for the esterification of fatty acid palm oil waste. Molecular Catalysis. 528:112504. https://doi.org/10.1016/j.mcat.2022.112504

Del Mondo A, Smerilli A, Sané E, Sansone C, Brunet CJMCF (2020) Challenging microalgal vitamins for human health. Microbial Cell Factories.19:1–23. https://doi.org/10.1186/s12934-020-01459-1

Devadas VV, Khoo KS, Chia WY, Chew KW, Munawaroh HSH, Lam M-K et al (2021) Algae biopolymer towards sustainable circular economy. Bioresource Technology. 325:124702. https://doi.org/10.1016/j.biortech.2021.124702

Devi S, Rani N, Sagar AJPA (2020) GC-MS analysis and antioxidant activity of two species of cyanobacteria isolated from Drang salt mine of district Mandi, Himachal Pradesh, India. Plant Archives. 20:7505–7510

Dikshit R, Tallapragada P (2018) Comparative study of natural and artificial flavoring agents and dyes. Natural and artificial flavoring agents and food dyes. Elsevier, pp 83–111.https://doi.org/10.1016/B978-0-12-811518-3.00003-X

Dineshkumar R, Kumaravel R, Gopalsamy J, Sikder MNA, Sampathkumar PJW, valorization b (2018) Microalgae as bio-fertilizers for rice growth and seed yield productivity. Waste and Biomass Valorization. 9:793–800. https://link.springer.com/article/10.1007/s12649-017-9873-5

Edelmann M., Aalto S., Chamlagain B., Kariluoto S., Piironen VJJoFC, Analysis (2019) Riboflavin, niacin, folate and vitamin B12 in commercial microalgae powders. Journal of Food Composition and Analysis. 82:103226. https://doi.org/10.1016/j.jfca.2019.05.009

Fabra MJ, Martínez-Sanz M, Gómez-Mascaraque LG, Gavara R, López-Rubio AJCp (2018) Structural and physicochemical characterization of thermoplastic corn starch films containing microalgae. Carbohydrate Polymers. 186:184–191. https://doi.org/10.1016/j.carbpol.2018.01.039

Fais G, Manca A, Bolognesi F, Borselli M, Concas A, Busutti M et al (2022) Wide range applications of *spirulina*: from earth to space missions. Marine Drugs. 20:299. https://doi.org/10.3390/md20050299

Fay PJMr (1992) Oxygen relations of nitrogen fixation in cyanobacteria. Microbiology and Molecular Biology Reviews 56:340–373. https://doi.org/10.1128/mr.56.2.340-373.1992

Ferreira J, Braga MQ, Gama RCNd, Magalhães IB, Marangon BB, Castro JdS et al (2024) Carotenoids from wastewater-grown microalgae biomass: life cycle assessment and techno-economical analysis. Journal of Cleaner Production. 434:140526. https://doi.org/10.1016/j.jclepro.2023.140526

Gaignard C, Laroche C, Pierre G, Dubessay P, Delattre C, Gardarin C et al (2019) Screening of marine microalgae: Investigation of new exopolysaccharide producers. Algal Research. 44:101711. https://doi.org/10.1016/j.algal.2019.101711

Gan F, Bryant DAJEm (2015) Adaptive and acclimative responses of cyanobacteria to far-red light. Environmental Microbiology. 17:34503465. https://doi.org/10.1111/1462-2920.12992

Garby TJ, Matys ED, Ongley SE, Salih A, Larkum AW, Walter MR et al (2017) Lack of methylated hopanoids renders the cyanobacterium *Nostoc punctiforme* sensitive to osmotic and pH stress. Applied and Environmental Microbiology. 83:e00777–17. https://doi.org/10.1128/AEM.00777-17

Gaysina LA, Saraf A, Singh P (2019) Cyanobacteria in diverse habitats. Cyanobacteria. Elsevier, pp 1–28. https://doi.org/10.1016/B978-0-12-814667-5.00001-5

Gerwick WH, Moore BSJC, biology (2012) Lessons from the past and charting the future of marine natural products drug discovery and chemical biology. Chemistry & Biology. 19:85–98. https://doi.org/10.1016/j.chembiol.2011.12.014

Gondi R, Kavitha S, Kannah RY, Karthikeyan OP, Kumar G, Tyagi VK et al (2022) Algal-based system for removal of emerging pollutants from wastewater: a review. Bioresource Technology. 344:126245. https://doi.org/10.1016/j.biortech.2021.126245

Guedes AC, Amaro HM, Malcata FXJMd (2011) Microalgae as sources of carotenoids. Marine Drugs. 9:625–644. https://doi.org/10.3390/md9040625
Guo, S., Wang, P., Wang, X., Zou, M., Liu, C., Hao, J. (2020). Microalgae as Biofertilizer in Modern Agriculture. In: Alam, M., Xu, JL., Wang, Z. (eds) Microalgae Biotechnology for Food, Health and High Value Products. Springer, Singapore. https://doi.org/10.1007/978-981-15-0169-2_12
Guskov A, Kern J, Gabdulkhakov A, Broser M, Zouni A, Saenger WJNs et al (2009) Cyanobacterial photosystem II at 2.9-Å resolution and the role of quinones, lipids, channels and chloride. Nature Structural & Molecular Biology. 16:334–342. https://doi.org/10.1038/nsmb.1559
Helliwell KE, Lawrence AD, Holzer A, Kudahl UJ, Sasso S, Kräutler B, et al (2016) Cyanobacteria and eukaryotic algae use different chemical variants of vitamin B12. Current Biology. 26:999–1008. https://doi.org/10.1016/j.cub.2016.02.041
Herrera A, D'Imporzano G, Acién Fernandez FG, Adani F (2021) Sustainable production of microalgae in raceways: nutrients and water management as key factors influencing environmental impacts. Journal of Cleaner Production. 287:125005. https://doi.org/10.1016/j.jclepro.2020.125005
Hossain N, Zaini J, Indra Mahlia TM (2019) Life cycle assessment, energy balance and sensitivity analysis of bioethanol production from microalgae in a tropical country. Renewable and Sustainable Energy Reviews. 115:109371. https://doi.org/10.1016/j.rser.2019.109371
Huisman J, Codd GA, Paerl HW, Ibelings BW, Verspagen JM, Visser PMJNRM (2018) Cyanobacterial blooms. Nature Reviews Microbiology. 16:471–483. https://doi.org/10.1038/s41579-018-0040-1
Hwang J-H, Church J, Lee S-J, Park J, Lee WHJEES (2016) Use of microalgae for advanced wastewater treatment and sustainable bioenergy generation. Environmental Engineering Science. 33:882–897. https://doi.org/10.1089/ees.2016.0132
Jahnke LL, Embaye T, Hope J, Turk KA, Van Zuilen M, Des Marais DJ et al (2004) Lipid biomarker and carbon isotopic signatures for stromatolite-forming, microbial mat communities and *Phormidium* cultures from Yellowstone National Park. Geobiology. 2:31–47. https://doi.org/10.1111/j.1472-4677.2004.00021.x
Jama A, Widiastuti DP, Gafur S, Davis JGJS (2023) Azolla biofertilizer is an effective replacement for urea fertilizer in vegetable crops. Sustainability. 15:6045. https://doi.org/10.3390/su15076045
Ji X, Verspagen JM, Van de Waal DB, Rost B, Huisman JJSa (2020) Phenotypic plasticity of carbon fixation stimulates cyanobacterial blooms at elevated CO2. Science Advances. https://doi.org/10.1126/sciadv.aax2926
Joliot, P., Joliot, A., Johnson, G. (2006). Cyclic Electron Transfer Around Photosystem I. In: Golbeck, J.H. (eds) Photosystem I. Advances in Photosynthesis and Respiration, vol 24. Springer, Dordrecht. https://doi.org/10.1007/978-1-4020-4256-0_37
Kumar N, Goel NJBr (2019) Phenolic acids: Natural versatile molecules with promising therapeutic applications. Biotechnology Reports. https://doi.org/10.1016/j.btre.2019.e00370
Li R, Narita R, Nishimura H, Marumoto S, Yamamoto SP, Ouda R et al (2018) Antiviral activity of phenolic derivatives in pyroligneous acid from hardwood, softwood, and bamboo. Biotechnology Reports. 6:119–126. https://doi.org/10.1021/acssuschemeng.7b01265
Lopes G, Clarinha D, Vasconcelos VJM (2020) Carotenoids from cyanobacteria: a biotechnological approach for the topical treatment of psoriasis. Microorganisms. https://doi.org/10.3390/microorganisms8020302
Luque R, Lovett JC, Datta B, Clancy J, Campelo JM, Romero AAJE et al (2010) Biodiesel as feasible petrol fuel replacement: a multidisciplinary overview. Energy & Environmental Science. 3:1706–1721. https://doi.org/10.1039/C0EE00085J
Madison LL, Huisman GWJM, reviews m.b (1999) Metabolic engineering of poly (3-hydroxyalkanoates): from DNA to plastic. Microbiology and Molecular Biology Reviews. 63:21–53. https://doi.org/10.1128/mmbr.63.1.21-53.1999

Magalhães IB, Ferreira J, Castro JdS, Assis LRd, Calijuri ML (2022) Agro-industrial wastewater-grown microalgae: a techno-environmental assessment of open and closed systems. Science of The Total Environment. 834:155282. https://doi.org/10.1016/j.scitotenv.2022.155282

Mahanty T, Bhattacharjee S, Goswami M, Bhattacharyya P, Das B, Ghosh A et al (2017) Biofertilizers: a potential approach for sustainable agriculture development. Environmental Science and Pollution Research. 24:3315–3335. https://doi.org/10.1007/s11356-016-8104-0

Martini F, Beghini G, Zanin L, Varanini Z, Zamboni A, Ballottari M (2021) The potential use of *Chlamydomonas reinhardtii* and *Chlorella sorokiniana* as biostimulants on maize plants. Algal Research 60:102515. https://doi.org/10.1016/j.algal.2021.102515

Mimuro M, Tsuchiya T, Inoue H, Sakuragi Y, Itoh Y, Gotoh T et al (2005) The secondary electron acceptor of photosystem I in Gloeobacter violaceus PCC 7421 is menaquinone-4 that is synthesized by a unique but unknown pathway. FEBS Letters. 579:3493–3496. https://doi.org/10.1016/j.febslet.2005.05.029

Monroe MBB, Easley AD, Grant K, Fletcher GK, Boyer C, Maitland DJJC (2018) Multifunctional shape-memory polymer foams with bio-inspired antimicrobials. ChemPhysChem. 19:1999–2008. https://doi.org/10.1002/cphc.201701015

Morone J, Lopes G, Preto M, Vasconcelos V, Martins RJMd (2020) Exploitation of filamentous and picoplanktonic cyanobacteria for cosmetic applications: Potential to improve skin structure and preserve dermal matrix components. Marine Drugs. 18:486. https://doi.org/10.3390/md18090486

Mourelle ML, Gómez CP, Legido JLJC (2017) The potential use of marine microalgae and cyanobacteria in cosmetics and thalassotherapy. Cosmetics. 4:46. https://doi.org/10.3390/cosmetics4040046

Nisar N, Li L, Lu S, Khin NC, Pogson BJJMp (2015) Carotenoid metabolism in plants. Molecular Plant. 8:68–82. https://doi.org/10.1016/j.molp.2014.12.007

Osman AI, Fang B, Zhang Y, Liu Y, Yu J, Farghali M et al (2024) Life cycle assessment and techno-economic analysis of sustainable bioenergy production: a review. Environmental Chemistry Letters. 22:1115. https://doi.org/10.1007/s10311-023-01694-z

Patipong T, Hibino T, Waditee-Sirisattha R, Kageyama HJNPC (2019) Induction of antioxidative activity and antioxidant molecules in the halotolerant cyanobacterium Halothece sp. PCC7418 by temperature shift. Natural Product Communications. http://dx.doi.org/10.1177/1934578X19865680

Rai P, Mehrotra S, Priya S, Gnansounou E, Sharma SKJBt (2021) Recent advances in the sustainable design and applications of biodegradable polymers. Bioresource Technology. 325:124739. https://doi.org/10.1016/j.biortech.2021.124739

Ray SS, Bousmina MJP (2005) Biodegradable polymers and their layered silicate nanocomposites: in greening the 21st century materials world. Progress in Materials Science. 50:962–1079. https://doi.org/10.1016/j.pmatsci.2005.05.002

Renuka N, Guldhe A, Prasanna R, Singh P, Bux FJBa (2018) Microalgae as multi-functional options in modern agriculture: current trends, prospects and challenges. Biotechnology Advances. 36:1255–1273. https://doi.org/10.1016/j.biotechadv.2018.04.004

Rippka R., Neilson A., Kunisawa R., Cohen-Bazire GJAfM (1971) Nitrogen fixation by unicellular blue-green algae. Archives of Microbiology. 76:341–348. https://doi.org/10.1007/BF00408530

Romagnoli F, Spaccini F, Boggia A, Paoli R, Feofilovs M, Ieviņa B et al (2024) Microalgae cultivation in a biogas plant: environmental and economic assessment through a life cycle approach. Biomass and Bioenergy 182:107116. https://doi.org/10.1016/j.biombioe.2024.107116

Rossi F, Li H, Liu Y, De Philippis RJE-SR (2017) Cyanobacterial inoculation (cyanobacterisation): perspectives for the development of a standardized multifunctional technology for soil fertilization and desertification reversal. Earth-Science Reviews. 171:28–43. https://doi.org/10.1016/j.earscirev.2017.05.006

Ruwanpathirana PS, Gnanavelrajah N, Ratnayake RR (2023) Potential of *Nostoc* sp. as a biofertilizer on growth and yield of paddy-*Oryza sativa*. In: Kadigamuwa CC (ed) 11th young scientists forum (YSF) symposium. National Science and Technology Comission

Sadvakasova AK, Kossalbayev BD, Zayadan BK, Kirbayeva DK, Alwasel S, Allakhverdiev SIJWJoM et al (2021) Potential of cyanobacteria in the conversion of wastewater to biofuels. World Journal of Microbiology and Biotechnology. 37:1–22. https://doi.org/10.1007/s11274-021-03107-1

Saha P, Aloui H, Yun JH, Kim HS, Kim BSJJoAPS (2021) Development of a novel composite film based on polyurethane and defatted Chlorella biomass: physical and functional characterization. Journal of Applied Polymer Science. 138:50152. https://doi.org/10.1002/app.50152

Saini DK, Pabbi S, Shukla PJF, toxicology c (2018) Cyanobacterial pigments: perspectives and biotechnological approaches. Food and Chemical Toxicology. 120:616–624. https://doi.org/10.1016/j.fct.2018.08.002

Santiago-Morales IS, Trujillo-Valle L, Márquez-Rocha FJ, Hernández JFLJAFB (2018) Tocopherols, phycocyanin and superoxide dismutase from microalgae: as potential food antioxidants. Applied Food Biotechnology. 5:19–27. https://doi.org/10.22037/afb.v5i1.17884

Schneider RCS, de Moura Lima M, Hoeltz M, de Farias Neves F, John DK, de Azevedo A (2018) Life cycle assessment of microalgae production in a raceway pond with alternative culture media. Algal Research. 32:280–292. https://doi.org/10.1016/j.algal.2018.04.012

Shahid A, Malik S, Alam MA, Nahid N, Mehmood MA (2019) The culture technology for freshwater and marine microalgae. In: Alam MA, Wang Z (eds) Microalgae biotechnology for development of biofuel and wastewater treatment. Springer Singapore, Singapore, pp 21–44

Shi B, Wideman G, Wang JHJJoP, Environment t (2012) A new approach of BioCO2 fixation by thermoplastic processing of microalgae. Journal of Polymers and the Environment. 20:124–131. https://doi.org/10.1007/s10924-011-0329-x

Silambarasan S, Logeswari P, Sivaramakrishnan R, Kamaraj B, Lan Chi NT, Cornejo P (2021) Cultivation of *Nostoc* sp. LS04 in municipal wastewater for biodiesel production and their deoiled biomass cellular extracts as biostimulants for Lactuca sativa growth improvement. Chemosphere 280:130644. https://doi.org/10.1016/j.chemosphere.2021.130644

Singh JS, Kumar A, Rai AN, Singh DPJFim (2016) Cyanobacteria: a precious bio-resource in agriculture, ecosystem, and environmental sustainability. Frontiers in Microbiology. 7:529. https://doi.org/10.3389/fmicb.2016.00529

Singh R, Parihar P, Singh M, Bajguz A, Kumar J, Singh S et al (2017) Uncovering potential applications of cyanobacteria and algal metabolites in biology, agriculture and medicine: current status and future prospects. Frontiers in Microbiology https://doi.org/10.3389/fmicb.2017.00515

Singh DP, Khattar JS, Rajput A, Chaudhary R, Singh RJPo (2019) High production of carotenoids by the green microalga Asterarcys quadricellulare PUMCC 51 1 under optimized culture conditions. PLOS ONE. 14:e0221930. https://doi.org/10.1371/journal.pone.0221930

Singh KB, Kaushalendra VS, Lalnunpuii R, Rajan JPJS (2023) Current issues and developments in cyanobacteria-derived biofuel as a potential source of energy for sustainable future. Sustainability 15:10439. https://doi.org/10.3390/su151310439

Sitther, V., Tabatabai, B., Fathabad, S. G., Gichuki, S., Chen, H., and Arumanayagam, A. C. S. (2020). Cyanobacteria as a biofuel source: advances and applications. In "Advances in Cyanobacterial Biology" (P. K. Singh, A. Kumar, V. K. Singh and A. K. Shrivastava, eds.), pp. 269-289. Academic Press. https://doi.org/10.1016/B978-0-12-819311-2.00018-8

Slocombe SP, Zúñiga-Burgos T, Chu L, Mehrshahi P, Davey MP, Smith AG et al (2023) Overexpression of PSR1 in *Chlamydomonas reinhardtii* induces luxury phosphorus uptake. Frontiers in Plant Science 14. https://doi.org/10.3389/fpls.2023.1208168

Sonani RR, Rastogi RP, Patel SN, Chaubey MG, Singh NK, Gupta GD et al (2019) Phylogenetic and crystallographic analysis of Nostoc phycocyanin having blue-shifted spectral properties. Scientific Reports. 9:9863. https://doi.org/10.1038/s41598-019-46288-4

Sylvander P., Häubner N., Snoeijs PJMe (2013) The thiamine content of phytoplankton cells is affected by abiotic stress and growth rate. Microbial Ecology. 65:566–577. https://doi.org/10.1007/s00248-012-0156-1

Tarento T.D., McClure D.D., Vasiljevski E., Schindeler A., Dehghani F., Kavanagh JMJAr (2018) Microalgae as a source of vitamin K1. Algal Research. 36:77–87. https://doi.org/10.1016/j.algal.2018.10.008

Teodoro GR, Ellepola K, Seneviratne CJ, Koga-Ito CYJFim (2015) Potential use of phenolic acids as anti-Candida agents: a review. Frontiers in Microbiology. 6:1420. https://doi.org/10.3389/fmicb.2015.01420

Tiwari ON, Bhunia B, Mondal A, Gopikrishna K, Indrama TJJoCP (2019) System metabolic engineering of exopolysaccharide-producing cyanobacteria in soil rehabilitation by inducing the formation of biological soil crusts: a review. Journal of Cleaner Production. 211:70–82. https://doi.org/10.1016/j.jclepro.2018.11.188

Hector Osorio, U., Lucy Belmar, B., and Mónica, V. (2016). Microalgae and Cyanobacteria as Green Molecular Factories: Tools and Perspectives. In "Algae" (T. Nooruddin and D. Dharumadurai, eds.), IntechOpen, Rijeka. DOI: 10.5772/63006

Villacreses-Freire D, Ketzer F, Rösch CJBR (2021) Advanced metabolic engineering approaches and renewable energy to improve environmental benefits of algal biofuels: LCA of large-scale biobutanol production with cyanobacteria Synechocystis. BioEnergy Research. PCC6803 1–16. https://doi.org/10.1007/s12155-021-10323-y

Villaró S, Acién G, Alarcón J, Ruiz Á, Rodríguez-Chikri L, Viviano E et al (2023) A zero-waste approach for the production and use of *Arthrospira platensis* as a protein source in foods and as a plant biostimulant in agriculture. Journal of Applied Phycology Publishing model 35:2619–2630. https://doi.org/10.1007/s10811-023-02993-2

Wang M, Luan G, Lu XJJob (2020) Engineering ethanol production in a marine cyanobacterium *Synechococcus* sp PCC7002 through simultaneously removing glycogen synthesis genes and introducing ethanolgenic cassettes. Journal of Biotechnology. 317:1–4. https://doi.org/10.1016/j.jbiotec.2020.04.002

Wang N, Xiang Y, Ma Y, Zhang P, Zhou X, Zhu H, Zhang Z, Li Z, Xiao X, He M, Mehmood MA, Zhu H (2024) Fermented *Spirulina platensis* alleviated DSS-induced ulcerative colitis by regulating gut microbiota and MyD88/TLR4 signaling pathway. Asian Journal of Agriculture & Biology. 1(1):2023232. https://doi.org/10.35495/ajab.2023.232

Widhalm JR, van Oostende C, Furt F, Basset GJJPotNAoS (2009) A dedicated thioesterase of the Hotdog-fold family is required for the biosynthesis of the naphthoquinone ring of vitamin K1. Biochemistry. 106:5599–5603. https://doi.org/10.1073/pnas.0900738106

Wuang SC, Khin MC, Chua PQD, Luo YD (2016) Use of *Spirulina* biomass produced from treatment of aquaculture wastewater as agricultural fertilizers. Algal Research. 15:59–64. https://doi.org/10.1016/j.algal.2016.02.009

Xu Y, Ibrahim IM, Wosu CI, Ben-Amotz A, Harvey PJJB (2018) Potential of new isolates of Dunaliella salina for natural β-carotene production. Biology. 7:14. https://doi.org/10.3390/biology7010014

Xu P, Li J, Qian J, Wang B, Liu J, Xu R et al (2023) Recent advances in CO_2 fixation by microalgae and its potential contribution to carbon neutrality. Chemosphere 319:137987. https://doi.org/10.1016/j.chemosphere.2023.137987

Xu Y, Wei C, Liu D, Li J, Tian B, Li Z et al (2024) Life-cycle and economic assessments of microalgae biogas production in suspension and biofilm cultivation systems. Bioresource Technology. 395:130381. https://doi.org/10.1016/j.biortech.2024.130381

Chapter 11
Bioprocessing of the Wastewater-Grown Cyanobacterial Biomass to Biofertilizers

Fatima Tahir, Mahwish Amin, Tanveer A. Tabish, Xin-Qing Zhao, and Chen-Guang Liu

Abstract A rapid increase in population has led to a significant rise in freshwater consumption, resulting in a huge amount of wastewater generation. Wastewater contains valuable resources like nitrogen, phosphorus, and organic compounds in bulk quantities, which makes it a suitable medium for algal cultivation. Cyanobacteria are recognized as a versatile biological resource with widespread applications in different industries. In agriculture, using wastewater-grown algal biomass as a biofertilizer enhances plant growth like traditional organic manures and fertilizers and reduces nutrient loss in the soil, offering a range of benefits to ensure nutrient availability, improve soil fertility, and minimize synthetic fertilizers. This sustainable approach maximizes plant nutrient utilization, contributing to efficient agricultural practices and environmental conservation. However, conventional chemical fertilizers pose significant long-term threats to ecosystems, contribute to food security challenges, and limit the potential for land reuse in agriculture. This chapter discusses suitable algal strains based on their biochemical composition and the methods of using cyanobacterial biomass as a biofertilizer. It also highlights the environmental and agricultural advantages of biofertilizers, emphasizing their role in nutrient management and sustainable farming.

Keywords Cyanobacteria · Biochemical composition · Biofertilizers · Sustainable farming · Nutrient management

---

F. Tahir · M. Amin
Bioenergy Research Center, Department of Bioinformatics and Biotechnology, Government College University Faisalabad, Faisalabad, Pakistan

T. A. Tabish
Division of Cardiovascular Medicine, Radcliffe Department of Medicine, University of Oxford, Oxford, UK

X.-Q. Zhao · C.-G. Liu (✉)
State Key Laboratory of Microbial Metabolism, Joint International Research Laboratory of Metabolic and Developmental Sciences, School of Life Sciences and Biotechnology, Shanghai Jiao Tong University, Shanghai, China
e-mail: cg.liu@sjtu.edu.cn

M. A. Mehmood et al. (eds.), *Cyanobacteria Biotechnology*,
https://doi.org/10.1007/978-3-031-70698-1_11

11.1 Introduction

Urbanization, high mobility, rising living standards, and increased industrialization have caused substantial utilization of natural resources such as agricultural land, freshwater for industry and homes, and excessive burning of fossil fuels to meet the demand (Shahid et al. 2021). As a result, there are significant greenhouse gas emissions and massive amounts of wastewater production, even though the situation is worse in heavily populated countries such as Bangladesh, Pakistan, China, India, Indonesia, and the Philippines. Human threats, such as resource shortages, water scarcity, food shortage, and environmental deterioration, stimulate a new way of managing wastewater and its treatment (Su 2021). This scenario necessitates the development of comprehensive procedures that intricately address these problems (Martinez-Hernandez and Samsatli 2017).

The daily discharge of wastewater is 61–2310 million m^3 globally (Renuka et al. 2021). Wastewater is rich in organic/inorganic carbon, nitrogen, and phosphorus forms. These nutrients make it potentially useful as a low-cost source of medium to produce sustainable bioresources (Kumar and Bera 2020). The nutrient in wastewater can be used for the cultivation of algae because algal biomass is a source of multiproduct and has various industrial applications as after cultivation the spent media can also be used for irrigation purposes. Algae are the potentially diverse, common, and dominant groups of photosynthetic organisms in nature, particularly cyanobacteria and microalgae (Deviram et al. 2020). Cyanobacteria are considered to be the earliest life forms responsible for oxygenating the early Earth's atmosphere.

Cyanobacteria are believed to be a feedstock of future multiproduct biorefineries because of their short cultivation period, stress-tolerant mechanism, highest CO_2 fixation rate, wastewater treatment efficiency, unique metabolite composition, and accumulation of valuable metabolites with industrial, environmental, and pharmaceutical significance (Shahid et al. 2021). Cyanobacteria consist of two main cell types: (1) heterocyst (responsible for nitrogen fixation) and (2) vegetative cells (exhibit reproductive growth and photosynthesis) (Chittora et al. 2020). Furthermore, cyanobacteria have shown great potential to be the most suitable feedstock for producing polyhydroxyalkanoates, pigments, medicines, biofuels, and biofertilizers (Encarnação et al. 2023).

Insufficient nutrient availability is the primary cause of low crop yields and is a significant agronomic problem across several regions of the world. The use of chemical fertilizers has been essential to agricultural development, but they also cause extensive nutrient pollution, and the deterioration of rivers, lakes, and coastlines (Al-Hazmi et al. 2023). Recent studies on the interactions between plants and rhizosphere-dwelling soil microbes have resulted in a worldwide increase in the production and application of microbial-based fertilizers, which are also known as biofertilizers (Kumar et al. 2023). In contrast to conventional fertilizers, which require energy-intensive manufacturing procedures and significantly increase emissions, biofertilizers work within the natural nutrient cycles. By lowering emissions from agriculture, using biofertilizers can contribute to reducing the carbon footprint in the struggle against climate change (Ferreira et al. 2021). Biofertilizers are considered eco-friendly and cost-effective products and are highly demanded, specifically in organic farming (Ronga et al. 2019). Wastewater-based grown algal biomass can be used as a biofertilizer. Algae-based biofertilizers assist in minimizing biotic and abiotic stresses while boosting plant growth, yield, quality, and nutrients. Several commercially available algae-based biofertilizers and biostimulants, such as AlgaFert Eco® and AlgaFert®, are commercialized by Biorizon Biotech SL (Almería, Spain).

11.2 Influence of Different Wastewaters on the Biochemical Composition of Biomass

The biochemical composition of algae is very dependent upon the composition of the media. Wastewater is used as an alternative nutrient source for cultivation, and it directly impacts the metabolite content of algae by influencing their growth and composition (Table 11.1). Microalgae and cyanobacteria utilize nutrients, including nitrates and phosphates, organic matter, pollutants and contaminants, and heavy metals from wastewater, which influences the production of protein, lipids, and carbohydrates in altered compositions or concentrations. So, there is a need to explore the potential of algae grown in different wastewaters while also considering the impact of different wastewaters on the concentration, composition, and quality of

Table 11.1 Influence of different types of wastewaters on final biomass composition

Algal species	Wastewater type	Biomass characteristics	References
Spirulina platensis	Internal desalination concentrate diluted in synthetic medium	Proteins 55%, lipids 4.8%, carbohydrates 12%	Matos et al. (2021)
Nostoc muscorum and *Chlorella minutissima*	Municipal wastewater	Higher levels of N were recorded for *chlorella* (5.87%) Higher P contents were found in *chlorella* and *Nostoc*	Khan et al. (2019)
Scenedesmus obliquus	Brewery effluent	Proteins 31%, carbohydrates 30–31%, lipids 17–18%	Ferreira et al. (2019)
Chlorella sp. and *Phormidium* sp.	Anaerobic-digested slurry (ADS) (high N and TSS) Livestock wastewater (LSW) (high N and P) Domestic gray water (DGW)	ADS: Proteins 44%, carbohydrates 13%, lipids 23% LSW: Proteins 32%, carbohydrates 22%, lipids 28% DGW: Proteins 43%, carbohydrates 24%, lipids 38%	Choudhary et al. (2017)
Chlorella pyrenoidosa	Rice mill wastewater	Lipids 34.15%, proteins 32.12%, carbohydrates 8.84%	Umamaheswari and Shanthakumar (2021)

metabolites produced, as this information is crucial for the selection of biomass applications such as biofuel, biofertilizer, and biopolymer.

11.2.1 Scenedesmus *sp.*

Scenedesmus sp. belongs to those genera of microalgae that have the potential to grow under autotrophic, heterotrophic, and mixotrophic cultivation conditions. Due to this ability, they grow well in wastewater by producing high biomass productivity, nutrient removal, and rapid growth. Shen et al. (2015) studied the metabolite composition and nutrient removal of *Scenedesmus obliquus* in secondary municipal wastewater. With regard to the growth and metabolite content, a notable increase in biomass and lipid productivity was observed approximately 577.6 and 16.7 mg L^{-1} day^{-1} with 5% CO_2 aeration and the nitrogen (N), phosphorus (P), and total organic compound (TOC) removal efficiencies were 97.8%, 95.6%, and 59.1%, respectively. Moreover, other metabolites, including proteins and carbohydrates, were 19% and 64%. In another study, to investigate the application of municipal wastewater-grown biomass of *S. obliquus,* kinetics of nitrogen and phosphorus elimination efficiency along with growth and metabolite content were assessed. The nitrogen and phosphorus removal efficiencies were 100% and 94.1%. Moreover, the biomass production, protein, lipid, and carbohydrate content were 0.88 g L^{-1}, 28.5%, 26.5%, and 27.5%, respectively (Ansari et al. 2019). There is plenty of research related to the cultivation of *Scenedesmus* sp. in wastewater (Daneshvar et al. 2018b; Oliveira et al. 2019; Jebali et al. 2018; Msanne et al. 2020) such as the cultivation of *S. quadricauda* in dairy wastewater clears the wastewater within eight days with 0.5 g L^{-1} biomass production. The nutrient uptake during the cultivation period, mainly N, P, and TOC, reached 86%, 90%, and 65%, respectively, and an increase in lipid content was observed. FTIR confirmed an increase in the concentration of saturated fatty acids in *S. quadricauda* (Msanne et al. 2020). Although algae-based wastewater treatment is oriented toward efficiently removing nutrients, they assimilate those nutrients as protein, lipid, and carbohydrate in their biomass and increase their susceptibility for vast applications.

11.2.2 Spirulina *sp.*

Spirulina is a spiral cyanobacterium, famous for its high protein content and vast applications in food supplements, cosmetics, therapeutics, biofuel, and animal feed production. In addition to protein, *Spirulina* is also rich in other high-value compounds such as phycocyanin, essential fatty acids, carbohydrates, phenolic compounds, minerals, and vitamins (Amin et al. 2023). Furthermore, *Spirulina* can remediate various wastewater, such as municipal and aquaculture wastewater, food processing effluents, and dairy and poultry industry wastewater. In a study, researchers grew *Spirulina platensis* in a media composed of Zarrouk's media and cheese whey wastewater at different concentrations (0%–10%). The highest biomass production of 2.98 g L^{-1} was obtained from media containing 5% whey wastewater. Moreover, wastewater-based cultivation produces variations in the metabolic profile of *Spirulina* (Usman et al. 2023) such as the cultivation of *Spirulina platensis* in 5% whey wastewater containing Zarrouk's media resulted in high carbohydrate and phenolic compounds 47.83% and 33.23 mg gallic acid g^{-1}, respectively (Pereira et al. 2019). In contrast, another study reported that cultivation of *Spirulina* sp. LEB18 in 25% aquaculture wastewater-based Zarrouk's media resulted in the production of the highest protein content, 65.73%, and the concentration of other metabolites, including phycocyanin, polyunsaturated fatty acids, and γ-linolenic, was 16.6 mg mL^{-1}, 38.2%, and 23.29%, respectively (Cardoso et al. 2021). So, the metabolic composition of the same species varies from strain to strain under the same cultivation conditions.

11.2.3 Chlorella *sp.*

Chlorella sp. shows great biorefinery potential due to its high lipid content such as omega-3 and omega-6 polyunsaturated fatty acids (PUFA), phenolic compounds, and some other high-value compounds including carotenoids and protein (Diprat et al. 2020). On the other hand, its ability to thrive in diverse environmental conditions and rapid growth makes it a promising candidate for nutrient removal from a variety of wastewater including aquaculture (Daneshvar et al. 2018a), cheese whey (Nazos et al. 2023), municipal (Pacheco et al. 2021), and industrial wastewater (Silva et al. 2022). Wastewater-based cultivation of algae gives dual benefits. Firstly, it absorbs and metabolizes the nutrients of wastewater and, secondly, after metabolization, leading to the production of a variety of metabolites. Research on the bioremediation capacity of *Chlorella vulgaris* highlighted its nutrient removal efficiency from wastewater. A study investigates the impact of secondary municipal wastewater on the biomass production and biodiesel production potential of *C. vulgaris*. Results revealed reasonable biomass productivity (39 mg L^{-1} day^{-1}) and triacylglycerides (TAG) 22.5%–41.3% (Sarpal et al. 2023). Another study reported the nutrient removal potential of *C. sorokiniana* from dairy wastewater (Hamidian and Zamani 2022). The results demonstrated the high biomass production (1.67 g L^{-1}), chlorophyll (20.64 mg g^{-1}), and carotenoid (4.74 mg g^{-1}) content (Hamidian and Zamani 2022). Moreover, the lipid and carbohydrate content was 0.433 gg^{-1} and 0.163 gg^{-1}, respectively (Hamidian and Zamani 2022). So, the extracted biomass and lipid content from *C. sorokiniana* holds potential applications in industries such as biofuel production, pharmaceuticals, nutraceuticals, and biofertilizers.

11.2.4 Haematococcus *sp.*

Haematococcus comprises solitary cells that appear ovoid to ellipsoidal, with a protoplast attached to the cell wall and thin branched cytoplasmic strands (Na 2015). Moreover, it comprises carotenoids (red accessory pigment), polyunsaturated fatty acids, carbohydrates (exopolysaccharides), phenolic compounds, and the least amount of protein (Mularczyk et al. 2020). On the other hand, it also has wastewater bioremediation potential. *Haematococcus pluvialis* SAG 34-1b has been cultivated in cassava wastewater to study its impact on the biochemical composition of microalgae and nutrient removal efficiency. The results showed that biomass production and pollutant removal efficiency of *H. pluvialis* SAG 34-1b were 1.79 g L^{-1} and 60.8% COD, 51.06% TN, and 54.68% phosphate, respectively (Sorgatto et al. 2021).

11.3 Revolutionizing Agriculture with Cyanobacteria

The revolution of cyanobacteria in agriculture as fertilizer creates a transformative paradigm shift, offering a sustainable and nutrient-rich solution to enhance soil health and crop productivity. The CO_2 sequestration ability of cyanobacteria not only reduces environmental pollution but also contributes to photosynthesis to produce biomass which is a potent biofertilizer. Packed with essential nutrients such as nitrogen, phosphorus, and micronutrients, algae-based fertilizers provide crops with a balanced and readily available nutrient source. Unlike conventional chemical fertilizers, algae-based fertilizers contribute a sustainable solution to soil, reduce nutrient runoff, and mitigate environmental impacts.

11.3.1 Conventional Versus Algal Biofertilizers

Fertilizer input is necessary for modern agriculture to produce food, fiber, and biofuel. However, many fertilizers are wasted, which raises agricultural costs, wastes energy, and pollutes the environment. Although phosphorus (P) is a crucial component of most fertilizers and is necessary for plant growth, only 20% of P is used efficiently (de Siqueira Castro et al. 2020). The use of fertilizers has increased globally by 30.8% (N), 31.4% (P_2O_5), and 61.3% (K_2O) as a result of their critical role in raising agricultural yields (Osorio-Reyes et al. 2023). Since almost 98% of ingested P ends up in wastewater and builds up in sewage sludge, it is a valuable resource worth recovering. Wastewater is a vast source of nutrients that can be extracted using various techniques and utilized to make fertilizers. Cultivating algae is the most effective method. This technique enables the recovery of N and P when the cultivation is done in wastewater, which produces algal biomass that is concentrated in nutrients and organic carbon. In addition to aiding in wastewater treatment, this technique produces algal biomass with various possible applications. In recent studies, it was reported that only 12% of algal biomass in triple superphosphate fertilizer mass can help in increasing phosphate uptake by plants (de Siqueira Castro et al. 2020).

Conventional Chemical Fertilizers

Inorganic (conventional) fertilizers are used primarily to increase crop yields and soil fertility; however, the sustained use of inorganic fertilizers causes a decrease in soil organic matter, increases acidity, degrades the physical and structural properties of the soil, contaminates groundwater and surface water, and has carcinogenic effects on human health. As a result, inorganic fertilizers have a detrimental long-term impact on the ecosystem, cause food security problems, and limit the reuse of land for agricultural production (Osorio-Reyes et al. 2023). The primary cause is the loss of basic cations from the soil, which results in acidification and alters the biodiversity of microorganisms, due to the overuse of nitrogenous fertilizers. A significant amount of reactive N_2 is produced worldwide by nitrogenous fertilizers, which can impact food security by rendering land unsuitable for cultivation (de Siqueira Castro et al. 2020). Numerous soil processes, including the decomposition of organic matter, the cycling of nutrients, and soil fertility, are carried out by microorganisms or bacteria. Pesticides and fertilizers applied in excess do not always boost yields proportionately, and over-fertilization can severely acidify farmland, causing metal accumulation and speeding up the loss of nutrients from the soil, weakening the soil's properties and rhizodeposition (Singh et al. 2020).

Carrots, cabbage, tomatoes, alfalfa, macadamia nuts, avocados, and bananas are among the crops that are particularly sensitive to pH variations in the soil. These crops would suffer from an incomplete, damaged, or very low-quality harvest because of pH changes, which would lead to additional fertilizers to improve the harvest (a never-ending cycle). Only 10–40% of chemical fertilizers are used by plants, according to a study; the remaining portion remains as insoluble inorganic salts and enters water bodies, where it fails to enter the biogeochemical cycle (Nosheen et al. 2021).

When nitrates are ingested, they have been degraded to nitrites and, therefore, to biologically active nitrogen oxides by bacterial strains in the oral cavity, catalyzed by the enzyme nitrate reductase (NO_S). Following ingestion, the gastric acid medium's nitrates and nitrites are converted into mitochondrial enzymes, non-enzymatic pH-dependent reduction, deoxy-hemoglobin, myoglobin, neuroglobin, xanthine oxidoreductase, aldehyde oxidase, and carbonic oxidase. After that, the small intestine absorbs the NO in the blood and enters the circulation, where it can spontaneously oxidize tissues to produce nitrites and nitrates. Some of these compounds accumulate in the salivary glands and are expelled as urine, which can lead to health problems like methemoglobinemia, diabetes, thyroid disease, and cancer (Camargo and Alonso 2006).

Algae-Based Fertilizers

Utilizing cyanobacteria as a biofertilizer or soil enhancer has several advantages for the environment, the soil's health, and the crops' eventual yield. When cyanobacterial biomass is in direct contact with the soil, the percentages of moisture, pH, and light encourage the viable cyanobacterial cells to become active and continue their metabolic activity, which helps fix nitrogen from the atmosphere. Along with providing additional macro- and micronutrients necessary for plant growth, this can even form an endo-symbiotic relationship between the roots of the crops and other soil-dwelling microbes. It has been demonstrated that the use of algae and other microorganisms, even in arid environments, increases soil fertility by improving the soil's ability to retain water, maintain stability, remove pollutants, and provide a better substrate for plants (Perera et al. 2018; Camargo and Alonso 2006).

In algae, photosynthesis and nitrogen fixation are natural processes that result in mineralization effects, the mobilization of organic and inorganic nutrients, and the production of various secondary metabolites with advantageous properties, such as growth hormones, polysaccharides, antimicrobial compounds, and many others. Applying live cyanobacteria and microalgae as biofertilizers improves crop growth and production yields. Because they can produce biocidal compounds like benzoic acid and majusculonic or hydrolytic enzymes that can control or inhibit the growth of pathogens like fungi, bacteria, and nematodes, algae are also referred to as biocontrol agents or biopesticides (Osorio-Reyes et al. 2023).

The most important element for the development of a plant is nitrogen (N), which plays an important role in the constitution of the plant structure and is present in chlorophyll. Soil microbiome, with the help of atmospheric nitrogen fixation, plays an essential role in nitrogen availability in the soil. So, with the increase in soil microbiota, plant growth increases because of increased nitrogen availability. While prokaryotic organisms such as *Bacillus*, *Clostridium*, and cyanobacteria are thought to be the only ones capable of N_2 fixation, adding eukaryotic microalgae can also facilitate this process. The fixation of bioavailable nitrogen in the soil beneath bean crops was demonstrated to increase upon the sprinkler application of *Chlorella vulgaris* IPPAS C-1 biomass, suggesting that using microalgae as a naturally occurring source of nitrogen for plants is a promising approach (Osorio-Reyes et al. 2023).

Phosphorus, found in DNA and making up the essential molecule ATP for all the plant's essential functions, is another chemical element that is significant to the development of the plant. The P that plants need to thrive is typically found in the soil, and while traditional fertilizers aim to enrich the soil, doing so can negatively affect water bodies and the soil itself. Instead, replacing the harmful P source with a less harmful one, such as cyanobacteria, is essential. Growing cyanobacteria in the contaminated waters and using the biomass they produce to treat the contaminated waters is one method of rerouting P pollution to the soil (Moreira et al. 2023).

Even though biofertilizers with bacteria as the main component have been studied extensively and shown great results for a wide variety of crops, fungi-based biofertilizers are also widespread. Algae have gained popularity as a versatile organism that can be used in a model of circular bio-economy in a wide range of industries, including their employment for the biocapture of CO_2, wastewater treatment, food, energy, secondary metabolite production, cosmetics, medicine, and even the manufacturing of biofertilizers (Moreira et al. 2023). There are various advantages and disadvantages of both chemical fertilizers and algae-based fertilizers as shown in Fig. 11.1.

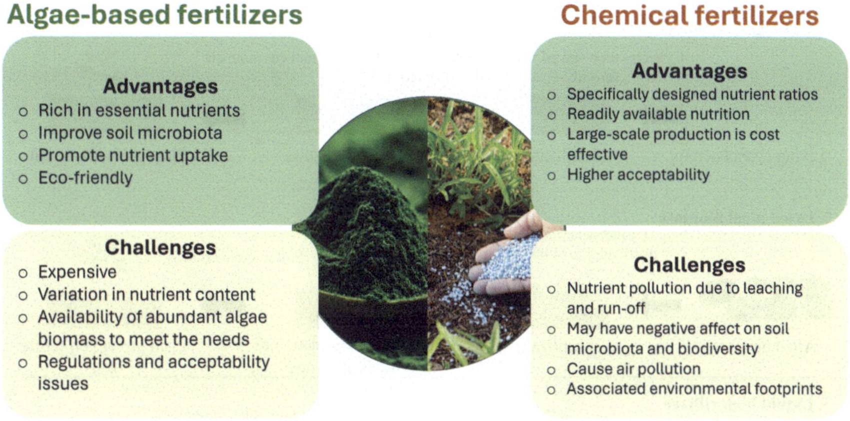

Fig. 11.1 Advantages and disadvantages of chemical and algae-based fertilizers

Using Cyanobacteria as Biofertilizer

Different methods exist for using cyanobacteria as a biofertilizer (Fig. 11.2). The simplest and least expensive method of using cyanobacteria as a biofertilizer is to employ the cells suspended in water or an extract of all the biomolecules present in them; this method is known as wet algae and may be used for both soil cultivation and seed germination. In one study, three different types of rice were germinated and cultured using a suspension of the cyanobacterium *Anabaena* sp. The researchers then compared the outcomes of the cultivation with the addition of blue-green algae versus the standard treatment with water. The researchers proposed using this cyanobacteria as a feasible biofertilizer because the data revealed that all three of the rice kinds employed had improved germination, and the plant development outperformed the control (Habibi et al. 2017).

Dried algal biomass is the second most frequently reported method for producing and using biofertilizers. The most accessible, most efficient, and least expensive method of producing solid biofertilizer is to recover the biomass by allowing it to air-dry directly in the sun. Lyophilization is another technique for drying algal cells; while this technique is good, its industrial applicability is improbable due to the high processing cost per sample. An investigation was carried out to assess the growth of *Urochloa brizantha* when treated with lyophilized microalgae (mostly *C. vulgaris*) based biofertilizer and commercial chemical fertilizer, compared to a control group. According to the results, microalgae biomass was an excellent choice for fertilizer because it produced plants with productivity comparable to those treated with chemical fertilizer (Lorentz et al. 2020).

Hydrothermal carbonization, which turns biomass into hydrocarbon, is performed in a liquid medium at high pressures (above the water vapor pressure) and temperatures (180–220 °C) to treat algal biomass. This technique appears to be effective in obtaining nutrients that are suspended in the liquid medium following the process, such as nitrogen, which is mainly found in the form of organic nitrogen,

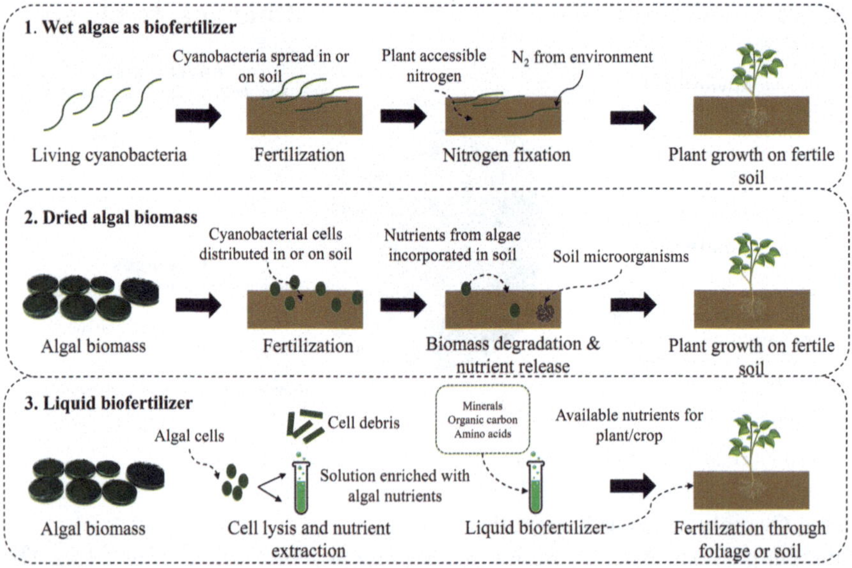

Fig. 11.2 Different modes of applying cyanobacteria as biofertilizer

nitrates, and ammonium, or orthophosphate, which is created from some polyphosphates present in the lignocellulosic raw material following the hydrolysis process (Osorio-Reyes et al. 2023).

There are reports of growth-promoting substances (phytohormones) produced by microalgae that play a critical role in the growth of plants, including cytokinins, auxins, gibberellins, and abscisic acid. It is more typical to discover the literature on microalgae extracts rather than crude biomass when these compounds are examined (Osorio-Reyes et al. 2023). Studies have demonstrated that a wide variety of cyanobacteria species can produce auxins and cytokinins, which are critical for regulating almost all aspects of plant physiology, such as the structure and growth of roots and shoots, as well as the formation of vascular networks and organs (Uniyal et al. 2022). It has been found that cyanobacteria, like *Arthrospira platensis*, produce gibberellin, which is involved in several biological processes, including the formation of fruit, the suppression of seed dormancy, stem lengthening, leaf expansion, early blooming, and sex expression (Han et al. 2018).

11.4 Impact of Cyanobacterial Extracts on Crop Production

A valuable source of biologically active substances that can significantly increase agricultural output is cyanobacteria. The wide range of compounds found in cyanobacteria and their different functions in crop production allow their use as crude algal compost or pure extract (Table 11.2). Additionally, the wide variety in metabolite composition impacts crop production differently. Some extracts are used to improve soil fertility to induce higher crop productivity, and some metabolites act directly on plant growth stimulation (Gonçalves 2021).

Table 11.2 Impact of cyanobacteria-based biofertilizer on crop production

Algal strain	Used as	Crops	Impacts	References
Arthrospira platensis, Porphyridium sp., and *Dunaliella salina*	Crude polysaccharide extract	*Solanum lycopersicum*	Improve the node number, increase pigment and protein content, and promote plant growth	Rachidi et al. (2020)
Spirulina platensis	Biomass	*Brassica rapa* ssp. chinensis, *Brassica oleracea alboglabra, Eruca sativa*	Improve the germination process and plant growth	Wuang et al. (2016)
Monoraphidium sp.	Dried biomass	*Solanum lycopersicum*	Nitrate leaching from the soil is reduced, and plant growth is equivalent to that of synthetic fertilizer	Jimenez et al. (2020)
Acutodesmus dimorphus	–	Roma tomato plant	Enhances floral production and plant growth	Ammar et al. (2022)
Anabaena cylindrica	–	Common bean	Increases plant growth, production, and nodulation	Ammar et al. (2022)
Aulosira fertilissima, Anabaena sphaerica, Nostoc hatei	–	Rice	Increase in grain and straw yield	Ammar et al. (2022)
Nostoc sp.	Biomass	–	Promote plant growth and enhance soil macro- and micronutrients	Khan et al. (2019)
Consortia of two *anabaena* and *Nostoc* strains	–	Rice	Enhance soil fertility and crop yield	Choudhary et al. (2024)
Anabaena spp., *Nostoc* spp., *Calothrix elenkinii*	–	Cotton	Increase the available nitrogen in soil which improves germination, microbial activity, and plant size	Ferreira et al. (2023)

11.4.1 Effect of Cyanobacterial Biomass on Microbial Community of the Soil

The three primary ways that cyanobacteria improve soil are by altering its physical properties, adding biomolecules, and increasing microbial activity (Kalyanasundaram et al. 2020). Soil conditioners are among the possible uses for cyanobacterial biomass, which, when added to the soil (as an algal soil amendment), can enhance physical characteristics such as soil structure and water retention. By increasing the supply or availability of nutrients to the plant through the activity of living microorganisms, cyanobacteria can also be employed as biofertilizers, which are microbial inoculants that promote plant development when applied to soil, seeds, or plant surfaces. These definitions are constantly changing.

When applying non-living cyanobacterial biomass, such as oven-dried biomass, cyanobacterial biomass as soil amendments that supply nutrients might generally be considered organic or organic fertilizers. Furthermore, cyanobacteria are being assessed as potential new sources of plant biostimulants. It is important to note that plant biostimulants are defined as compounds, blends, or microbes that enhance plant growth, nutrient uptake efficiency, resistance to abiotic stress, quality characteristics, and the availability of confined nutrients in the rhizosphere or soil, regardless of the nutrients' actual content (Alvarez et al. 2021).

11.4.2 Soil Reclamation and Biocrust Formation

Certain types of microalgae and cyanobacteria and their extracts have positive effects on soil (Table 11.3). Cyanobacteria are considered ecological engineers because of their ability to alter soil ecological processes and create habitats for other organisms (Rocha et al. 2020). As an indicator of fertility, soil structure is determined by the distribution of pore spaces and soil aggregates, which are the fundamental units of soil structure. This distribution affects water drainage, water-holding capacity, soil aeration, and soil's ability to hold water and nutrients for crop growth (Ramakrishnan et al. 2023). Maintaining the right amounts of organic matter and soil structure is essential for successful agriculture to obtain high crop yield and maintain soil fertility. Synthetic fertilizers are typically administered to maintain soil nutrition. However, it is anticipated that these fertilizers may alter the soil's native microbiota and may not be used as efficiently as they could, harming soil nutrition (Dhawi 2023). The algal extract has well-known functions such as (1) fixing atmospheric nitrogen into nitrogen species that plants can readily absorb, (2) supplying vital nutrients to plants, and (3) enhancing the chemical and physical characteristics of the soil.

Cyanobacteria found in biocrusts can be divided into three categories: (1) filamentous cyanobacteria, like *Microcoleus*, which stabilize soils by adhering to extracellular matrix and forming soil aggregates; (2) unicellular or filamentous

Table 11.3 Impact of cyanobacterial biomass as a biofertilizer on soil remediation (Ferreira et al. 2023)

Algal strain	Target crop/soil	Conclusion
Oscillatoria sp., *Nostoc* spp., *Scytonema* sp.	Burned soils	Improvement in microbial activity and biocrust formation
Anabaena spp., *Nostoc* spp.	Maize rhizosphere	Promote soil functional activities
Chlorella sp.	Agricultural soil	Increase the activities of heterotrophic microorganisms in the soil
Chlorella vulgaris	Maize and wheat	Improve water retention capacity and the amount of soil organic matter
Arthrospira platensis	Biodiesel-contaminated soils	The phycocyanin extracts are effective in biodiesel removal (88.8%)
Lyngbya sp., *Nostoc* sp., *Oscillatoria* sp.	Erosion-prone soils	Increase the total organic carbon, nitrogen, and soil organic matter

cyanobacteria, like Chroococcidiopsis, Scytonema, and Stigonema, which contribute to the cycling of carbon and nitrogen; and (3) cyanobacteria that occur randomly. Algae use a variety of survival techniques, such as the production of intracellular antioxidants (such as the carotenoid lutein in microalgae) and antioxidative enzymes, the secretion of extracellular polymeric substances (EPSs) and cellular membranes (such as trehalose and water stress protein), and the development of microbial consortia with bacteria and fungi to strengthen their cell walls. These survival techniques significantly influence the ecological roles of biocrusts (Ramakrishnan et al. 2023). Microbial species adopt various strategies, such as forming EPS and biofilms, to withstand stressful environments, such as salinity. Additionally, cyanobacteria can restore desert soils and mitigate stressful conditions in deserts by biocrust formation (Karimi et al. 2023).

11.4.3 Nitrogen Fixation

Cyanobacteria are nitrogen-fixing microorganisms; inorganic nitrogen (N_2) from the atmosphere is transformed by cyanobacteria into organic nitrogen that easily gets absorbed by higher plants (Kumar 2016). They can fix <10 kg of N ha^{-1} $year^{-1}$, although annually ~10–30 kg of N ha^{-1} is fixed by thick mats of cyanobacteria. The primary means of biological N_2 fixation used in tropical rice cultivation is cyanobacteria, naturally occurring in paddy fields. Rice is the oldest cultivated and most important crop. It is the staple crop for most of the world's population (Choudhary et al. 2024). Approximately 32 Tg of nitrogen is fixed annually in these cultivated agricultural systems by biological nitrogen fixers, and cyanobacteria add about 20–30 kg fixed nitrogen ha^{-1} to the paddy fields along with the organic matter. Many heterocystous cyanobacterial genera such as *Nostoc*, *Chlorogloea*, *Mastigocladus*, *Gloeotrichia*, and *Anabaena* are efficient N_2 fixers (Pathak et al. 2024).

In different countries, especially India and Chile, efforts have been made to use cyanobacterial strains to promote rice growth. It has been demonstrated that rice paddles can accumulate nitrogen more efficiently when local cyanobacterial strains from Chile are used. It is recommended that the symbiotic relationship of Anabaena and Azolla (water fern) develop an N_2 fixing complex that is considered self-renewable natural nitrogen resources to minimize inorganic N requirements (Vaishampayan et al. 2001).

11.4.4 Crop Protection

Plant diseases and pests regularly jeopardize the availability and security of plants for the nourishment of humans and animals. The use of synthetic items in agriculture is currently restricted by legislation due to the detrimental effects of these products. A sustainable way is to increase agricultural productivity with cyanobacterial-based biofertilizers for both crop protection and production because of their biostimulating and biofertilizing potential. Cyanobacteria play a significant role in crop protection because they produce bioactive compounds that trigger plant development and activate the defense mechanism in response to various stressors (Table 11.4) (Sithole et al. 2023). Algae are included in the class of "plant stimulants." These plant stimulants attract interest in modern agriculture to increase plant/crop performance, nutrient use efficiency, and persistence to environmental stress (Colla and Rouphael 2020). The most dominating species of algae with biostimulatory action are *Spirulina maxima*, *Scenedesmus* sp., *Calothrix elenkinii*, *Chlorella vulgaris*, and *Acutodesmus dimorphus*. The biostimulatory action of algal extracts has been connected with the content of primary metabolites such as proteins, carbohydrates and lipids, key amino acids (tryptophan and arginine), polysaccharides (β-glucan), and vitamins (Chiaiese et al. 2018). Several algal strains have phytohormone-like activities, including gibberellins, auxins, abscisic acid, and cytokinins, which are considered essential factors in enhancing plant yield and growth (Ronga et al. 2019).

11.4.5 Enhancing Nutrient Availability and Plant Productivity

The N- and P-rich algal biomass works as a slow-release biofertilizer in soil. Furthermore, it has been demonstrated that the amount of nitrogen algae release does not surpass the nutrients plants need, reducing the eutrophication of ground waters and surfaces caused by the improper use of traditional fertilizers (Khan et al. 2023). Moreover, it has been demonstrated that wastewater-grown NPK-rich algal biomass substantially enhanced baby spinach and corn crop yields, as it contains NPK quantities more significant than those found in de-oiled cakes, vermicompost, and waste slurries from biogas plants (Sharma et al. 2021). By secreting bioactive

Table 11.4 Bioactive compounds from cyanobacteria: potential agricultural benefits (Gonçalves 2021; Gonçalves et al. 2023)

Bioactive compounds	Examples	Biological activity	Potential applications	Source/Strain
Phytohormones	Abscisic acid, cytokinins, gibberellins auxins	Chemical messengers	Regulation of cellular activities in crops, response to stress conditions, and stimulation of plant growth	*Phormidium* and *Arthrospira*
Phenolic compounds	Phenylpropanoids, phenolic acids, polyphenols, flavonoids	Antibacterial, antifungal, and antioxidant	Crop protection against abiotic and biotic stress conditions, or other pathogens	*Phaeodactylum tricornutum*, *Skeletonema costatum,* and *Tetraselmis suecica*
Free fatty acids	Unsaturated and saturated fatty acids	Antiviral, antioxidant, antifungal, and anticarcinogenic	Protect against pathogens and various biotic and abiotic stressors	*Anabaena, Porphyridium,* and *Spirulina*
Carotenoids	Astaxanthin, alpha-carotene, lutein, zeaxanthin, beta-carotene, lycopene	Anti-inflammatory, anticancer, and antioxidant	Crop protection against insects, bacteria, and various biotic and abiotic stress conditions. Soil fertilization, fortification, and bioremediation	*Spirulina* sp., *Muriellopsis* sp., *Chlorella zofingiensis*, and *Haematococcus pluvialis*
Terpenoids	Monoterpenes, triterpenes, sesquiterpenes, hemiterpenes	Antioxidant, anticarcinogenic, and antibacterial	Stimulate the development and growth of plants, attract pollinators, and protect against insects and bacteria	*Oscillatoria perornata, Synechocystis sp., Hypnea pannosa,* and *Chondrococcus hornemannii*
Polysaccharides	Extracellular polysaccharides, energy-storage polysaccharides, structural polysaccharides	Anticoagulant, antioxidant, anticancer, antibacterial, and anti-inflammatory	Promote plant growth and protection against biotic and abiotic stress conditions. Improve soil quality	*Arthrospira, Chlamydomonas, Porphyridium,* and *Aphanothece*

compounds and exopolysaccharides (EPS), cyanobacterial biomass helps restore soil nutrients. This biomass could potentially mobilize insoluble inorganic phosphate forms. Additionally, algalization has been used to lower the exchangeable salt

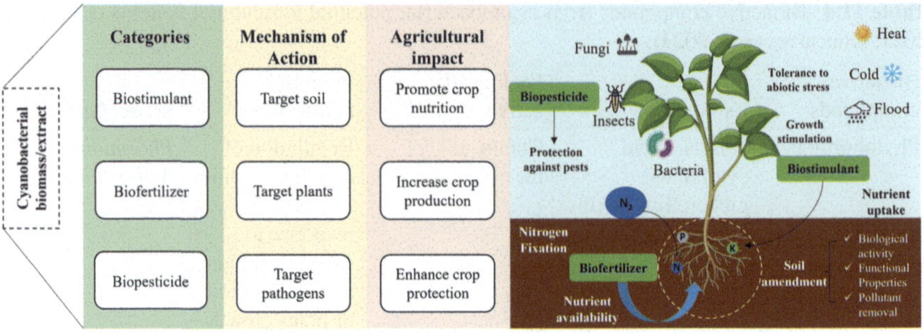

Fig. 11.3 Impact of cyanobacterial biomass as a biofertilizer for agricultural applications

content, changing the pH of the soil and leading to the reclamation of sodic soils (Chatterjee et al. 2017). Cyanobacteria play a pivotal role as contributors to organic matter within the agroecosystem. These organic entities also enhance the soil's organic carbon content and promote the development of other micro-flora and fauna (Krings et al. 2009).

Cyanobacteria are unique, organic creatures well known for their capacity to thrive under various abiotic stressors, which tend to be the primary challenge to agricultural productivity. By quickly invading the rhizosphere and using processes like nitrogen fixation, mineralization, and rock phosphate solubility to transform inaccessible mineral forms into necessary nutrients, they promote plant growth and development (Fig. 11.3) (Ammar et al. 2022).

11.5 Conclusion and Prospects

Research on applying algal biomass to produce biostimulants and fertilizers has gained attention in recent years. The process streams of conventional wastewater treatment facilities contain varying concentrations of nitrogenous, phosphorus, and organic compounds that lead to eutrophication. Different urban wastewaters have varying N/P ratios appropriate for growing various algal species and transforming eutrophying N and P molecules into biomass rich in protein, carbohydrates, and fats. The cyanobacterial biomass from wastewater has a greater NPK content than other organic manures, which increases agricultural yield and lowers the risk of eutrophication caused by conventional inorganic fertilizers. A sustainable agricultural technique, "biofertilization" involves using biofertilizers to improve the soil's nutrient content and organic matter to increase productivity. Cyanobacteria are eco-friendly bio-based fertilizers for pollution-free agricultural practices. By reducing dependency on chemical fertilizers and mitigating environmental degradation, this comprehensive approach targets long-term sustainability goals in addition to the crucial agricultural requirements for higher yields.

References

Al-Hazmi HE, Mohammadi A, Hejna A et al (2023) Wastewater reuse in agriculture: prospects and challenges. Environ Res 236:116711. https://doi.org/10.1016/j.envres.2023.116711

Alvarez AL, Weyers SL, Goemann HM et al (2021) Microalgae, soil and plants: a critical review of microalgae as renewable resources for agriculture. Algal Res 54:102200. https://doi.org/10.1016/j.algal.2021.102200

Amin M, Haq MU, Xu JR et al (2023) Advances in green chemistry. Green Chem 1:23002

Ammar EE, AaA A, Elesawy AE et al (2022) Algae as bio-fertilizers: between current situation and future prospective. Saudi J Biol Sci 29:3083–3096. https://doi.org/10.1016/j.sjbs.2022.03.020

Ansari FA, Ravindran B, Gupta SK et al (2019) Techno-economic estimation of wastewater phycoremediation and environmental benefits using *Scenedesmus obliquus* microalgae. J Environ Manag 240:293–302. https://doi.org/10.1016/j.jenvman.2019.03.123

Camargo JA, Alonso Á (2006) Ecological and toxicological effects of inorganic nitrogen pollution in aquatic ecosystems: a global assessment. Environ Int 32:831–849. https://doi.org/10.1016/j.envint.2006.05.002

Cardoso LG, Duarte JH, Costa JV et al (2021) *Spirulina* sp. as a bioremediation agent for aquaculture wastewater: production of high added value compounds and estimation of theoretical biodiesel. Bioenergy Res 14:254–264. https://doi.org/10.1007/s12155-020-10153-4

Chatterjee A, Singh S, Agrawal C et al (2017) Role of algae as a biofertilizer. In: Rastogi RP, Madamwar D, Pandey A (eds) Algal green chemistry. Elsevier, Amsterdam. https://doi.org/10.1016/B978-0-444-63784-0.00010-2

Chiaiese P, Corrado G, Colla G et al (2018) Renewable sources of plant biostimulation: microalgae as a sustainable means to improve crop performance. Front Plant Sci 9. https://doi.org/10.3389/fpls.2018.01782

Chittora D, Meena M, Barupal T et al (2020) Cyanobacteria as a source of biofertilizers for sustainable agriculture. Biochem Biophys Rep 22:100737. https://doi.org/10.1016/j.bbrep.2020.100737

Choudhary P, Prajapati SK, Kumar P et al (2017) Development and performance evaluation of an algal biofilm reactor for treatment of multiple wastewaters and characterization of biomass for diverse applications. Bioresour Technol 224:276–284. https://doi.org/10.1016/j.biortech.2016.10.078

Choudhary N, Tripathi A, Singh PK et al (2024) Application of algae for enhanced plant growth and food productivity. Syst Microbiol Biomanuf 4:564–574. https://doi.org/10.1007/s43393-024-00233-3

Colla G, Rouphael Y (2020) Microalgae: new source of plant biostimulants. Agronomy [Online] 10. https://doi.org/10.3390/agronomy10091240

Daneshvar E, Antikainen L, Koutra E et al (2018a) Investigation on the feasibility of *Chlorella vulgaris* cultivation in a mixture of pulp and aquaculture effluents: treatment of wastewater and lipid extraction. Bioresour Technol 255:104–110. https://doi.org/10.1016/j.biortech.2018.01.101

Daneshvar E, Zarrinmehr MJ, Hashtjin AM et al (2018b) Versatile applications of freshwater and marine water microalgae in dairy wastewater treatment, lipid extraction and tetracycline biosorption. Bioresour Technol 268:523–530. https://doi.org/10.1016/j.biortech.2018.08.032

De Siqueira CJ, Calijuri ML, Ferreira J et al (2020) Microalgae based biofertilizer: a life cycle approach. Sci Total Environ 724:138138. https://doi.org/10.1016/j.scitotenv.2020.138138

Deviram G, Mathimani T, Anto S et al (2020) Applications of microalgal and cyanobacterial biomass on a way to safe, cleaner and a sustainable environment. J Clean Prod 253:119770. https://doi.org/10.1016/j.jclepro.2019.119770

Dhawi F (2023) How can we stabilize soil using microbial communities and mitigate desertification? Sustainability [Online] 15. https://doi.org/10.3390/su15010863

Diprat AB, Silveira Thys RC, Rodrigues E et al (2020) *Chlorella sorokiniana*: a new alternative source of carotenoids and proteins for gluten-free bread. LWT 134:109974. https://doi.org/10.1016/j.lwt.2020.109974

Encarnação T, Ramos P, Mohammed D et al (2023) Bioremediation using microalgae and cyanobacteria and biomass valorisation. Marine organisms: a solution to environmental pollution? Uses in bioremediation and in biorefinery. Springer. https://doi.org/10.1007/978-3-031-17226-7_2

Ferreira A, Ribeiro B, Ferreira AF et al (2019) *Scenedesmus obliquus* microalga-based biorefinery–from brewery effluent to bioactive compounds, biofuels and biofertilizers–aiming at a circular bioeconomy. Biofuels Bioprod Biorefin 13:1169–1186. https://doi.org/10.1002/bbb.2032

Ferreira A, Melkonyan L, Carapinha S et al (2021) Biostimulant and biopesticide potential of microalgae growing in piggery wastewater. Environ Adv 4:100062. https://doi.org/10.1016/j.envadv.2021.100062

Ferreira A, Bastos CRV, Marques-Dos-Santos C et al (2023) Algaeculture for agriculture: from past to future. Front Agron 5:1064041. https://doi.org/10.3389/fagro.2023.1064041

Gonçalves AL (2021) The use of microalgae and cyanobacteria in the improvement of agricultural practices: a review on their Biofertilising, biostimulating and biopesticide roles. Appl Sci [Online] 11. https://doi.org/10.3390/app11020871

Gonçalves J, Freitas J, Fernandes I et al (2023) Microalgae as biofertilizers: a sustainable way to improve soil fertility and plant growth. Sustainability [Online] 15. https://doi.org/10.3390/su151612413

Habibi A, Zayadan B, Baizhigitova A et al (2017) The effect of isolated cyanobacteria on rice seed germination and growth. Eurasian J Ecol 51:114–121

Hamidian N, Zamani H (2022) Biomass production and nutritional properties of *Chlorella sorokiniana* grown on dairy wastewater. J Water Process Eng 47:102760. https://doi.org/10.1016/j.jwpe.2022.102760

Han X, Zeng H, Bartocci P et al (2018) Phytohormones and effects on growth and metabolites of microalgae: a review. Fermentation 4:25. https://doi.org/10.3390/fermentation4020025

Jebali A, Acién FG, Rodriguez Barradas E et al (2018) Pilot-scale outdoor production of *Scenedesmus* sp. in raceways using flue gases and centrate from anaerobic digestion as the sole culture medium. Bioresour Technol 262:1–8. https://doi.org/10.1016/j.biortech.2018.04.057

Jimenez R, Markou G, Tayibi S et al (2020) Production of microalgal slow-release fertilizer by valorizing liquid agricultural digestate: growth experiments with tomatoes. Applied Sciences [Online] 10. https://doi.org/10.3390/app10113890

Kalyanasundaram GT, Ramasamy A, Rakesh S et al (2020) Microalgae and cyanobacteria: role and applications in agriculture. Applied algal biotechnology. Nova Science Publishers, Hauppauge, NY

Karimi A, Tahmourespour A, Hoodaji M (2023) The formation of biocrust and improvement of soil properties by the exopolysaccharide-producing cyanobacterium: a biogeotechnological study. Biomass Convers Biorefinery 13:15489–15499. https://doi.org/10.1007/s13399-022-02336-0

Khan SA, Sharma GK, Malla FA et al (2019) Microalgae based biofertilizers: a biorefinery approach to phycoremediate wastewater and harvest biodiesel and manure. J Clean Prod 211:1412–1419. https://doi.org/10.1016/j.jclepro.2018.11.281

Khan S, Thaher M, Abdulquadir M et al (2023) Utilization of microalgae for urban wastewater treatment and valorization of treated wastewater and biomass for biofertilizer applications. Sustainability [Online] 15. https://doi.org/10.3390/su152216019

Krings M, Hass H, Kerp H et al (2009) Endophytic cyanobacteria in a 400-million-yr-old land plant: a scenario for the origin of a symbiosis? Rev Palaeobot Palynol 153:62–69. https://doi.org/10.1016/j.revpalbo.2008.06.006

Kumar N (2016) Effect of algal bio-fertilizer on the Vigna radiata: a critical review. Int J Eng Res Appl 6:85–94

Kumar A, Bera S (2020) Revisiting nitrogen utilization in algae: a review on the process of regulation and assimilation. Bioresour Technol Rep 12:100584. https://doi.org/10.1016/j.biteb.2020.100584

Kumar D, Singh SK, Arya SK et al (2023) Multifunctional growth-promoting microbial consortium-based biofertilizers and their techno-commercial feasibility for sustainable agriculture. In: Parray JA, Shameem N, Egamberdieva D, Sayyed RZ (eds) Rhizobiome. Academic Press. https://doi.org/10.1016/B978-0-443-16030-1.00010-9

Lorentz JF, Calijuri ML, Assemany PP et al (2020) Microalgal biomass as a biofertilizer for pasture cultivation: plant productivity and chemical composition. J Clean Prod 276:124130. https://doi.org/10.1016/j.jclepro.2020.124130

Martinez-Hernandez E, Samsatli S (2017) Biorefineries and the food, energy, water nexus—towards a whole systems approach to design and planning. Curr Opin Chem Eng 18:16–22. https://doi.org/10.1016/j.coche.2017.08.003

Matos ÂP, Da Silva T, Sant'anna ES (2021) The feasibility of using inland desalination concentrate (DC) as an alternative substrate for *Spirulina platensis* mass cultivation. Waste Biomass Valorization 12:3193–3203. https://doi.org/10.1007/s12649-020-01233-9

Moreira JB, Santos TD, Duarte JH et al (2023) Role of microalgae in circular bioeconomy: from waste treatment to biofuel production. Clean Techn Environ Policy 25:427–437. https://doi.org/10.1007/s10098-021-02149-1

Msanne J, Polle J, Starkenburg S (2020) An assessment of heterotrophy and mixotrophy in *Scenedesmus* and its utilization in wastewater treatment. Algal Res 48:101911. https://doi.org/10.1016/j.algal.2020.101911

Mularczyk M, Michalak I, Marycz K (2020) Astaxanthin and other nutrients from *Haematococcus pluvialis*—multifunctional applications. Mar Drugs 18:459. https://doi.org/10.3390/md18090459

Na N (2015) The algae. Springer. https://doi.org/10.1007/978-94-017-7321-8

Nazos TT, Stratigakis NC, Spantidaki M et al (2023) Characterization of cheese whey effluents and investigation of their potential to be used as a nutrient substrate for *Chlorella* biomass production. Waste Biomass Valorization 14:3643–3655. https://doi.org/10.1007/s12649-023-02081-z

Nosheen S, Ajmal I, Song Y (2021) Microbes as biofertilizers, a potential approach for sustainable crop production. Sustain For 13:1868. https://doi.org/10.3390/su13041868

Oliveira AC, Barata A, Batista AP et al (2019) *Scenedesmus obliquus* in poultry wastewater bioremediation. Environ Technol 40:3735–3744. https://doi.org/10.1080/09593330.2018.1488003

Osorio-Reyes JG, Valenzuela-Amaro HM, Pizaña-Aranda JJP et al (2023) Microalgae-based biotechnology as alternative biofertilizers for soil enhancement and carbon footprint reduction: advantages and implications. Mar Drugs 21:93. https://doi.org/10.3390/md21020093

Pacheco D, Rocha ACS, Garcia A et al (2021) Municipal wastewater: a sustainable source for the green microalgae *Chlorella vulgaris* biomass production. Appl Sci 11:2207. https://doi.org/10.3390/app11052207

Pathak J, Jaiswal J, Shukla RK et al (2024) Cyanobacterial/algal biofertilizers as plant growth stimulants for green sustainable agriculture. Crops 42:43. https://doi.org/10.22271/plants.2024.v12.i1a.1622

Pereira MIB, Chagas BME, Sassi R et al (2019) Mixotrophic cultivation of *Spirulina platensis* in dairy wastewater: effects on the production of biomass, biochemical composition and antioxidant capacity. PLoS One 14:e0224294. https://doi.org/10.1371/journal.pone.0224294

Perera I, Subashchandrabose SR, Venkateswarlu K et al (2018) Consorţia of cyanobacteria/microalgae and bacteria in desert soils: an underexplored microbiota. Appl Microbiol Biotechnol 102:7351–7363. https://doi.org/10.1007/s00253-018-9192-1

Rachidi F, Benhima R, Sbabou L et al (2020) Microalgae polysaccharides bio-stimulating effect on tomato plants: growth and metabolic distribution. Biotechnol Rep 25:e00426. https://doi.org/10.1016/j.btre.2020.e00426

Ramakrishnan B, Maddela NR, Venkateswarlu K et al (2023) Potential of microalgae and cyanobacteria to improve soil health and agricultural productivity: a critical view. Environ Sci: Adv 2:586–611. https://doi.org/10.1039/D2VA00158F

Renuka N, Ratha SK, Kader F et al (2021) Insights into the potential impact of algae-mediated wastewater beneficiation for the circular bioeconomy: a global perspective. J Environ Manag 297:113257. https://doi.org/10.1016/j.jenvman.2021.113257

Rocha F, Esteban Lucas-Borja M, Pereira P et al (2020) Cyanobacteria as a nature-based biotechnological tool for restoring salt-affected soils. Agronomy 10:1321. https://doi.org/10.3390/agronomy10091321

Ronga D, Biazzi E, Parati K et al 2019) Microalgal biostimulants and biofertilisers in crop productions. Agronomy [Online] 9. https://doi.org/10.3390/agronomy9040192

Sarpal AS, Teixeira CM, Costa IC (2023) Cultivation of *Chlorella vulgaris* in wastewater: biodiesel potential and wastewater remediation. https://doi.org/10.21203/rs.3.rs-3177891/v1

Shahid A, Usman M, Atta Z et al (2021) Impact of wastewater cultivation on pollutant removal, biomass production, metabolite biosynthesis, and carbon dioxide fixation of newly isolated cyanobacteria in a multiproduct biorefinery paradigm. Bioresour Technol 333:125194. https://doi.org/10.1016/j.biortech.2021.125194

Sharma GK, Khan SA, Shrivastava M et al (2021) Circular economy fertilization: Phycoremediated algal biomass as biofertilizers for sustainable crop production. J Environ Manag 287:112295. https://doi.org/10.1016/j.jenvman.2021.112295

Shen Q-H, Jiang J-W, Chen L-P et al (2015) Effect of carbon source on biomass growth and nutrients removal of *Scenedesmus obliquus* for wastewater advanced treatment and lipid production. Bioresour Technol 190:257–263. https://doi.org/10.1016/j.biortech.2015.04.053

Silva DA, Cardoso LG, De Jesus Silva JS et al (2022) Strategy for the cultivation of *Chlorella vulgaris* with high biomass production and biofuel potential in wastewater from the oil industry. Environ Technol Innov 25:102204. https://doi.org/10.1016/j.eti.2021.102204

Singh D, Singh SK, Modi A et al (2020) Impacts of agrochemicals on soil microbiology and food quality. Agrochemicals detection, treatment and remediation. Elsevier. https://doi.org/10.1016/B978-0-08-103017-2.00004-0

Sithole N, Gupta S, Dube Z et al (2023) Algae and cyanobacteria-based biostimulants in controlling plant-parasitic nematodes: a sustainable approach for crop protection. Phytoparasitica 51:803–813. https://doi.org/10.1007/s12600-023-01094-7

Sorgatto VG, Soccol CR, Molina-Aulestia DT et al (2021) Mixotrophic cultivation of microalgae in cassava processing wastewater for simultaneous treatment and production of lipid-rich biomass. Fuels 2:521–532. https://doi.org/10.3390/fuels2040030

Su Y (2021) Revisiting carbon, nitrogen, and phosphorus metabolisms in microalgae for wastewater treatment. Sci Total Environ 762:144590. https://doi.org/10.1016/j.scitotenv.2020.144590

Umamaheswari J, Shanthakumar S (2021) Paddy-soaked rice mill wastewater treatment by phycoremediation and feasibility study on use of algal biomass as biofertilizer. J Chem Technol Biotechnol 96:394–403. https://doi.org/10.1002/jctb.6551

Uniyal S, Bhandari M, Singh P et al (2022) Cytokinin biosynthesis in cyanobacteria: insights for crop improvement. Front Genet 13:933226. https://doi.org/10.3389/fgene.2022.933226

Usman M, Amin M, Kamal I et al (2023) Algae-mediated resource recovery from urban wastewater. Curr Pollut Rep 9:243–258. https://doi.org/10.1007/s40726-023-00254-x

Vaishampayan A, Sinha RP, Hader DP et al (2001) Cyanobacterial biofertilizers in rice agriculture. Bot Rev 67:453–516. https://doi.org/10.1007/BF02857893

Wuang SC, Khin MC, Chua PQD et al (2016) Use of *Spirulina* biomass produced from treatment of aquaculture wastewater as agricultural fertilizers. Algal Res 15:59–64. https://doi.org/10.1016/j.algal.2016.02.009

Chapter 12
Cyanobacteria for the Global Space Biology Program: Challenges and Opportunities

Fiaz Ahmad and Da-Chuan Yin

Abstract Space microbiological research, with a particular emphasis on cyanobacteria and microalgae, has garnered significant attention over the past five decades. The quest for suitable microbes is pivotal in facilitating human habitation on both Luna and the Martian surface, given the inhospitable atmospheric conditions, including low temperature, atmospheric pressure, and the imperative need for fundamental life-sustaining resources such as oxygen and water. This chapter elucidates the pivotal role of these microorganisms in supporting manned missions to Mars, examining the current landscape of research successes and failures in establishing a sustainable bioregenerative life support system (LSS). It explores ongoing endeavors to unravel mysteries and identifies prospective avenues for further investigation. Furthermore, it probes the potential of these microbes to instigate oxygen, nitrogen, and carbon cycles within closed-loop systems, thereby diminishing reliance on resource storage or resupply missions and, consequently, mitigating the overall cost of space missions. The chapter also scrutinizes the technical challenges encountered in cyanobacteria and microalgae space experiments, particularly in their cultivation on Luna and Mars. Nevertheless, current research remains confined to a limited array of microbes, leaving a plethora of unresolved issues, including the imperative task of large-scale cultivation of these organisms amidst the synergistic effects of substantial stresses such as weightlessness, cosmic radiation, elevated CO_2 levels, and low atmospheric pressure, all crucial for both lunar and Martian manned space missions.

Keywords Space · Cyanobacteria · Microalgae · Life support system · Crewed deep-space missions

---

F. Ahmad (✉) · D.-C. Yin (✉)
Key Laboratory for Space Biosciences and Biotechnology, School of Life Sciences (SLS), Northwestern Polytechnical University (NPU), Xi'an, People's Republic of China
e-mail: fiaz.a@mail.nwpu.edu.cn; yindc@nwpu.edu.cn

M. A. Mehmood et al. (eds.), *Cyanobacteria Biotechnology*,
https://doi.org/10.1007/978-3-031-70698-1_12

12.1 Introduction

In recent times, discussions pertaining to space exploration, planets, and satellites have gained significant traction. Consequently, there has been a surge in studies focusing on developing crewed space missions and addressing the challenges associated with prolonged stays in space, particularly during deep space voyages to Mars. Concurrently, ensuring the safety and well-being of spacecraft crews, along with maintaining an optimal working environment, has become paramount. Several noteworthy projects are currently underway for deep-space missions, including the Lunar Orbital Platform-Gateway (Haws et al. 2019), the Artemis Lunar program (Von Ehrenfried 2020), the Tiangong space station (Xuedong Li et al. 2024), and the Axiom commercial space station (Maender et al. 2022). Furthermore, the involvement of the private sector in the space race is noteworthy, with recent initiatives from companies such as Blue Origin and Sierra Space leading to the development of fully functional commercial space stations (https://www.blueorigin.com/news/orbital-reef-commercial-space-station, accessed April 2024). Notably, SpaceX is actively developing the Starship spacecraft for crewed flights to Mars. Under the Artemis program, the first Starship test mission could entail a manned flight to the moon, with funding already secured from NASA (Heldmann et al. 2021). Despite the rapid development of space technology for deep-space crewed missions, transporting consumables such as oxygen, water, food, and fuel from Earth still incurs considerable costs.

According to various estimates, the cost of delivering one kilogram of cargo from Earth to Low Earth Orbits (LEO) stands at around $10,000, while the cost increases significantly to approximately $300,000 for delivery to the Martian surface (Jones 2015; Massa et al. 2006; Musk 2017; Lopez et al. 2015). Despite advancements leading to reduced spacecraft launch costs, these figures underscore the considerable financial barriers associated with space exploration. Moreover, the duration of a one-way journey to Mars and the timing constraints imposed by "launch windows" represent critical factors demanding careful consideration. These launch windows, occurring every 26 months, determine the optimal orbital energy required for missions from Earth to Mars. The absence of effective safety measures against cosmic radiation poses a significant challenge (Musk 2017; Chen and Li 2022; Mapstone et al. 2022; Saghamanesh et al. 2024). Consequently, planning deep-space crewed expeditions to Mars carries inherent risks to the health and lives of astronauts (Bahadori 2024; Abadie and Francisco 2024; Cucinotta et al. 2013). One potential solution to address these challenges involves implementing a closed-loop life support system (LSS). Such a system must prioritize reliability to maximize payload reuse, thereby allowing for increased allocation to scientific endeavors and other mission objectives.

The air regeneration system within the LSS relies on well-established processes such as the Sabatier (Ridley et al. 2023), Bosch, and water–gas shift reactions (Greenwood et al. 2018). These processes, including carbon dioxide (CO_2) scrubbing and electrolysis of H_2O (Ridley et al. 2023), have been utilized on the

International Space Station (ISS) (Ridley et al. 2023; Winton et al. 2016). However, they are not without limitations. For instance, carbon produced during these reactions either accumulates on catalysts, decreasing efficiency or is transformed into non-recyclable methane, posing challenges for long-term sustainability (Sargusingh et al. 2015; Shaw et al. 2021). Furthermore, these physicochemical methods necessitate high-energy inputs to maintain elevated temperatures for optimal catalytic activity and process efficiency (Abney et al. 2013). Moreover, none of these approaches offer a seamless solution for sustaining a continuous food supply or complete recycling of essential elements such as hydrogen, oxygen, carbon, nitrogen, and phosphorus from inorganic waste (Shaw et al. 2021). Consequently, deep-space missions remain reliant on continuous resource replenishment from Earth, highlighting the need for further innovation in resource recycling technologies.

Integrating biological components within LSS offers a promising pathway toward achieving comprehensive resource recycling, atmospheric revitalization, and water purification (Nangle et al. 2020; Maity and Saxena 2024). These hybrid LSSs, which combine biological and physicochemical elements, are positioned to enhance efficiency, reliability, and self-sufficiency, addressing the limitations of individual systems and ensuring seamless operation even in emergency situations (Dominoni 2024; Revellame et al. 2021). Despite their higher launch mass compared to purely physicochemical LSS counterparts, these hybrid configurations offer significant advantages, including reduced reliance on long-term resupply missions, energy conservation, and on-site synthesis of critical biomolecules (Belz et al. 2013; Greenman et al. 2024). This shift toward internal resource generation, encompassing essentials like food supplements and medicines, markedly decreases dependence on supplies from Earth (Çelekli and Zariç 2024). Furthermore, incorporating bioregenerative subsystems helps minimize carbon and hydrogen losses, thereby reducing the necessity for resupply missions and releasing crucial elements into space (Revellame et al. 2021; Fais et al. 2024). Current research efforts are focused on identifying optimal flora and microorganisms suitable for sustained space travel. Cyanobacteria and Chlorophyta are emerging as leading candidates due to their easy cultivation, high productivity, and efficient photosynthetic capabilities essential for oxygen and sustenance production, carbon dioxide absorption, and water purification (Verseux et al. 2016; Keller et al. 2023; Walker and Granjou 2017; Revellame et al. 2024).

When designing bioregenerative life support systems and selecting suitable organisms, it's vital to consider daily human needs and waste management (Fig. 12.1) (Revellame et al. 2021). Organisms within closed life support systems should ensure oxygen and water recirculation at a minimum, strive for complete mineral component reuse, and ideally facilitate a full carbon cycle (Kerner et al. 2024). Creating such systems for long-term crewed space missions and projected settlements on the Moon or Mars necessitates analyzing existing data and identifying avenues for further theoretical and practical research. This entails addressing challenges such as: (1) Enhancing the closed-loop functionality of LSS and optimizing waste recycling across compartments (Dominoni 2024; Yin et al. 2024). (2) Ensuring the enduring stability of all LSS components, with a specific emphasis on

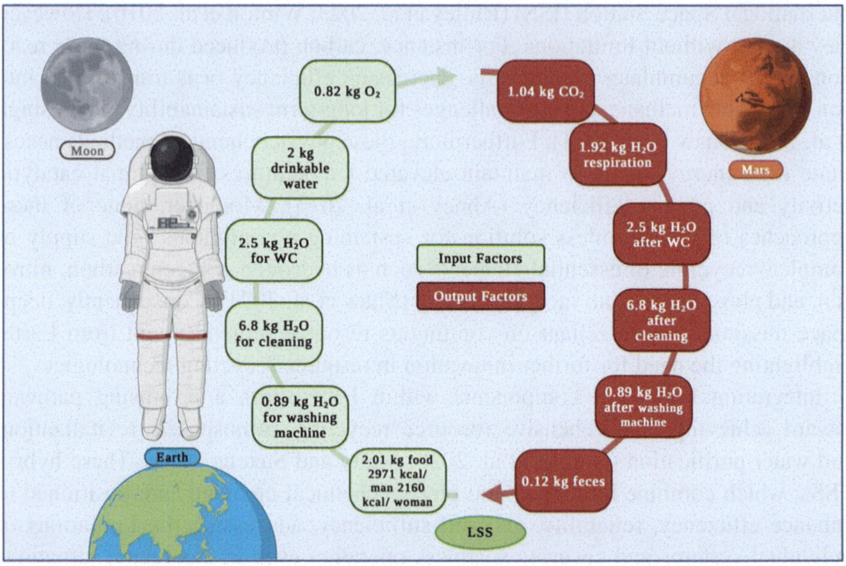

Fig. 12.1 Daily intake requirements and outputs of astronauts during a space mission to space

biological compartments (Bahadori 2024). (3) Applying stringent constraints on the weight and dimensions of system components (Casula et al. 2024a). (4) Investigating the potential effects of microgravity and weightlessness on gas distribution within the cabin and bioreactors (Casula et al. 2024b). (5) Assessing the solar-induced impact of cosmic radiation on the entirety of the life support system, including microorganisms, plants, and crew (II Brown et al. 2010). (6) Exploring the feasibility of In-Situ Resource Utilization (ISRU) to leverage external resources such as regolith, atmospheric gases, water, and ice, where available (Rapp and Inglezakis 2024; II Brown et al. 2010). (7) Investigating the possibility of complete automation of the life support system to streamline maintenance and reduce complexity (Detrell 2021a). (8) Maintaining low inertia amidst significant parameter fluctuations, including air leakage, water loss, and cultivated biomass volume, while upholding sustainability (Fahrion et al. 2023). It's imperative to comprehensively assess the impact of long-term spaceflight factors on the biomass of phototrophic microorganisms, whether cultivated in photobioreactors or reserved for future use in the life support system in emergencies. Following this, we delve into contemporary strategies for harnessing phototrophic microorganisms in life support systems, while addressing challenges associated with their extended cultivation in conditions influenced by cosmic radiation, microgravity, low atmospheric pressure, and heightened CO_2 levels.

12.2 Implications of Cyanobacteria and Microalgae in Deep-Space Missions

Currently, cyanobacteria and microalgae are considered to be the most promising candidates for their use in LSS, as they can recycle wastes like CO_2 and water and produce food and oxygen (O_2) (Verseux et al. 2016; Çelekli and Zariç 2024). It has been demonstrated that certain types of cyanobacteria and microalgae exhibit growth rates and biomass productivities (per unit surface area) up to ten times higher than terrestrial plants on Earth. Additionally, the photosynthetic efficiency of cyanobacteria and microalgae surpasses that of terrestrial plants by several folds (Sharafi laleh et al. 2023; Brown and Zeiler 1993; Brennan and Owende 2010). For instance, the photosynthetic efficiency of widely studied phototrophic microorganisms like *Chlorella vulgaris* and *Arthrospira platensis* can be up to 5–7 times greater than that of higher plants (Kerner et al. 2024; Revellame et al. 2021). This observation is presumed to hold true under terrestrial conditions and in spaceflight settings, although comparative studies with higher plants are lacking (Verseux et al. 2016). Some reports suggest that biomass productivity, including the volume of oxygen released per unit volume per unit time, is comparable on the ISS and Earth (Poughon et al. 2020). However, productivity is influenced not only by the microbial strain but also by the photobioreactor design (Ibrahim et al. 2020; Wang et al. 2024). Biofilm cultivation offers advantages over suspension cultivation, such as efficient biomass collection (in certain biofilm reactor designs), reduced nutrient medium volume requirements, high biomass concentration, and productivity. However, the efficiency of biofilm cultivation of phototrophic microorganisms under spaceflight conditions has not been thoroughly studied. Moreover, scaling such PBRs can be challenging due to limitations on biofilm thickness under low light conditions, complexities in moistening, and nutrient distribution gradients along the biofilm direction (Li et al. 2017a). Beyond just the photobioreactor design, several factors exert significant influence on oxygen production. These include the rate and concentration of CO_2 supply, the growth phase of the culture, agitation rate, luminescence intensity, light penetration depth into the biomass layers, pH of the cultivation medium, temperature, and the techniques employed for continuous cultivation (Sánchez-Bayo et al. 2020; Gonçalves et al. 2016; Yin et al. 2024).

In LSS, cyanobacteria and microalgae offer multifaceted utility beyond mere oxygen production and air purification. They serve roles such as generating nutritious edible biomass, treating human organic waste and wastewater, providing nutrients for other organisms within LSS, and producing fuel and clean energy like hydrogen (Keller et al. 2023; Verseux et al. 2016). Moreover, the potential utilization of microalgae capable of withstanding sudden temperature fluctuations from 4 to 30 °C for spacecraft cabin thermoregulation has been proposed (Matula and Nabity 2021a). These strategies present viable means to recycle limited resources in space environments and enhance payload management across various space missions.

Moreover, to diminish Earth's reliance and maximize spacecraft payload efficiency, the concept of utilizing resources on the Moon and Mars through ISRU to support cyanobacteria and microalgae cultivation has gained traction (Verseux et al. 2016; Keller et al. 2023). Consequently, the lunar and Martian regolith could be harnessed to formulate nutrient-rich cultivation mediums for these organisms. Several studies have explored the use of regolith in cultivation mediums for various cyanobacteria species, such as *Arthrospira platensis* (Macário et al. 2022), *Haematococcus pluvialis* (Casula et al. 2024a), *Anabaena* sp. PCC 7938 (Ramalho et al. 2022), *Chroococcidiopsis* sp. (Olsson-Francis and Cockell 2010), *Synechococcus nidulans* (Casulà et al. 2024b), *Anabaena cylindrica* (Olsson-Francis and Cockell 2010), *Nostoc muscorum* (Macário et al. 2022), several species of Siderophilic cyanobacteria isolated from iron-depositing hot springs (Brown et al. 2010), and *Nostoc* sp. (Ye et al. 2021), as well as microalgae like *Chlorella vulgaris* (Macário et al. 2022). However, these investigations have revealed suboptimal growth rates and biomass production for the studied microbes, underscoring the necessity for further research into the integration of regolith into cultivation mediums.

Beyond regolith, alternative resources such as atmospheric gases (Verseux et al. 2021), water ice (Rapp and Inglezakis 2024; Detrell 2021a), and water from hydrated minerals and atmospheric water vapor (Rapp and Inglezakis 2024) could also be considered for utilization. Notably, the Martian atmospheric water content is less than 30 ppm, rendering practical atmospheric water harvesting unfeasible (Rapp and Inglezakis 2024). Consequently, the most viable approach entails the utilization of buried water ice and minerals hydrated with water contents of up to 10% (Rapp and Inglezakis 2024).

The primary function of incorporating organisms into LSS is to produce food or pharmaceuticals for astronauts. Cyanobacteria and microalgae, due to their edible nature, represent self-sustaining and renewable resource rich in essential nutrients such as micronutrients, macronutrients, antioxidants, unsaturated fatty acids, vitamins, and essential amino acids (Fahrion et al. 2023; Dominoni 2024). However, alongside these advantages, using cyanobacteria and microalgae in diets comes with certain drawbacks. For instance, they often contain less fiber compared to edible plant biomass, lack essential oils, and possess a low carbohydrate-to-protein ratio (Way et al. 2011). Conversely, in some cases, the content of nucleic acids and minerals in these microbes may be excessively high, posing potential health risks for astronauts (Revellame et al. 2021). Additionally, the cellulose nature of microalgal biomass may impede its digestibility (Revellame et al. 2021). Ultimately, the biomass yield of these microbes may only amount to approximately 10 g person^{-1} day^{-1}, significantly below the 50% protein requirements of an individual (Revellame et al. 2021). Nonetheless, these challenges can be addressed through the application of cutting-edge bioengineering techniques (Llorente et al. 2022).

More recently, there has been a growing interest in the utilization of axenic (denoting a pure, singular strain of cyanobacteria or microalgae) and non-axenic (comprising a blend of at least two or more cyanobacteria or microalgae strains) cultures within LSS. Currently, there is a notable lack of documented instances demonstrating the extended-term (e.g., spanning several years) utilization of axenic cultures in LSS. Research has highlighted the utilization of cyanobacteria-microalgae, bacteria-microalgae, and microalgae-microalgal symbiosis to bolster biomass production and enhance survival in extreme environmental conditions on Earth (Xu et al. 2024; Ramanan et al. 2016). The advancement of technologies aimed at exploring the potential application of non-axenic cultures holds the promise of making this approach practically viable in LSS, facilitating the creation of sustainable, long-lasting renewable resources for the synthesis of valuable products in extraterrestrial environments (Casula et al. 2024a).

12.3 Prospects and Challenges

12.3.1 Deep-Space Missions

Driven by the insatiable desire to uncover resources beyond our home planet, extensive efforts have been dedicated to investigating the surface conditions and environments of the Moon. To this end, a multitude of satellites—including Luna, Apollo, Surveyor, the Chang'e program, Lunar Orbiter, Ranger, and various other missions—have been meticulously crafted and launched into space. In a parallel endeavor, the Martian landscape and atmosphere have also been meticulously examined through a series of space missions, featuring the likes of Mars Odyssey, the Mars Express Orbiter, the Interplanetary Stations of the Mars Series (USSR), the Mars Global Surveyor, and a fleet of landers and rovers such as Spirit, InSight, Viking-1 and 2, ExoMars, Mars Pathfinder, and Perseverance, each meticulously engineered and launched into the cosmos. Consequently, armed with an array of sophisticated instruments, these space missions have yielded invaluable data for assessing the physical, geological, and atmospheric conditions of celestial bodies both proximal to Earth and in the depths of space. This data corpus underscores the presence of microorganisms in these environments, highlighting their potential role in bolstering life support systems. These microbial entities serve as pivotal subjects for elucidating stress thresholds that do not induce significant alterations to microbial genomes and metabolomes.

12.3.2 Influence of High CO_2 Concentration and Low Atmospheric Pressure

Given the lack of atmosphere on the Moon (Holcomb et al. 2023), this section focuses on microbial cultivation within spacecraft cabins and under Mars-like environmental conditions. As illustrated in Fig. 12.2 the conditions on Earth starkly contrast those on Mars. Mars exhibits pressures 100–200 times lower than Earth's,

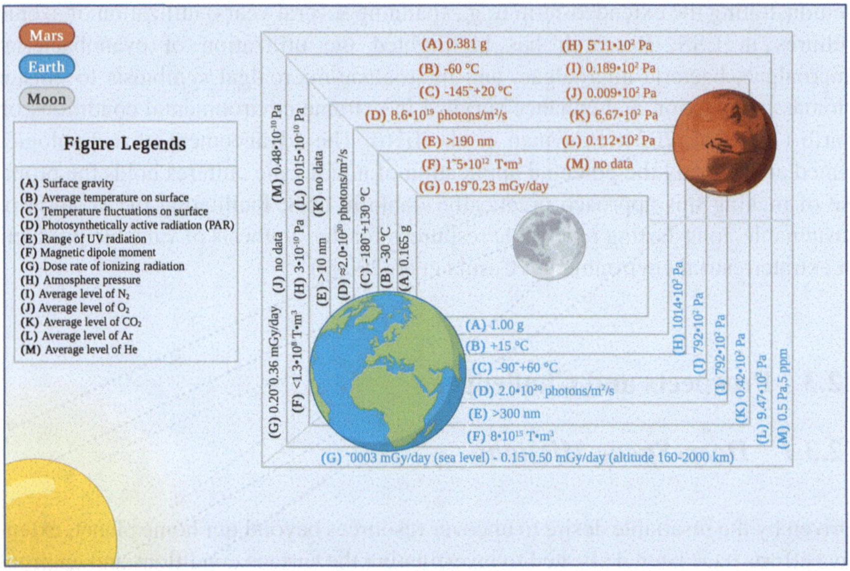

Fig. 12.2 Comparative presentation of atmospheric conditions on the surface of Earth, Moon, and Mars

with CO_2 dominating the atmosphere at 95%. Consequently, it appears unlikely that cyanobacteria and/or microalgae strains thriving in Earth's conditions, conducive to high biomass and oxygen production, would fare well on Mars. Currently, engineering a physical LSS mimicking terrestrial conditions for Mars could address microbial cultivation challenges and support human settlements. However, such LSS would necessitate maintenance, optimization, and intricate technological design, entailing significant resource and energy consumption (Ewert et al. 2022). Nevertheless, such facilities might still be viable for small-scale microbial cultivation.

Alternatively, identifying cyanobacteria and microalgae adaptable to Mars-like conditions could present another avenue. Nonetheless, prevailing conditions must support efficient photosynthesis and biomass production and likely influence microbial biochemical properties (Verseux et al. 2021). For instance, decreased pressure required in photobioreactors (PBRs) for cyanobacteria and microalgae cultivation necessitates reconsidering material reliability limits, reducing mass, and enhancing PBR sustainability. However, it's imperative to note the physical constraints on pressure application for specific organisms. This constraint arises from the need to maintain the liquid phase at growth-permissive temperatures. Lower pressure experiments typically involve saturating the air above the liquid to prevent nutrient medium evaporation (Rautela et al. 2024; Kanervo et al. 2005; Battistuzzi et al. 2024). Cyanobacteria and microalgal cultivation fluctuations in CO_2 concentration in surroundings and the cultivation medium modulate the functionality of carbonic anhydrase enzyme involved in the regulation of carbon sequestration mechanisms of these microbes (Miyachi et al. 2003). Also, literature reported the impact of depleted nutrient medium and elevated concentrations of CO_2 on the rate of photosynthesis and biomass production (Kryvenda et al. 2023).

Additionally, the utilization of mutagenesis, facilitated by ultraviolet (UV) light, presents a promising avenue for research aimed at developing microalgal strains with enhanced tolerance to elevated concentrations of CO_2 in cultivation mediums lacking in nutrients. An illustrative example is the mutant strain of *Chlorella vulgaris*-SDEC-3M, which demonstrated heightened tolerance to elevated CO_2 levels, alongside improvements in biomass production, photosynthetic rate, and CO_2 fixation (Qi et al. 2016; Priyadharsini et al. 2022). Furthermore, genetic stability assessments conducted over five consecutive generations revealed that this mutant strain exhibited no propensity for further genetic alterations. Similarly, research conducted on *Scenedesmus obliquus*, and its mutant variant designated as SDEC-1M (Qi et al. 2018), yielded comparable results. These findings collectively suggest that microalgal strains possess promising potential for integration into LSS.

On the contrary, cyanobacteria have shown significant potential for proliferation in environments with elevated CO_2 concentrations and low atmospheric pressure. For instance, a study detailed the cultivation of *Synechocystis* sp. PCC 6803 in freshwater under conditions where the supplied air was saturated with CO_2 at 20%, at an atmospheric pressure of 0.1 atm. This study demonstrated enhanced growth compared to control experiments conducted under ambient conditions (Kanervo et al. 2005). In another investigation, researchers found that certain cyanobacterial strains could thrive in CO_2 concentrations as high as 40%. Species such as *Anabaena* sp., *Plectonema boryanum*, and *Synechococcus* sp., after a period of adaptation, could grow well in conditions where the supplied air was completely saturated with CO_2 at normal atmospheric pressure (Thomas et al. 2005). Conversely, specific strains of cyanobacteria, like *Arthrospira platensis* and *Synechocystis* sp., exhibited better proliferation in an atmosphere saturated with CO_2 at a pressure of 0.05 atm than those cultivated under ambient conditions. These findings strongly advocate for the cultivation of cyanobacteria on Mars due to the potentially lower operational and maintenance costs involved.

In the preceding paragraphs, it has been discussed that not all cyanobacteria and microalgae possess the capability to proliferate and survive under conditions divergent from their natural ambient environment. For instance, certain strains of green microalgae, like *Chlamydomonas reinhardtii*, exhibited reduced growth and altered CO_2 absorption when cultured under conditions where CO_2 constituted 4.8% of the atmosphere at a pressure of 0.7 atm (Wagner and Posten 2017). Similarly, studies have documented adverse effects on the growth of various cyanobacterial strains, such as *Merismopedia* sp., *Anabaena* sp., and *Microcystis aeruginosa*, in response to reduced atmospheric pressure, even without alterations in air composition (Qin et al. 2014). Further investigations have explored the impact of both reduced atmospheric pressure and modified air composition on the culture of *Anabaena* sp. in PBR. In one study, *Anabaena* sp. was subjected to a constant pressure of 0.1 atm, with air compositions of 4% CO_2 and 96% nitrogen, while ambient air served as the control. Results indicated no statistically significant differences in biomass production, carbohydrate content, or chlorophyll levels (Verseux et al. 2021). Moreover, another study demonstrated the production of biomass and clean energy sources, such as H_2, by cyanobacteria under simulated Martian-like atmospheric conditions enriched with approximately 100% CO_2 at reduced concentrations of nitrogen and varying levels of atmospheric pressures (Murukesan et al. 2016). Additionally, the pH level of the cultivation medium has been identified as a significant factor influencing biomass production and CO_2 sequestration in cyanobacteria and microalgae (Caparroz et al. 2024). These investigations underscore the intricate dependencies of photosynthetic microbe proliferation on not only CO_2 levels but also pH levels in the cultivation medium. Given the nitrogen-limited atmosphere of Mars, certain cyanobacteria, such as those belonging to the genus *Anabaena*, have the ability to fix atmospheric nitrogen into the cultivation medium, potentially benefiting other organisms within an LSS (Subramanian and Shanmugasundaram 1986; Fais et al. 2024). Overall, these findings suggest that leveraging cyanobacteria could enhance the viability of deep-space expeditions while facilitating sustainable human settlement in the Martian atmosphere.

The primary obstacle to nitrogen fixation in the Martian atmosphere is the low nitrogen pressure (pN_2), which stands at 0.2 hPa. Recent studies have identified certain cyanobacteria capable of nitrogen fixation at a pN_2 of 5 hPa (Maity and Saxena 2024; McKay and Marinova 2001). However, a notable reduction in biomass production has been observed in specific cyanobacterial species, such as *Anabaena variabilis* and *Anabaena cylindrica*, at a pressure of 500 hPa (Nicola et al. 2024; Silverman et al. 2019). Conversely, *Anabaena* sp. PCC 7938 demonstrated consistent biomass accumulation over ten consecutive days when cultivated at a pN_2 of 96 hPa (Verseux et al. 2021). These findings indicate that different cyanobacterial and microalgal strains require varying pN_2 levels and cultivation conditions for optimal growth and proliferation. Thus, targeted research is necessary to establish a definitive pN_2 threshold that can satisfy the basic requirements of microbial cultivation, potentially obtainable from the Martian atmosphere. One approach could involve the extraction of N_2 and CO_2 from the atmosphere to create a tailored air mixture suitable for efficient cyanobacterial or microalgal cultivation on Mars (Verseux et al. 2016). Alternatively, identifying naturally occurring strains of photosynthetic microbes or engineering specific strains with desired characteristics through advanced synthetic biology methods could enable cultivation under natural or partially controlled Martian conditions. For effective growth and cultivation of cyanobacteria and microalgae at low atmospheric pressure, it is crucial to increase the partial pressure of CO_2. This facilitates the dissolution of CO_2 in the cultivation medium, providing a readily available carbon source for photosynthetic microbes. However, it is essential to consider the optimal partial pressure of CO_2 for different microbial genera (Huang et al. 2022) and even strains within the same species (Cheng et al. 2006; Qi et al. 2016). Additionally, cyanobacteria require nitrogen for growth in cultivation media, necessitating the provision of nitrogen sources such as nitrates, urea, and/or ammonium.

12.3.3 Impact of Microgravity

The cultivation of cyanobacteria and microalgae in spacecraft and on the surface of celestial bodies such as the Moon and Mars involves exposure to complete or partial weightlessness. Previous theoretical studies suggested that microgravity might not directly affect cells smaller than 10 μm due to the relative homogeneity of the intracellular matrix and the diminutive size of these cells (Pollard 1965; Horneck et al. 2010). Subsequent experiments aimed to confirm this theory, yet many of these investigations do not provide unequivocal evidence (Sychev et al. 2001; Nickerson Cheryl et al. 2000; Su et al. 2023). Simulated microgravity experiments have revealed disruptions in growth patterns, cell concentration, and the expression of stress-responsive genes in various bacteria such as *Escherichia coli*, *Salmonella typhimurium*, and *Bacillus subtilis* (Benoit and Klaus 2007; Acres et al. 2021). Conversely, species like *Streptomyces clavuligerus* and *Streptomyces hygroscopicus*, which develop hyphae with a diameter of approximately 2 μm, exhibit decreased growth rates and altered secondary metabolite production under microgravity conditions (Bing Huang et al. 2018; Tirumalai et al. 2017; Chavez et al. 2024). Similar investigations have been conducted on cyanobacteria. For instance, increased growth of *Nostoc sphaeroides* has been observed (Wang et al. 2004), while retarded growth associated with the production of toxic compounds by *Microcystis aeruginosa* has also been documented (Xiao et al. 2010). Furthermore, exposure of the green microalga *Euglena gracilis* to microgravity resulted in the modulation of gene expression related to stress responses (e.g., oxidative stress and heat shock protein production), cell cycle regulation, and cell division (Nasir et al. 2014).

Verseux and his colleagues proposed that microbial mobility is inversely correlated with the impact of microgravity (Verseux et al. 2016). However, the precise mechanism by which microgravity influences microbial growth and proliferation still needs to be understood. It could be speculated that this deleterious effect stems from various factors, including the sluggish movement of the cultivation medium, hindered nutrient access, and the accumulation of cellular exudates, which may become toxic to the cell itself, thereby altering the microenvironment surrounding the cell and resulting in inhibited cellular growth (Marra et al. 2023; Horneck et al. 2010).

Cellular architecture plays a crucial role in mitigating stress induced by microgravity. For instance, cells with rigid cytoskeletons are better equipped to withstand microgravity-induced stress than those with more flexible structures (Nasir et al. 2014). Nevertheless, studies have shown that microgravity does not impede the photosynthesis and nitrogen fixation processes of *Anabaena siamensis* when cultivated in a photobioreactor (PBR) (Gaohong Wang et al. 2006). Therefore, the engineering of appropriate PBRs equipped with suitable devices to ensure a sustainable supply of light, gas exchange, and mixing of the cultivation medium could eliminate or at least mitigate the negative effects of microgravity. This hypothesis has been supported by the successful cultivation of *Arthrospira*-B onboard the International Space Station (ISS), accompanied by measurements of biomass generation and oxygen production (Poughon et al. 2020). Additionally, studies have demonstrated that cultivating *Limnospira indica* in a photobioreactor can yield biomass and oxygen production equivalent to that achieved on Earth (Nowicka-Krawczyk et al. 2019).

Recently, various types of PBRs, such as membrane PBRs composed of materials like polytetrafluoroethylene, polyethylene, silicone polymers, and polyvinylidene, have been designed and tested for the cultivation of cyanobacteria and microalgae in microgravity environments (Oncel and Sukan 2008; Cogne et al. 2005). Typically, membrane PBRs are employed to facilitate gas exchange between the aqueous cultivation medium and the air, thereby preventing the formation of large air bubbles (Drexler and Yeh 2014). Alternative methods, such as the Karman reactor, can also be utilized for microbial cultivation under microgravity conditions, creating low-intensity homogenized turbulence to mitigate the formation of large air bubbles (Detrell 2021b).

12.3.4 Impact of Cosmic Ionizing Radiations

Cosmic ionizing radiations (CIR) encompass a blend of high-energy solar particles, gamma rays, ultraviolet (UV) and X-rays, along with galactic cosmic rays (GCR). These rays have the potential to induce ionization, posing risks to astronauts, onboard organisms such as microbes within Life Support Systems (LSS), and the electronic equipment within spacecraft. If left unmitigated, they could also have ramifications for life on Earth unless they are deflected or intercepted by Earth's magnetic field (DeVanzo and Hayes 2020). Moreover, the cultivation of aquatic microbial cultures in LSS is being explored as a protective measure against the harmful effects of ionizing cosmic rays. These microbes are expected to serve as a shield, safeguarding the spacecraft's interior. Hence, it's imperative for the microbes bred within LSS to demonstrate resilience to such ionizing radiations while maintaining high rates of oxygen production, biomass, and the synthesis of value-added compounds (Elsaesser et al. 2023; Cottin et al. 2017).

Cosmic radiation has detrimental effects on the cells of both higher organisms and microbes, disrupting the normal functions of essential macromolecules such as DNA, RNA, and proteins. This harm occurs through direct radiation impact, where energy is transferred directly to these molecules, as well as indirect radiation impact, where energy is transferred to the aqueous solutions housing these molecules. Consequently, a variety of reactive oxygen species (ROS) and radicals are produced, leading to the alteration of the normal functionality of crucial macromolecules within living cells (Riccio 2023; Fagliarone et al. 2023; Moeller et al. 2017; Cottin et al. 2017). To survive under such conditions, cells adapt by either synthesizing specific molecules (both macro and micro) or succumbing to death. Cyanobacteria and microalgae, for instance, produce various protective molecules, including but not limited to antioxidants such as vitamins C and E, phycobiliproteins, carotenoids, and polyamines. Additionally, they utilize enzymes, notably superoxide dismutase

(SOD), ascorbate peroxidase, and glutathione reductase, to regulate ROS concentrations (Rastogi et al. 2020; Eregie et al. 2024; Amin et al. 2024).

On average, the distance between the Sun and Mars is approximately 1.5 times greater than the distance from Earth to the Sun. Consequently, solar radiation reaching Mars is roughly half of that which reaches Earth (Fig. 12.2). However, these radiations are either unchanged or manifest as secondary rays, as Mars lacks a magnetic field and its atmospheric density is significantly lower compared to Earth's (Smith et al. 1965; Acuña et al. 2001). While challenges arise when contemplating human settlement on the lunar surface, as the atmospheric density and magnetic forces on the Moon are even lower than those on Mars (Fig. 12.2). Thus, the impact of cosmic radiation on cyanobacteria and microalgae cultivated in Life Support Systems (LSS) on the lunar surface will be akin to that experienced in open space environments, except for radiation impeded by the Moon itself. UV light wavelengths reaching Mars range from medium to far bands, distinct from those on Earth and the Moon, and are considered more harmful to humans, cyanobacteria, and microalgae. Data from the Curiosity rover on the Martian surface indicates a UV ray flux of about 20 W/m^2, with UV spectra ranging between 200 and 380 nm. However, the wavelength range varies depending on the local Martian season and surface location (Gómez-Elvira et al. 2014). Nevertheless, this data collected by the Curiosity rover can serve as a valuable reference for conducting research to develop resources for Mars implementation.

Numerous studies have investigated the impact of electromagnetic infrared (EMIR) rays on cyanobacteria and microalgae, focusing particularly on their effects on photosystems I and II activity. The findings reveal a significant decrease in the performance of photosystem II, albeit with variations depending on the microbial type and strains. Conversely, changes in the performance of photosystem-I appear to be minor. Moreover, alterations have been observed in the structure of the chlorophyll-a antenna, protein accumulation in photosystem II, thylakoid structure, and the composition of pigment-protein complexes. The predominant effects of UV rays at cellular and molecular levels encompass various phenomena, including but not limited to delays in cell differentiation, decreased or inhibited nitrogen fixation, hindered cellular growth, reduced total protein content and oxygen production, changes in the thylakoid and cell wall composition, and alterations in plasma membrane structure due to lipid unsaturation. These effects also entail membrane leakage, impacting organelle integrity and overall cell viability, as well as a reduction in the efficiency of the Ribulose-1,5-bisphosphate carboxylase/oxygenase (RUBISCO) enzyme activity (Rastogi et al. 2020; Amin et al. 2024).

To date, numerous studies have demonstrated that a thin layer of Mars's soil can effectively mitigate the penetration of UV rays (Patel et al. 2003; Rontó et al. 2003; Cockell and Raven 2004). For example, *Bacillus subtilis* spores covered with just 1 mm of Martian soil retained viability for germination, whereas spores exposed directly to UV rays experienced a viability reduction of over 99% in just 15 s. Similarly, investigations into the survivability of *Chroococcidiopsis* sp. under UV exposure revealed a loss of viability of up to 99% within 5 min, with complete cellular viability loss occurring within 30 min of irradiation under optimal growth conditions (Cockell et al. 2005). However, applying UV filters that allow the passage of photosynthetically active radiation (PAR) could present an alternative method for cultivating these microbes in the Martian environment. Given the detrimental impact of UV rays, exploring the aforementioned suggestions or alternative strategies to mitigate this issue is imperative.

Numerous studies have investigated the tolerance of various strains of cyanobacteria and microalgae to electromagnetic infrared (EMIR) radiation. For example, filamentous Anabaena spp. cells have demonstrated resilience, surviving exposure to 15 kGy of EMIR radiation, with no apparent effects on growth parameters and survival observed at 5 kGy (Singh et al. 2013). Mechanisms enabling survival under such harsh conditions have also been elucidated (Srivastava et al. 2023). Certain cyanobacterial strains, such as *Chroococcidiopsis* spp., have exhibited viability even after exposure to EMIR radiation exceeding 15 kGy. However, DNA damage was observed in exposed cells, although recovered within 24 h post-exposure (Billi et al. 2000). To date, the most resistant photosynthetic microorganism known, capable of withstanding up to 20 kGy of radiation, is *Coccomyxa actinabiotis*, originally isolated from a nuclear reactor's cooling pool. Remarkably, this microalga can restore its highest growth rate just 14 days after the initial exposure (Rivasseau et al. 2013).

The growth and proliferation of cyanobacteria and microalgae are significantly influenced by cosmic radiation, including galactic cosmic radiation (GCR) and solar energetic particles (SEP). Studies have demonstrated the absorption rates of radiation doses in Martian soil, ranging from 50 to 150 mGy annually (Pavlov et al. 2012). The rover Curiosity further confirmed the rate of ray absorption on the Martian surface to be 210 ± 15 μGy/day, equivalent to 76 mGy/year (Hassler et al. 2014). Additional data from the Mars orbiter's MARIE instrument revealed an average flux of absorbed rays at 213.6 μGy/day. During July 2002, this flux increased to 1682.8 μGy/day due to a surge in solar energetic particles (Hassler et al. 2014). Scientists have also researched the Martian surface to assess whether Mars's regolith can shield against cosmic radiation. Their findings indicate that at a depth of 3 m, the radiation levels approximate 1 μGy/day, consistent with background radiation levels on both Earth and Mars (Hassler et al. 2014).

Similar findings have been documented regarding human settlement on the Moon, indicating that lunar regolith possesses the capability to shield against radioactive rays (Naser and Chehab 2018). Research suggests that at a depth of 0.5 m within the Moon's regolith, which boasts an approximate density ranging from 1.0 to 1.4 g cm^{-3}, protection from cosmic radiation is attainable (Miller et al. 2009). However, this safeguarding mechanism doesn't extend to neutron rays, which possess high penetration energy, necessitating a relatively greater mass for adequate protection. Consequently, a separate study aimed to ascertain whether the Moon's regolith can effectively shield against neutron radiation. The results indicated that the protective properties of regolith resemble those of aluminum. To shield against SPE, a thickness of a few dozen centimeters of regolith with a density of 1 g cm^{-3} is required, whereas protection against GCR mandates a similar regolith depth of 2 m (Meurisse et al. 2020).

Relatively fewer studies have explored the impact of corpuscular ionizing radiation on the survival and proliferation of cyanobacteria and microalgae. However, a comparative investigation has been undertaken to assess the effects of various linear energy transfer (LET) rays on genetic alteration in *Chroococcidiopsis* spp. (Verseux et al. 2017). The results revealed that Fe, Si, and He ions shielded cyanobacteria from genotoxic and mutagenic effects, while γ-rays with a flux rate exceeding 24 kGy disrupted the normal growth behavior of these microorganisms. This study elucidated that corpuscular radiation exerts more damage compared to electromagnetic rays. Moreover, the combined deleterious effects of cosmic ionizing radiation and microgravity have been documented (Yatagai et al. 2019). Typically, such synergistic impacts of these two parameters have been employed to investigate the behavior of human cells (Adamopoulos et al. 2024; Fu et al. 2020). Conversely, studies focusing on cyanobacteria and microalgae are relatively scarce, and often conducted aboard spacecraft in Earth's orbit (Yatagai et al. 2019). These investigations operate under the assumption that space radiations in Earth's orbit are generally analogous to interplanetary space radiation conditions (Janssen et al. 2014).

12.3.5 Synergistic Effect of Stresses on Cyanobacteria and Microalgae

Researchers often grapple with the challenge of understanding how microgravity and cosmic radiation interact. However, they frequently find themselves relying on simplified models of ionizing radiation flux, which may not fully capture the complexities of actual spaceflight conditions (Heinicke et al. 2021). Recreating the precise fluxes of photons, electrons, protons, and high-energy ionizing heavy particles that comprise real cosmic radiation poses a significant challenge (Cockell et al. 2011). Achieving a precise balance of energy particles and distribution is necessary to simulate conditions closely resembling space conditions. Yet, engineering a sustainable long-term life support system is possible by comprehending the synergistic effects of various stress-inducing factors, including ionizing radiation, on cyanobacteria and microalgae.

Hence, it is urgent to understand the combined impact of all potential stressors present in the space environment on cyanobacteria and microalgae. Promising venues for conducting such research include the Moon's Orbital Platform-Gateway and prospective space-based laboratories situated on the lunar surface (Heinicke et al. 2021). These facilities offer the advantage of conducting research in conditions closely resembling those found in space, thus providing valuable insights into how organisms cope with stress factors like high CO_2 levels and low atmospheric pressure, crucial for the success of crewed missions to Mars. However, as these facilities are currently unavailable, similar research has been carried out aboard the ISS and other related spacecraft (Wassmann et al. 2012; Billi et al. 2019).

Numerous studies have investigated the ecological behavior of cyanobacteria and microalgae and higher plants and animals. For instance, Foton satellites in low Earth orbit were utilized to conduct experiments with Aquacells and OmegaHab, involving microalgae *Euglena gracilis* and larvae of *Oreochromis mossambicus* over a period of less than 21 days. Results revealed larval mortality attributed to insufficient oxygen production by *E. gracilis* (Lebert et al. 1995; Häder et al. 2018). Similarly, research on the Shenzhou 8 spacecraft involved two microalgae species, *Chlorella pyrenoidosa* and *E. gracilis*, along with three snails, *Bulinus australianus*, during a 17.5-day mission. Comparative analysis with similar experiments conducted on Earth revealed a decline in microalgal cell density, total soluble carbohydrates, and proteins, as well as inhibited nitrogen and phosphate fixation and reduced oxygen production due to the suppression of the photosynthetic process in microalgae (Xiaoyan Li et al. 2017b).

In 2018, the Eu:CROPIS satellite was launched to conduct a scaled-up experiment similar to one planned for two greenhouses under simulated Moon and Mars gravity setups (Fig. 12.2). This mission, intended to last one year, aimed to employ synthetic urine as a bacterial growth medium, utilizing bacteria such as *Nitrosomonas* and *Nitrobacter* to initiate nitrification. Additionally, the green microalgae *Euglena gracilis* was introduced to generate oxygen, facilitating bacterial-mediated nitrification. Tomato plants were used as the final consumers of nitrates. Regrettably, the mission failed due to onboard software malfunctions (Hauslage et al. 2018). Nevertheless, a successful experiment was conducted using a secondary payload on the same satellite (https://www.dlr.de/en/latest/news/2020/01/20200113_farewell-to-the-eucropis-mission (accessed May 9, 2024). This experiment explored the synergistic effect of genetically modified photosynthetic cyanobacteria *Anabaena*, which extracellularly produces sucrose, and a sucrose-consuming bacterium, *Bacillus subtilis*, under varying levels of gravity exposure (McCutcheon et al. 2016). The focus was on establishing a miniature microbial ecology, its integration into spacecraft, and understanding its stability as a model system. In other studies, specific cyanobacteria and microalgae were investigated for targeted purposes under controlled spacecraft conditions and simulated environments like on different planets. The ADAPT I experiment, conducted on the EXPOSE-E facility of the ISS, involved placing *Bacillus subtilis* spores and a community of epilithic microbes, supplemented with layers of cyanobacteria *Anabaena cylindrica* and cells of *Nostoc commune* and *Chroococcidiopsis*, outside the ISS for 1.5 years. Upon revival, it was discovered that *Nostoc commune* samples were deceased, cyanobacteria *Anabaena cylindrica* survived only under Martian-like conditions with reduced UV flux and approximately 103 Pa CO_2. Intriguingly, samples of cyanobacteria *Chroococcidiopsis* were able to grow and proliferate under almost all tested conditions (Wassmann et al. 2012; Cockell et al. 2011).

The discoveries have sparked significant interest in utilizing *Chroococcidiopsis* to study its resilience to space conditions for the betterment of humanity. Consequently, a hypothesis was formulated suggesting that multilayered dried plankton of the cyanobacteria *Chroococcidiopsis* would exhibit lesser resistance to space conditions and the Martian environment than dried biofilms of the same cyanobacteria. This hypothesis underwent evaluation in a space test mission conducted aboard the ISS. The results of the mission confirmed the hypothesis, indicating that pure cyanobacterial biofilms demonstrated tolerance to nearly all space conditions when contrasted with layers of cyanobacteria associated with planktonic microbes. These findings were rooted in observations of growth behavior in exposed cultures, identification of unviable cells within different layers, and DNA damage in cyanobacterial cells. The authors provided further elucidation by noting that cyanobacterial cells within the biofilm exhibited more significant metabolic activity and physiological differences compared to cells in planktonic cultures (Billi et al. 2019). However, to develop durable and sustainable bioregenerative LSS, it is crucial to validate microorganisms in their viable form rather than dormant or inactive state. Therefore, there is considerable merit in conducting experiments related to the cultivation and scaling up of cyanobacteria and microalgae in actual space environments. In pursuit of this objective, successful research on Arthrospira-B was conducted aboard the ISS from 2017 to 2018, demonstrating the regeneration of the atmosphere in microgravity and the suitability of using *Limnospira indica* for this purpose (Poughon et al. 2020). Moreover, the European Space Agency (ESA) has initiated the BIORAT-1 program, designed to continue the cultivation of *Limnospira indica* on the ISS in relatively large volumes (e.g., 2.6 L in a Photobioreactor (PBR)), compared to its previous cultivation in only 60 ml during the Arthrospira-B project. This endeavor is facilitated by the provision of air from a crewed chamber as a source of CO_2 for the cyanobacteria (Poughon et al. 2020). In another initiative aimed at evaluating microalgae onboard, the PBR@LSR project was established in May 2019 to investigate the cultivation of *Chlorella vulgaris* in a microgravity photobioreactor (μgPBR) in conjunction with a CO_2 concentrator aboard the ISS. However, this experiment was not completed due to technological issues (Detrell et al. 2019).

12.4 Discussion

Based on the current data, the creation of enduring, self-sufficient, dependable, automated, and energy-efficient bioregenerative Life Support Systems (LSS) still needs to be discovered. To realize this objective, extensive research is imperative to identify cyanobacteria or green microalgae, either naturally occurring or engineered through synthetic biology, capable of thriving efficiently in conditions closely resembling the actual space environment (Fig. 12.3). Nonetheless, existing data strongly advocate for the potential utilization of cyanobacteria and microalgae within LSS to sustain astronauts during prolonged space missions. Notably, certain microalgae and cyanobacteria genera, such as *Chlorella* and *Arthrospira*, respectively, have emerged as promising candidates for space programs. They contribute to maintaining a clean environment by generating oxygen and consuming carbon dioxide for their metabolic processes. Moreover, they produce edible biomass rich in various valuable compounds essential for astronaut consumption. However, it is

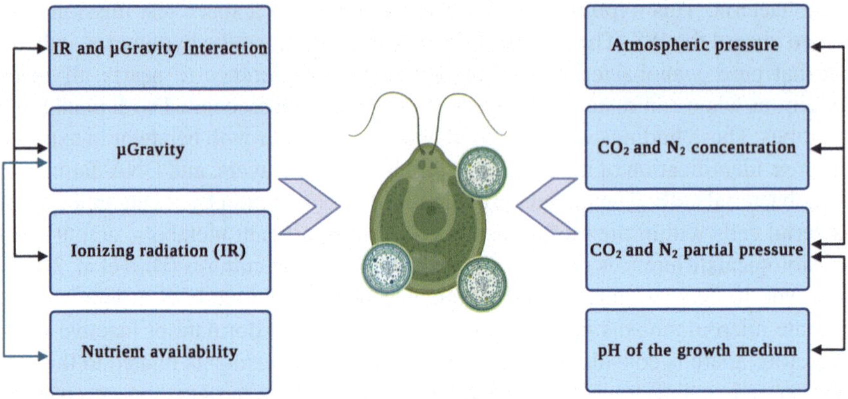

Fig. 12.3 Different stress factors affecting the growth and metabolic activities of cyanobacteria and microalgae in spacecraft-based Life Support Systems (LSS) and on the surfaces of the Moon and Mars

pertinent to acknowledge that these microorganisms necessitate specific cultivation conditions, including an appropriate growth medium, nutrient availability, optimal pH levels, adjusted atmospheric pressure, and controlled temperature, for successful cultivation on extraterrestrial bodies such as the Moon and Mars (Murukesan et al. 2016; de Morais and Costa 2007).

Conducting comprehensive research to identify cyanobacteria and microalgae capable of enduring synergistic stress exposure akin to space conditions is paramount. Rather than assessing individual stressors in isolation, it is imperative to prioritize studying these organisms' resilience to multiple stress factors simultaneously. The following stressors warrant significant attention during experimental investigations:

CO_2 Concentration: Cyanobacteria and microalgae intended for space programs must tolerate higher CO_2 concentrations in their environment. Studies on photosynthesis rates, oxygen production, and biomass generation are crucial. Insights gained from such research can aid in developing photobioreactors (PBRs) with improved microbial cultivation control, optimal growth parameter selection, and utilization of resources onboard spacecraft and on extraterrestrial surfaces.

Low Atmospheric Pressure: Atmospheric pressure and CO_2 concentrations synergistically affect cyanobacteria and microalgae cultivation. Understanding the combined impact of these parameters is essential for establishing sustainable bioregenerative Life Support Systems (LSS) for deep-space missions.

Microgravity: Microbes intended for space use will experience stress due to microgravity. Therefore, careful consideration should be given to incorporating this stress factor when selecting cyanobacteria and microalgae. Understanding how microgravity affects microbial cultivation, photosynthesis rates, and biomass production within PBRs is crucial for designing reliable processes for sustainable LSS.

Cosmic Radiation: Experimentally reproducing cosmic radiation in laboratories is nearly impossible (Matula and Nabity 2021b). While various studies have investigated cyanobacteria and microalgae under different types of ionizing radiation, comparing results is challenging due to variations in experimental conditions. Understanding the effects of cosmic radiation on microorganisms can enhance the reliability and sustainability of LSS and potentially mitigate its impact on spacecraft crews in the future.

Utilization of Resources in Lunar and Martian Regolith: In-situ resource utilization (ISRU) is essential for the success of deep-space missions. Establishing methods to utilize water resources available on the Moon and Mars, whether in the form of ice, vapor, or regolith, is critical. Few studies have successfully explored certain cyanobacterial strains that can utilize optimal atmospheric pressure of CO_2 and N_2.

Synergistic Effects of Stress Factors: Evaluating potential microbes under the collective impact of stress factors over extended periods, equivalent to real-time space programs, is essential. This comprehensive understanding of selected microbes is crucial for developing sustainable bioregenerative LSS. Key objectives include upscaling PBRs and assessing growth rates, biomass production, and oxygen generation under synergistic exposure to various stressors, including radiation and microgravity. Scaling PBRs is essential, as the current culture volume is insufficient to produce enough oxygen for even one spacecraft crew member. Investigating the combined effects of all stressors is vital for optimizing microbial performance in space environments.

12.5 Conclusion and Prospects

This chapter delved into the critical factors influencing cyanobacteria and microalgae within bioregenerative Life Support Systems (LSS) during near-Earth orbit and deep-space missions. An extensive examination of the literature shows an urgent need for research to cultivate cyanobacteria and microalgae on a large scale to meet the demands of bioregenerative and hybrid LSS for onboard crew members. Concurrently, exploring the nuanced effects of stressors on these microbes is essential to gain a comprehensive understanding of their behavior, thus aiding in selecting the most suitable candidates for space missions. Lunar orbit stations emerge as promising environments for such investigations. Carrying out these studies on Earth poses challenges in replicating characteristic cosmic radiation and microgravity, while Low Earth Orbit presents unsuitable cosmic radiation conditions.

References

Abadie LJ, Francisco DR (2024) Human health and performance: keeping astronauts safe & productive on a mission to Mars

Abney MB, Mansell JM, DuMez S, Thomas J, Cooper C, Long D (2013) Ongoing development of a series Bosch reactor system. In: 43rd international conference on environmental systems, p 3512

Acres JM, Youngapelian MJ, Nadeau J (2021) The influence of spaceflight and simulated microgravity on bacterial motility and chemotaxis. npj Microgravity 7(1):7. https://doi.org/10.1038/s41526-021-00135-x

Acuña MH, Connerney JEP, Wasilewski P, Lin RP, Mitchell D, Anderson KA et al (2001) Magnetic field of Mars: summary of results from the aerobraking and mapping orbits. J Geophys Res: Planets 106(E10):23403–23417. https://doi.org/10.1029/2000JE001404

Adamopoulos KI, Sanders LM, Costes SV (2024) NASA GeneLab derived microarray studies of Mus musculus and Homo sapiens organisms in altered gravitational conditions. npj Microgravity 10(1):49. https://doi.org/10.1038/s41526-024-00392-6

Amin N, Sinha RP, Kannaujiya VK (2024) Effects of ultraviolet and photosynthetically active radiation on morphogenesis, antioxidants and photoprotective defense mechanism in a hot-spring cyanobacterium Nostoc sp. strain VKB02. Res Microbiol 104180:104180. https://doi.org/10.1016/j.resmic.2024.104180

Bahadori AA (2024) Space radiation protection in the modern era: new approaches to familiar challenges. Radiat Phys Chem 111764:111764. https://doi.org/10.1016/j.radphyschem.2024.111764

Battistuzzi M, Morlino MS, Cocola L, Trainotti L, Treu L, Campanaro S et al (2024) Transcriptomic and photosynthetic analyses of Synechocystis sp. PCC6803 and Chlorogloeopsis fritschii sp. PCC6912 exposed to an M-dwarf spectrum under an anoxic atmosphere (original research). Front Plant Sci 14. https://doi.org/10.3389/fpls.2023.1322052

Belz S, Ganzer B, Messerschmid E, Friedrich KA, Schmid-Staiger U (2013) Hybrid life support systems with integrated fuel cells and photobioreactors for a lunar base. Aerosp Sci Technol 24(1):169–176. https://doi.org/10.1016/j.ast.2011.11.004

Benoit MR, Klaus DM (2007) Microgravity, bacteria, and the influence of motility. Adv Space Res 39(7):1225–1232. https://doi.org/10.1016/j.asr.2006.10.009

Billi D, Friedmann EI, Hofer Kurt G, Caiola Maria G, Ocampo-Friedmann R (2000) Ionizing-radiation resistance in the desiccation-tolerant cyanobacterium Chroococcidiopsis. Appl Environ Microbiol 66(4):1489–1492. https://doi.org/10.1128/AEM.66.4.1489-1492.2000

Billi D, Staibano C, Verseux C, Fagliarone C, Mosca C, Baqué M et al (2019) Dried biofilms of desert strains of Chroococcidiopsis survived prolonged exposure to space and Mars-like conditions in low Earth orbit. Astrobiology 19(8):1008–1017. https://doi.org/10.1089/ast.2018.1900

Brennan L, Owende P (2010) Biofuels from microalgae—a review of technologies for production, processing, and extractions of biofuels and co-products. Renew Sust Energ Rev 14(2):557–577. https://doi.org/10.1016/j.rser.2009.10.009

Brown LM, Zeiler KG (1993) Aquatic biomass and carbon dioxide trapping. Energy Convers Manag 34(9):1005–1013. https://doi.org/10.1016/0196-8904(93)90048-F

Brown I, Allen C, Garrison D, Sarkisova S, Galindo C, Mckay DS (2010) Developing biological ISRU: implications for life support and space exploration. No. JSC-CN-19923. 2010

Caparroz M, Guzmán JL, Berenguel M, Acien FG (2024) A novel data-driven model for prediction and adaptive control of pH in raceway reactor for microalgae cultivation. New Biotechnol 82:1–13. https://doi.org/10.1016/j.nbt.2024.04.001

Casula M, Caboni P, Fais G, Dessì D, Scano P, Lai N et al (2024a) In-situ resource utilization to produce Haematococcus pluvialis biomass in simulated Martian environment. Algal Res 79:103489. https://doi.org/10.1016/j.algal.2024.103489

Casula M, Fais G, Sidorowicz A, Caboni P, Cao G, Concas A (2024b) Effects of a novel bioprocess for the cultivation Synechococcus nidulans on Mars on its biochemical composition:

focus on the lipidome. Biotechnol Bioprocess Eng 29(2):361–376. https://doi.org/10.1007/s12257-024-00019-3
Çelekli A, Zariç ÖE (2024) Breathing life into Mars: terraforming and the pivotal role of algae in atmospheric genesis. Life Sci Space Res 41:181–190. https://doi.org/10.1016/j.lssr.2024.03.001
Chavez A, Topolski C, Hicks J, Villafania M, Baez N, Burke M et al (2024) Phenotypical changes in Escherichia coli K12 after prolonged exposure to simulated microgravity (brief research report). Front Space Technol 4. https://doi.org/10.3389/frspt.2023.1282850
Chen L, Li J (2022) Optimization of Earth-Mars transfer trajectories with launch constraints. Astrophys Space Sci 367(1):12. https://doi.org/10.1007/s10509-021-04025-2
Cheng L, Zhang L, Chen H, Gao C (2006) Carbon dioxide removal from air by microalgae cultured in a membrane-photobioreactor. Sep Purif Technol 50(3):324–329. https://doi.org/10.1016/j.seppur.2005.12.006
Cockell CS, Raven JA (2004) Zones of photosynthetic potential on Mars and the early Earth. Icarus 169(2):300–310. https://doi.org/10.1016/j.icarus.2003.12.024
Cockell, C. S., Schuerger, A. C., Billi, D., Friedmann, E. I., & Panitz, C. (2005). Effects of a simulated Martian UV flux on the cyanobacterium, Chroococcidiopsis sp. 029. Astrobiology, 5(2), 127–140. https://doi.org/10.1089/ast.2005.5.127
Cockell CS, Rettberg P, Rabbow E, Olsson-Francis K (2011) Exposure of phototrophs to 548 days in low Earth orbit: microbial selection pressures in outer space and on early Earth. ISME J 5(10):1671–1682. https://doi.org/10.1038/ismej.2011.46
Cogne G, Cornet J-F, Gros J-B (2005) Design, operation, and modeling of a membrane photobioreactor to study the growth of the cyanobacterium Arthrospira platensis in space conditions. Biotechnol Prog 21(3):741–750. https://doi.org/10.1021/bp0495926
Cottin H, Kotler JM, Billi D, Cockell C, Demets R, Ehrenfreund P et al (2017) Space as a tool for astrobiology: review and recommendations for experimentations in earth orbit and beyond. Space Sci Rev 209(1):83–181. https://doi.org/10.1007/s11214-017-0365-5
Cucinotta FA, Kim M-HY, Chappell LJ, Huff JL (2013) How safe is safe enough? Radiation risk for a human mission to Mars. PLoS One 8(10):e74988. https://doi.org/10.1371/journal.pone.0074988
de Morais MG, Costa JAV (2007) Biofixation of carbon dioxide by Spirulina sp. and Scenedesmus obliquus cultivated in a three-stage serial tubular photobioreactor. J Biotechnol 129(3):439–445. https://doi.org/10.1016/j.jbiotec.2007.01.009
Detrell G (2021a) Chlorella vulgaris photobioreactor for oxygen and food production on a Moon base—potential and challenges (review). Front Astronomy Space Sci 8. https://doi.org/10.3389/fspas.2021.700579
Detrell G (2021b) Microalgae-based hybrid life support system from simulations to flight experiment. In: 50th international conference on environmental systems. https://hdl.handle.net/2346/87146
Detrell G, Helisch H, Keppler J, Martin J, Angerer O, Adrian A et al (2019) PBR@ LSR: the algae-based Photobioreactor experiment at the ISS–configuration and operations. In: 49th international conference on environmental systems. https://hdl.handle.net/2346/84416
DeVanzo M, Hayes RB (2020) Ionizing radiation shielding properties of metal oxide impregnated conformal coatings. Radiat Phys Chem 171:108685. https://doi.org/10.1016/j.radphyschem.2020.108685
Dominoni A (2024) Design for sustainability in space: future nature. In: Dominoni A (ed) Design for sustainability and inclusion in space: how new European Bauhaus principles drive nature & parastronauts projects. Springer International Publishing, Cham, pp 25–48. https://doi.org/10.1007/978-3-031-56004-0_2
Drexler ILC, Yeh DH (2014) Membrane applications for microalgae cultivation and harvesting: a review. Rev Environ Sci Biotechnol 13(4):487–504. https://doi.org/10.1007/s11157-014-9350-6
Elsaesser A, Burr DJ, Mabey P, Urso RG, Billi D, Cockell C et al (2023) Future space experiment platforms for astrobiology and astrochemistry research. npj Microgravity 9(1):43. https://doi.org/10.1038/s41526-023-00292-1

Eregie SB, Sanusi IA, Kana GEB, Olaniran AO (2024) Effect of ultra-violet light radiation on Scenedesmus vacuolatus growth kinetics, metabolic performance, and preliminary biodegradation study. Biodegradation 35(1):71–86. https://doi.org/10.1007/s10532-023-10029-2

Ewert MK, Chen TT, Powell CD (2022) Life support baseline values and assumptions document

Fagliarone C, Mosca C, Di Stefano G, Leuko S, Moeller R, Rabbow E et al (2023) Enabling deep-space experimentations on cyanobacteria by monitoring cell division resumption in dried Chroococcidiopsis sp. 029 with accumulated DNA damage (original research). Front Microbiol. https://doi.org/10.3389/fmicb.2023.1150224

Fahrion J, Dussap CG, Leys N (2023) Assessment of batch culture conditions for cyanobacterial propagation for a bioreactor in space (original research). Front Astronomy Space Sci. https://doi.org/10.3389/fspas.2023.1178332

Fais G, Casula M, Sidorowicz A, Manca A, Margarita V, Fiori PL et al (2024) Cultivation of Chroococcidiopsis thermalis using available in situ resources to sustain life on Mars. Life 14(2):251. https://doi.org/10.3390/life14020251

Fu H, Su F, Zhu J, Zheng X, Ge C (2020) Effect of simulated microgravity and ionizing radiation on expression profiles of miRNA, lncRNA, and mRNA in human lymphoblastoid cells. Life Sci Space Res 24:1–8. https://doi.org/10.1016/j.lssr.2019.10.009

Gómez-Elvira J, Armiens C, Carrasco I, Genzer M, Gómez F, Haberle R et al (2014) Curiosity's rover environmental monitoring station: overview of the first 100 sols. J Geophys Res: Planets 119(7):1680–1688. https://doi.org/10.1002/2013JE004576

Gonçalves AL, Rodrigues CM, Pires JCM, Simões M (2016) The effect of increasing CO2 concentrations on its capture, biomass production and wastewater bioremediation by microalgae and cyanobacteria. Algal Res 14:127–136. https://doi.org/10.1016/j.algal.2016.01.008

Greenman J, Thorn R, Willey N, Ieropoulos I (2024) Energy harvesting from plants using hybrid microbial fuel cells; potential applications and future exploitation. Front Bioeng Biotechnol 12:1276176. https://doi.org/10.3389/fbioe.2024.1276176

Greenwood ZW, Abney MB, Brown BR, Fox ET, Stanley C (2018) State of NASA oxygen recovery. In: 48th international conference on environmental systems, pp 8–12

Häder D-P, Braun M, Hemmersbach R (2018) Bioregenerative life support systems in space research. In: Braun M, Böhmer M, Häder D-P, Hemmersbach R, Palme K (eds) Gravitational biology I: gravity sensing and graviorientation in microorganisms and plants. Springer International Publishing, Cham, pp 113–122. https://doi.org/10.1007/978-3-319-93894-3_8

Hassler DM, Zeitlin C, Wimmer-Schweingruber RF, Ehresmann B, Rafkin S, Eigenbrode JL et al (2014) Mars' surface radiation environment measured with the Mars science laboratory's curiosity rover. Science 343(6169):1244797. https://doi.org/10.1126/science.1244797

Hauslage J, Strauch SM, Eßmann O, Haag FWM, Richter P, Krüger J et al (2018) Eu:CROPIS – "Euglena gracilis: combined regenerative organic-food production in space" – a space experiment testing biological life support systems under lunar and Martian gravity. Microgravity Sci Technol 30(6):933–942. https://doi.org/10.1007/s12217-018-9654-1

Haws TD, Zimmerman JS, Fuller ME (2019) SLS, the gateway, and a lunar outpost in the early 2030s. In: 2019 IEEE aerospace conference. 2–9 March 2019, pp 1–15

Heinicke C, Adeli S, Baqué M, Correale G, Fateri M, Jaret S et al (2021) Equipping an extraterrestrial laboratory: overview of open research questions and recommended instrumentation for the moon. Adv Space Res 68(6):2565–2599. https://doi.org/10.1016/j.asr.2021.04.047

Heldmann JL, Marinova MM, Lim DSS, Wilson D, Carrato P, Kennedy K et al (2021) Mission architecture using the SpaceX Starship vehicle to enable a sustained human presence on Mars. New Space 10(3):259–273. https://doi.org/10.1089/space.2020.0058

Holcomb JA, O'Leary B, Darrin AG, Mandel RD, Kling C, Wegmann KW (2023) Planetary geoarchaeology as a new frontier in archaeological science: evaluating site formation processes on Earth's moon. Geoarchaeology 38(5):513–533. https://doi.org/10.1002/gea.21966

Horneck G, Klaus David M, Mancinelli Rocco L (2010) Space microbiology. Microbiol Mol Biol Rev 74(1):121–156. https://doi.org/10.1128/mmbr.00016-09

Huang B, Li D-G, Huang Y, Liu C-T (2018) Effects of spaceflight and simulated microgravity on microbial growth and secondary metabolism. Mil Med Res 5(1):18. https://doi.org/10.1186/s40779-018-0162-9

Huang G, Han Y, Li W, Xue Z, He D, He H (2022) Rapid screening of microalgae by a 96-hole air-flowing device. Bioprocess Biosyst Eng 45(5):943–953. https://doi.org/10.1007/s00449-022-02714-y

Ibrahim FG, Torre RM, Moya BL, de Godos Crespo I (2020) Carbon dioxide capture from carbon dioxide–rich gases by microalgae. In: Soreanu G, Dumont É (eds) From biofiltration to promising options in gaseous fluxes biotreatment. Elsevier, pp 373–396. https://doi.org/10.1016/B978-0-12-819064-7.00018-2

Janssen PJD, Lambreva MD, Plumeré N, Bartolucci C, Antonacci A, Buonasera K et al (2014) Photosynthesis at the forefront of a sustainable life (review). Front Chem 2. https://doi.org/10.3389/fchem.2014.00036

Jones HW (2015) Estimating the life cycle cost of space systems. In: International conference on environmental systems. ARC-E-DAA-TN24212

Kanervo E, Lehto K, Ståhle K, Lehto H, Mäenpää P (2005) Characterization of growth and photosynthesis of Synechocystis sp. PCC 6803 cultures under reduced atmospheric pressures and enhanced CO2 levels. Int J Astrobiol 4(1):97–100. https://doi.org/10.1017/S1473550405002466

Keller R, Goli K, Porter W, Alrabaa A, Jones JA (2023) Cyanobacteria and algal-based biological life support system (BLSS) and planetary surface atmospheric revitalizing bioreactor brief concept review. Life 13(3):816. https://doi.org/10.3390/life13030816

Kerner M, Wolff T, Brinkmann T (2024) Efficient supply with carbon dioxide from flue gas during large scale production of microalgae: a novel approach for bioenergy facades. Bioresour Technol 391:129917. https://doi.org/10.1016/j.biortech.2023.129917

Kryvenda A, Tischner R, Steudel B, Griehl C, Armon R, Friedl T (2023) Testing for terrestrial and freshwater microalgae productivity under elevated CO2 conditions and nutrient limitation. BMC Plant Biol 23(1):27. https://doi.org/10.1186/s12870-023-04042-z

Lebert M, Porst M, Häder D (1995) Long-term culture of Euglena gracilis: an AQUARACK progress report. In: Proceedings of the CEBAS workshops annual, pp 223–231 11

Li T, Strous M, Melkonian M (2017a) Biofilm-based photobioreactors: their design and improving productivity through efficient supply of dissolved inorganic carbon. FEMS Microbiol Lett 364(24):fnx218. https://doi.org/10.1093/femsle/fnx218

Li X, Richter PR, Hao Z, An Y, Wang G, Li D et al (2017b) Operation of an enclosed aquatic ecosystem in the Shenzhou-8 mission. Acta Astronaut 134:17–22. https://doi.org/10.1016/j.actaastro.2017.01.014

Li X, Xie Y, Tian Y, An F (2024) A study on the design and implementation technologies of EVA at the China Space Station. Aerospace 11(4):264. https://www.mdpi.com/2226-4310/11/4/264

Llorente B, Williams TC, Goold HD, Pretorius IS, Paulsen IT (2022) Harnessing bioengineered microbes as a versatile platform for space nutrition. Nat Commun 13(1):6177. https://doi.org/10.1038/s41467-022-33974-7

Lopez P, Schultz E, Mattfeld B, Stromgren C, Goodliff K (2015) Logistics needs for potential deep space mission scenarios post asteroid redirect crewed mission. In: 2015 IEEE aerospace conference. IEEE, pp 1–10. https://doi.org/10.1109/AERO.2015.7119161

Macário IPE, Veloso T, Frankenbach S, Serôdio J, Passos H, Sousa C et al (2022) Cyanobacteria as candidates to support mars colonization: growth and biofertilization potential using Mars regolith as a resource (original research). Front Microbiol 13. https://doi.org/10.3389/fmicb.2022.840098

Maender C, Aspiotis J, Clemens R, Gupta A, Panchanathan D, Zuniga D et al (2022) Beyond the ISS : the world's first commercial space station. In-Space manufacturing and resources, pp 319–337. https://doi.org/10.1002/9783527830909.ch17

Maity T, Saxena A (2024) Challenges and innovations in food and water availability for a sustainable Mars colonization. Life Sci Space Res 42:27–36. https://doi.org/10.1016/j.lssr.2024.03.008

Mapstone LJ, Leite MN, Purton S, Crawford IA, Dartnell L (2022) Cyanobacteria and microalgae in supporting human habitation on Mars. Biotechnol Adv 59:107946. https://doi.org/10.1016/j.biotechadv.2022.107946
Marra D, Karapantsios T, Caserta S, Secchi E, Holynska M, Labarthe S et al (2023) Migration of surface-associated microbial communities in spaceflight habitats. Biofilms 5:100109. https://doi.org/10.1016/j.bioflm.2023.100109
Massa GD, Emmerich JC, Morrow RC, Bourget CM, Mitchell CA (2006) Plant-growth lighting for space life support: a review. Gravitational and space biology 19(2):19–30
Matula EE, Nabity DJA (2021a) Metabolic response of Chlorella vulgaris to a transient thermal environment for supporting simultaneous air revitalization and thermal control in a crewed habitat. Acta Astronaut 187:406–415. https://doi.org/10.1016/j.actaastro.2021.07.003
Matula EE, Nabity JA (2021b) Effects of stepwise changes in dissolved carbon dioxide concentrations on metabolic activity in Chlorella for spaceflight applications. Life Sci Space Res 29:73–84. https://doi.org/10.1016/j.lssr.2021.03.005
McCutcheon G, Kent R, Paulino-Lima I, Pless E, Ricco A, Mazmanian E et al (2016) PowerCells payload on EuCROPIS-measuring synthetic biology in space. AIAA/USU Conference on Small Satellites SSC-16-XI-04
McKay CP, Marinova MM (2001) The physics, biology, and environmental ethics of making Mars habitable. Astrobiology 1(1):89–109. https://doi.org/10.1089/153110701750137477
Meurisse A, Cazzaniga C, Frost C, Barnes A, Makaya A, Sperl M (2020) Neutron radiation shielding with sintered lunar regolith. Radiat Meas 132:106247. https://doi.org/10.1016/j.radmeas.2020.106247
Miller J, Taylor L, Zeitlin C, Heilbronn L, Guetersloh S, DiGiuseppe M et al (2009) Lunar soil as shielding against space radiation. Radiat Meas 44(2):163–167. https://doi.org/10.1016/j.radmeas.2009.01.010
Miyachi S, Iwasaki I, Shiraiwa Y (2003) Historical perspective on microalgal and cyanobacterial acclimation to low- and extremely high-CO2 conditions. Photosynth Res 77(2):139–153. https://doi.org/10.1023/A:1025817616865
Moeller R, Raguse M, Leuko S, Berger T, Hellweg CE, Fujimori A et al (2017) STARLIFE—an international campaign to study the role of galactic cosmic radiation in astrobiological model systems. Astrobiology 17(2):101–109. https://doi.org/10.1089/ast.2016.1571
Murukesan G, Leino H, Mäenpää P, Ståhle K, Raksajit W, Lehto HJ et al (2016) Pressurized Martian-like pure CO2 atmosphere supports strong growth of cyanobacteria, and causes significant changes in their metabolism. Origins Life Evol Biospheres 46(1):119–131. https://doi.org/10.1007/s11084-015-9458-x
Musk E (2017) Making humans a multi-planetary species. New Space 5(2):46–61. https://doi.org/10.1089/space.2017.29009.emu
Nangle SN, Wolfson MY, Hartsough L, Ma NJ, Mason CE, Merighi M et al (2020) The case for biotech on Mars. Nat Biotechnol 38(4):401–407. https://doi.org/10.1038/s41587-020-0485-4
Naser MZ, Chehab AI (2018) Materials and design concepts for space-resilient structures. Prog Aerosp Sci 98:74–90. https://doi.org/10.1016/j.paerosci.2018.03.004
Nasir A, Strauch SM, Becker I, Sperling A, Schuster M, Richter PR et al (2014) The influence of microgravity on Euglena gracilis as studied on Shenzhou 8. Plant Biol 16(s1):113–119. https://doi.org/10.1111/plb.12067
Nickerson Cheryl A, Ott CM, Mister Sarah J, Morrow Brian J, Burns-Keliher L, Pierson Duane L (2000) Microgravity as a novel environmental signal affecting salmonella enterica Serovar typhimurium virulence. Infect Immun 68(6):3147–3152. https://doi.org/10.1128/iai.68.6.3147-3152.2000
Nicola W, Eva ES, Thorsten B, Michelle MG (2024) Exploring the influence of atmospheric CO2 and O2 levels on the utility of nitrogen isotopes as proxy for biological N2 fixation. bioRxiv, 2024.03.28.587259. https://doi.org/10.1101/2024.03.28.587259

Nowicka-Krawczyk P, Mühlsteinová R, Hauer T (2019) Detailed characterization of the Arthrospira type species separating commercially grown taxa into the new genus Limnospira (cyanobacteria). Sci Rep 9(1):694. https://doi.org/10.1038/s41598-018-36831-0

Olsson-Francis K, Cockell CS (2010) Use of cyanobacteria for in-situ resource use in space applications. Planet Space Sci 58(10):1279–1285. https://doi.org/10.1016/j.pss.2010.05.005

Oncel S, Sukan FV (2008) Comparison of two different pneumatically mixed column photobioreactors for the cultivation of Artrospira platensis (Spirulina platensis). Bioresour Technol 99(11):4755–4760. https://doi.org/10.1016/j.biortech.2007.09.068

Patel MR, Bérces A, Kolb C, Lammer H, Rettberg P, Zarnecki JC et al (2003) Seasonal and diurnal variations in Martian surface ultraviolet irradiation: biological and chemical implications for the Martian regolith. Int J Astrobiol 2(1):21–34. https://doi.org/10.1017/S1473550402001180

Pavlov AA, Vasilyev G, Ostryakov VM, Pavlov AK, Mahaffy P (2012) Degradation of the organic molecules in the shallow subsurface of Mars due to irradiation by cosmic rays. Geophys Res Lett 39(13). https://doi.org/10.1029/2012GL052166

Pollard EC (1965) Theoretical studies on living systems in the absence of mechanical stress. J Theor Biol 8(1):113–123. https://doi.org/10.1016/0022-5193(65)90097-4

Poughon L, Laroche C, Creuly C, Dussap C-G, Paille C, Lasseur C et al (2020) Limnospira indica PCC8005 growth in photobioreactor: model and simulation of the ISS and ground experiments. Life Sci Space Res 25:53–65. https://doi.org/10.1016/j.lssr.2020.03.002

Priyadharsini P, Nirmala N, Dawn SS, Baskaran A, SundarRajan P, Gopinath KP et al (2022) Genetic improvement of microalgae for enhanced carbon dioxide sequestration and enriched biomass productivity: review on CO2 bio-fixation pathways modifications. Algal Res 66:102810. https://doi.org/10.1016/j.algal.2022.102810

Qi F, Pei H, Hu W, Mu R, Zhang S (2016) Characterization of a microalgal mutant for CO2 biofixation and biofuel production. Energy Convers Manag 122:344–349. https://doi.org/10.1016/j.enconman.2016.05.093

Qi F, Wu D, Mu R, Zhang S, Xu X (2018) Characterization of a microalgal UV mutant for CO2 biofixation and biomass production. Biomed Res Int 2018:4375170. https://doi.org/10.1155/2018/4375170

Qin L, Yu Q, Ai W, Tang Y, Ren J, Guo S (2014) Response of cyanobacteria to low atmospheric pressure. Life Sci Space Res 3:55–62. https://doi.org/10.1016/j.lssr.2014.09.001

Ramalho TP, Chopin G, Salman L, Baumgartner V, Heinicke C, Verseux C (2022) On the growth dynamics of the cyanobacterium anabaena sp. PCC 7938 in Martian regolith. npj Microgravity 8(1):43. https://doi.org/10.1038/s41526-022-00240-5

Ramanan R, Kim B-H, Cho D-H, Oh H-M, Kim H-S (2016) Algae–bacteria interactions: evolution, ecology and emerging applications. Biotechnol Adv 34(1):14–29. https://doi.org/10.1016/j.biotechadv.2015.12.003

Rapp D, Inglezakis VJ (2024) Mars in situ resource utilization (ISRU) with focus on atmospheric processing for near-term application—a historical review and appraisal. Applied Sci 14(2):653. https://doi.org/10.3390/app14020653

Rastogi RP, Madamwar D, Nakamoto H, Incharoensakdi A (2020) Resilience and self-regulation processes of microalgae under UV radiation stress. J Photochem Photobiol C: Photochem Rev 43:100322. https://doi.org/10.1016/j.jphotochemrev.2019.100322

Rautela A, Yadav I, Gangwar A, Chatterjee R, Kumar S (2024) Photosynthetic production of α-farnesene by engineered Synechococcus elongatus UTEX 2973 from carbon dioxide. Bioresour Technol 396:130432. https://doi.org/10.1016/j.biortech.2024.130432

Revellame ED, Aguda R, Chistoserdov A, Fortela DL, Hernandez RA, Zappi ME (2021) Microalgae cultivation for space exploration: assessing the potential for a new generation of waste to human life-support system for long duration space travel and planetary human habitation. Algal Res 55:102258. https://doi.org/10.1016/j.algal.2021.102258

Revellame ED, Aguda R, Gatdula KM, Holmes W, Fortela DL, Sharp W et al (2024) Microalgae in bioregenerative life support systems for space applications. Algal Res 77:103332. https://doi.org/10.1016/j.algal.2023.103332

Riccio GE (2023) Toward risk-informed decision making in situated systems medicine: metascience for smart homes on earth and in space. Systems medicine for human spaceflight. World Scientific, pp 199–229. https://doi.org/10.1142/9789811287695_0008
Ridley A, Brown C, Garr J, Gavin L, Hornyak D, Toon K et al (2023) International Space Station as a testbed for exploration environmental control and life support systems–2023 status. In: 53rd international conference on environmental systems. No. ICES-2024-316
Rivasseau C, Farhi E, Atteia A, Couté A, Gromova M, de Gouvion Saint Cyr D et al (2013) An extremely radioresistant green eukaryote for radionuclide bio-decontamination in the nuclear industry (10.1039/C2EE23129H). Energy Environ Sci 6(4):1230–1239. https://doi.org/10.1039/C2EE23129H
Rontó G, Bérces A, Lammer H, Cockell CS, Molina-Cuberos GJ, Patel MR et al (2003) Solar UV irradiation conditions on the surface of Mars. Photochem Photobiol 77(1):34–40. https://doi.org/10.1562/0031-8655(2003)0770034SUICOT2.0.CO2
Saghamanesh M, Hu W, Rastinasab V (2024) Low-thrust trajectory optimization for Earth-Mars-Jupiter gravity-assist spacecraft mission. Preprints. 2024/01/15. Preprints. https://doi.org/10.20944/preprints202401.1077.v1
Sánchez-Bayo A, Morales V, Rodríguez R, Vicente G, Bautista LF (2020) Cultivation of microalgae and cyanobacteria: effect of operating conditions on growth and biomass composition. Molecules 25(12): 2834
Sargusingh M, Perry J, Howard D, Toomarian N (2015) Notional environmental control and life support system architectures for human exploration beyond low-earth orbit. AIAA SPACE 2015 Conference and exposition: American Institute of Aeronautics and Astronautics. https://doi.org/10.2514/6.2015-4456
Sharafi laleh S, Gharamaleki FP, Alavi SF, Soltani S, Mahmoudi SMS, Rosen MA (2023) A novel sustainable biomass-fueled cogeneration cycle integrated with carbon dioxide capture utilizing LNG regasification and green hydrogen production via PEM electrolysis: thermodynamic assessment. J Clean Prod 421:138529. https://doi.org/10.1016/j.jclepro.2023.138529
Shaw L, Garr J, Gavin L, Hornyak D, Matty C, Ridley A et al (2021) International space station as a testbed for exploration environmental control and life support systems 2021 status. In: 50th international conference on environmental systems
Silverman SN, Kopf SH, Bebout BM, Gordon R, Som SM (2019) Morphological and isotopic changes of heterocystous cyanobacteria in response to N2 partial pressure. Geobiology 17(1):60–75. https://doi.org/10.1111/gbi.12312
Singh H, Anurag K, Apte SK (2013) High radiation and desiccation tolerance of nitrogen-fixing cultures of the cyanobacterium anabaena sp. strain PCC 7120 emanates from genome/proteome repair capabilities. Photosynth Res 118(1):71–81. https://doi.org/10.1007/s11120-013-9936-9
Smith EJ, Davis L, Coleman PJ, Jones DE (1965) Magnetic field measurements near Mars. Science 149(3689):1241–1242. https://doi.org/10.1126/science.149.3689.1241
Srivastava A, Kumar A, Biswas S, Kumar R, Srivastava V, Rajaram H et al (2023) Gamma (γ)-radiation stress response of the cyanobacterium anabaena sp. PCC7120: regulatory role of LexA and photophysiological changes. Plant Sci 326:111529. https://doi.org/10.1016/j.plantsci.2022.111529
Su X, Fang T, Fang L, Wang D, Jiang X, Liu C et al (2023) Effects of short-term exposure to simulated microgravity on the physiology of Bacillus subtilis and multiomic analysis. Can J Microbiol 69(12):464–478. https://doi.org/10.1139/cjm-2023-0061
Subramanian G, Shanmugasundaram S (1986) Uninduced ammonia release by the nitrogen-fixing cyanobacterium anabaena. FEMS Microbiol Lett 37(2):151–154. https://doi.org/10.1111/j.1574-6968.1986.tb01784.x
Sychev VN, Shepelev EY, Meleshko GI, Gurieva TS, Levinskikh MA, Podolsky IG et al (2001) Main characteristics of biological components of developing life support system observed during the experiments aboard orbital complex MIR. Adv Space Res 27(9):1529–1534. https://doi.org/10.1016/S0273-1177(01)00245-9
Thomas DJ, Sullivan SL, Price AL, Zimmerman SM (2005) Common freshwater cyanobacteria grow in 100% CO2. Astrobiology 5(1):66–74. https://doi.org/10.1089/ast.2005.5.66

Tirumalai MR, Karouia F, Tran Q, Stepanov VG, Bruce RJ, Ott CM et al (2017) The adaptation of Escherichia coli cells grown in simulated microgravity for an extended period is both phenotypic and genomic. npj Microgravity 3(1):15. https://doi.org/10.1038/s41526-017-0020-1

Verseux C, Baqué M, Lehto K, de Vera J-PP, Rothschild LJ, Billi D (2016) Sustainable life support on Mars – the potential roles of cyanobacteria. Int J Astrobiol 15(1):65–92. https://doi.org/10.1017/S147355041500021X

Verseux C, Baqué M, Cifariello R, Fagliarone C, Raguse M, Moeller R et al (2017) Evaluation of the resistance of Chroococcidiopsis spp. to sparsely and densely ionizing irradiation. Astrobiology 17(2):118–125. https://doi.org/10.1089/ast.2015.1450

Verseux C, Heinicke C, Ramalho TP, Determann J, Duckhorn M, Smagin M et al (2021) A low-pressure, N2/CO2 atmosphere is suitable for cyanobacterium-based life-support systems on Mars (original research). Front Microbiol 12. https://doi.org/10.3389/fmicb.2021.611798

Von Ehrenfried MD (2020) The Artemis lunar program overview. In: von Ehrenfried MD (ed) The Artemis lunar program: returning people to the moon. Springer International Publishing, Cham, pp 7–47. https://doi.org/10.1007/978-3-030-38513-2_2

Wagner I, Posten C (2017) Pressure reduction affects growth and morphology of Chlamydomonas reinhardtii. Eng Life Sci 17(5):552–560. https://doi.org/10.1002/elsc.201600131

Walker J, Granjou C (2017) MELiSSA the minimal biosphere: human life, waste and refuge in deep space. Futures 92:59–69. https://doi.org/10.1016/j.futures.2016.12.001

Wang GH, Li GB, Li DH, Liu YD, Song LR, Tong GH et al (2004) Real-time studies on microalgae under microgravity. Acta Astronaut 55(2):131–137. https://doi.org/10.1016/j.actaastro.2004.02.005

Wang G, Chen H, Li G, Chen L, Li D, Hu C et al (2006) Population growth and physiological characteristics of microalgae in a miniaturized bioreactor during space flight. Acta Astronaut 58(5):264–269. https://doi.org/10.1016/j.actaastro.2005.11.001

Wang R, Wang X, Zhu T (2024) Research progress and application of carbon sequestration in industrial flue gas by microalgae: a review. J Environ Sci 152:14. https://doi.org/10.1016/j.jes.2024.04.018

Wassmann M, Moeller R, Rabbow E, Panitz C, Horneck G, Reitz G et al (2012) Survival of spores of the UV-resistant Bacillus subtilis strain MW01 after exposure to low-Earth orbit and simulated Martian conditions: data from the space experiment ADAPT on EXPOSE-E. Astrobiology 12(5):498–507. https://doi.org/10.1089/ast.2011.0772

Way JC, Silver PA, Howard RJ (2011) Sun-driven microbial synthesis of chemicals in space. Int J Astrobiol 10(4):359–364. https://doi.org/10.1017/S1473550411000218

Winton D, Isobe J, Henson P, MacKnight A, Yates S, Schuck D (2016) Carbon dioxide removal technologies for space vehicles-past, present, and future. In: 46th international conference on environmental systems. http://hdl.handle.net/2346/67727

Xiao Y, Liu Y, Wang G, Hao Z, An Y (2010) Simulated microgravity alters growth and microcystin production in Microcystis aeruginosa (cyanophyta). Toxicon 56(1):1–7. https://doi.org/10.1016/j.toxicon.2010.01.026

Xu Z, Theodoropoulos C, Pittman JK (2024) Optimization of a Chlorella–Saccharomyces co-culture system for enhanced metabolite productivity. Algal Res 79:103455. https://doi.org/10.1016/j.algal.2024.103455

Yatagai F, Honma M, Dohmae N, Ishioka N (2019) Biological effects of space environmental factors: a possible interaction between space radiation and microgravity. Life Sci Space Res 20:113–123. https://doi.org/10.1016/j.lssr.2018.10.004

Ye T, Wang B, Li C, Bian P, Chen L, Wang G (2021) Exposure of cyanobacterium Nostoc sp. to the Mars-like stratosphere environment. J Photochem Photobiol B Biol 224:112307. https://doi.org/10.1016/j.jphotobiol.2021.112307

Yin S, Jin W, Xi T, Zhou X, He Z, Meng X et al (2024) Factors affect the oxygen production of Chlorella pyrenoidosa in a bacterial-algal symbiotic system: light intensity, temperature, pH and static magnetic field. Process Saf Environ Prot 184:492–501. https://doi.org/10.1016/j.psep.2024.02.004

SPRINGER NATURE

GPSR Compliance

The European Union's (EU) General Product Safety Regulation (GPSR) is a set of rules that requires consumer products to be safe and our obligations to ensure this.

If you have any concerns about our products, you can contact us on ProductSafety@springernature.com

In case Publisher is established outside the EU, the EU authorized representative is:

Springer Nature Customer Service Center GmbH
Europaplatz 3
69115 Heidelberg, Germany

The manufacturer's authorised representative in the EU is Springer Nature Customer Service Centre GmbH, Europaplatz 3, 69115 Heidelberg, Germany. If you have any concerns regarding our products, please contact ProductSafety@springernature.com

Printed and bound by CPI Group (UK) Ltd, Croydon, CR0 4YY
30/09/2026
02225567-0015